London Mathematical Society Student Texts 79

Introduction to Compact Riemann Surfaces and Dessins d'Enfants

ERNESTO GIRONDO
Universidad Autónoma de Madrid

GABINO GONZÁLEZ-DIEZ
Universidad Autónoma de Madrid

CAMBRIDGE
UNIVERSITY PRESS

CAMBRIDGE UNIVERSITY PRESS
Cambridge, New York, Melbourne, Madrid, Cape Town,
Singapore, São Paulo, Delhi, Tokyo, Mexico City

Cambridge University Press
The Edinburgh Building, Cambridge CB2 8RU, UK

Published in the United States of America by Cambridge University Press, New York

www.cambridge.org
Information on this title: www.cambridge.org/9780521519632

First published 2012

Printed in the United Kingdom at the University Press, Cambridge

A catalogue record for this publication is available from the British Library

ISBN 978-0-521-51963-2 Hardback
ISBN 978-0-521-74022-7 Paperback

LONDON MATHEMATICAL SOCIETY STUDENT TEXTS

Managing Editor

Professor D. Benson, Department of Mathematics, University of Aberdeen, UK

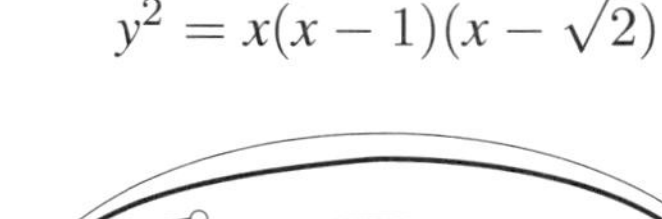

$$y^2 = x(x-1)(x-\sqrt{2}) \qquad\qquad y^2 = x(x-1)(x+\sqrt{2})$$

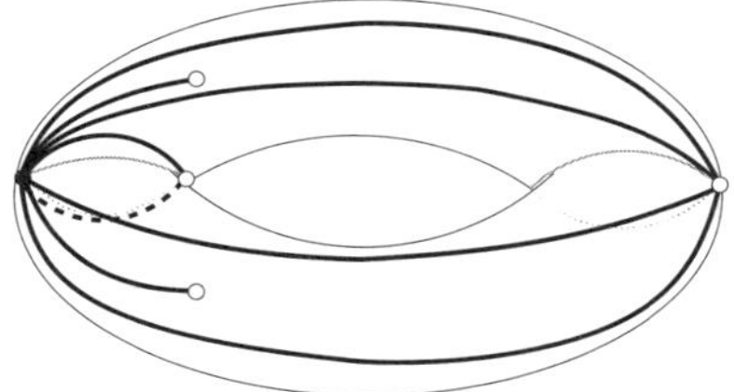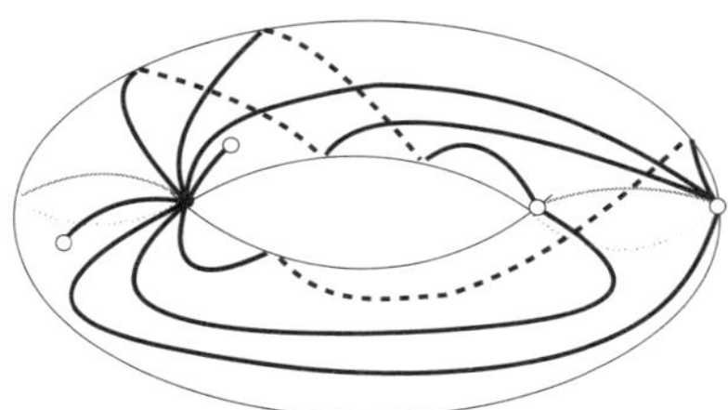

For our families

Emi and Álvaro

and

Isabel, Jorge and David

Contents

Preface

The present text is an expanded version of the lecture notes for a course on Riemann surfaces and dessins d'enfants which the authors have taught for several years to students of the masters degree in mathematics at the Universidad Autónoma de Madrid.

Riemann surfaces are an ideal meeting ground for several branches of mathematics. For example, a student taking a course like this will encounter concepts of algebraic topology (fundamental group, theory of covering spaces, monodromy), elements of Riemannian geometry (geodesics, isometries, tessellations), objects belonging to algebra and algebraic geometry (field extensions, algebraic curves, valuations), definitions belonging to arithmetic geometry (fields of moduli and definition of an algebraic variety), some elementary graph theory (dessins d'enfants), tools of (complex) analysis (Weierstrass functions and Poincaré series) and some of the most relevant groups in analytic number theory (principal congruence subgroups).

One of the main features of the theory of Riemann surfaces is that there is a bijective correspondence between isomorphism classes of compact Riemann surfaces and isomorphism classes of complex algebraic curves. Establishing this correspondence requires proving first that a Riemann surface has enough meromorphic functions to separate its points. This can be done by either applying the Riemann–Roch Theorem or using the Uniformization Theorem to construct these functions by means of Poincaré series (Weierstrass functions, in the genus one case). In this book we have chosen the second option, thereby introducing Fuchsian groups, the third member of this trinity of equivalent objects. The Uniformization Theorem is the only result we assume without proof. On the one hand it has a very simple statement, and on the other the students at this level have become familiar with the result for open sets of the complex plane in their first course on complex analysis, in which they probably have had to find Riemann mappings between different simply connected regions of the plane, and at the same time learnt that, due to Liouville's theorem, this is impossible if one of the regions is the whole plane. Otherwise the prerequisites have been kept to a minimum. An undergraduate student who has taken courses on vector calculus, point-set topology, field theory and complex analysis should be in a position to follow a course based in this book. Whenever we thought that an example could help to understand the theory we included one. About one third of the pages of the book are devoted to worked out examples and illustrative pictures.

The text is divided in two parts, consisting of two chapters each. In the first part we give an elementary introduction to the theory of compact Riemann surfaces. The main goal of these two first chapters is to establish the equivalence between compact Riemann surfaces, compact algebraic curves and Fuchsian groups of finite type (or lattices of $\mathbb{C}$, in the genus 1 case). We have made an effort to work out in detail the Riemann surface structure associated to a number of particularly interesting curves (hyperelliptic, Fermat, Klein, etc.) and Fuchsian groups (triangle groups, groups with special symmetric fundamental domains, etc.). When possible we have shown the link between these objects. This first part could serve by itself as a textbook for an elementary introduction to the theory of compact Riemann surfaces from the point of view of algebraic curves and Fuchsian groups. The results in it are therefore very classical as they go back to Riemann, Weierstrass, Hurwitz, Poincaré, Klein, etc.

On the contrary, the theory presented in the second part (third and fourth chapters) is much more modern. It was launched by Grothendieck in the 1980s, in his now famous *Equisse d'un programme*, following the disclosure of Belyi's celebrated theorem, whose amazingly simple proof seems to have impressed him deeply (in his own words: *'jamais sans doute un résultat profond et déroutant ne fut démontré en si peu de lignes!'*†). Here we focus on those Riemann surfaces whose corresponding algebraic equa-

---

† Never, without a doubt, was such a deep and disconcerting result proved in so few lines!

tion has coefficients in $\overline{\mathbb{Q}}$, the field of algebraic numbers. This is an extraordinarily beautiful theory. Without needing to invoke all the far-reaching ideas and proposals raised by Grothendieck in the Equisse, some of them beyond the scope of this book (Grothendieck–Teichmüller Theory, etc), it is impossible to resist the attraction of Grothendieck's correspondence:

$$\left\{ \begin{array}{c} \text{graphs dividing an} \\ \text{orientable surface} \\ \text{into a disjoint} \\ \text{union of cells} \end{array} \right\} \longleftrightarrow \left\{ \begin{array}{c} \text{algebraic curves } C \text{ endowed} \\ \text{with a function } f \text{ ramified} \\ \text{over three values, both with} \\ \text{coefficients in } \overline{\mathbb{Q}} \end{array} \right\}$$

whose proof is the second part's main goal. We stress the fact that the objects on the left-hand side, called dessins d'enfants (child's drawings) by Grothendieck, are purely topological, whereas those on the right-hand side have an arithmetic nature. It turns out that this correspondence is a consequence of Belyi's correspondence:

$$\left\{ \begin{array}{c} \text{algebraic curves} \\ \text{with coefficients in } \overline{\mathbb{Q}} \end{array} \right\} \longleftrightarrow \left\{ \begin{array}{c} \text{Riemann surfaces with} \\ \text{a meromorphic function} \\ \text{ramified over three values} \end{array} \right\}$$

The discovery of this theory seems to have made a big impact on Grothendieck:

Cette découverte, qui techniquement se réduit à si peu de choses, a fait sur moi une impression très forte, et elle représente un tournant décisif dans le cours de mes réflexions, un déplacement notamment de mon centre d'intérêt en mathématique, qui soudain s'est trouvé fortement localisé. Je ne crois pas qu'un fait mathématique m'ait jamais autant frappé que celui-là, et ait eu un impact psychologique comparable.†

Let us observe that via the natural action of the absolute Galois group $\mathrm{Gal}(\overline{\mathbb{Q}}/\mathbb{Q})$ on pairs (C, f), as above, Grothendieck's correspondence implies an action of $\mathrm{Gal}(\overline{\mathbb{Q}}/\mathbb{Q})$ on dessins. Thus, this theory can be applied to understand the structure of the group $\mathrm{Gal}(\overline{\mathbb{Q}}/\mathbb{Q})$, which one can regard as embodying the whole of classical Galois theory over $\mathbb{Q}$.

The proof of Belyi's correspondence is given in Chapter 3. In

---

† This discovery, which is technically so simple, made a very strong impression on me, and it represents a decisive turning point in the course of my reflections, a shift in particular of my centre of interest in mathematics, which suddenly found itself strongly focused. I do not believe that a mathematical fact has ever struck me quite so strongly as this one, nor had a comparable psychological impact (translation by Schneps and Lochak).

one direction (algebraic curves defined over $\overline{\mathbb{Q}}$ admit functions with only three branching values) the result follows from a surprisingly simple construction due to Belyi. For the opposite direction he invokes a criterion of rationality due to Weil. However, the machinery used by Weil in the proof of this criterion exceeds by far the elementary level we want to retain in this book. So, with the tools we have at our disposal, we work out a (much) weaker criterion which nevertheless satisfies our needs. Grothendieck's correspondence itself is proved in Chapter 4. In it we also study the first properties of the action of $\mathrm{Gal}(\overline{\mathbb{Q}}/\mathbb{Q})$. A good handful of examples are explicitly described. All in all this chapter leads readers to the boundary of current research in the subject.

Those who wish to pursue this theory are adviced to consult the book by Lando and Zvonkin, the conference proceedings edited by Schneps and Lochak, the survey articles by Jones–Singerman, Shabat–Voevodsky, Wolfart, Jones, Cohen–Itzykson–Wolfart and Lochak, and, from a different point of view, the monograph by Bowers–Stephenson, which are all included in the references to this book.

As for readers wishing to study the subject of Riemann surfaces beyond or complementing the introduction given in the first part of this book, there are many excellent references available. Among them are the books by Jones–Singerman, Beardon, Farkas–Kra, Siegel, Jost, Forster, Miranda, Buser and Kirwan.

Acknowledgment: In the process of writing this book the authors have benefited from the following research projects:
- MTM2006-01859 (MEC, Spain)
- MTM2006-28257-E (MEC, Spain)
- CCG08-UAM/ESP-4145 (UAM-C. de Madrid, Spain)
- MTM2009-08213-E (MICINN, Spain).

The second author also wishes to thank the Departamento de Matemáticas of the Universidad Autónoma de Madrid for a sabbatical leave during the academic year 2009/2010.

1

Compact Riemann surfaces and algebraic curves

1.1 Basic definitions

1.1.1 Riemann surfaces – examples

Definition 1.1 A *topological surface* X is a Hausdorff topological space provided with a collection $\{\varphi_i : U_i \longrightarrow \varphi_i(U_i)\}$ of homeomorphisms (called *charts*) from open subsets $U_i \subset X$ (called *coordinate neighbourhoods*) to open subsets $\varphi_i(U_i) \subset \mathbb{C}$ such that:

(i) the union $\bigcup_i U_i$ covers the whole space X; and

(ii) whenever $U_i \cap U_j \neq \emptyset$, the *transition function*

$$\varphi_j \circ \varphi_i^{-1} : \varphi_i(U_i \cap U_j) \longrightarrow \varphi_j(U_i \cap U_j)$$

is a homeomorphism (Figure 1.1).

A collection of charts fulfilling these properties is called a (topological) *atlas*, and the inverse φ_i^{-1} of a chart is called a *parametrization*.

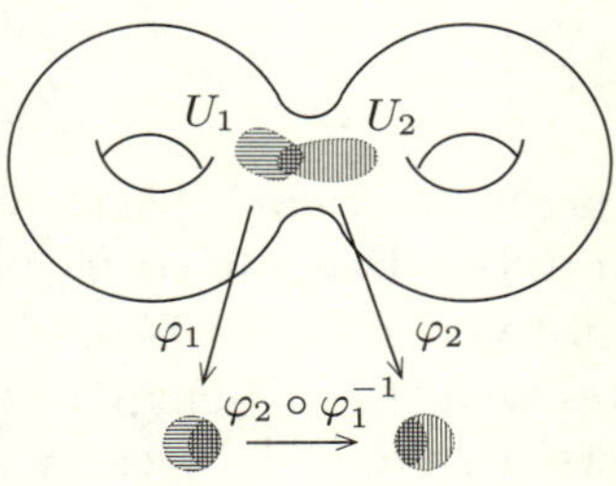

Fig. 1.1. The transition function between two coordinate charts.

Definition 1.2 A *Riemann surface* is a connected topological surface such that the transition functions of the atlas are holomorphic mappings between open subsets of the complex plane $\mathbb{C}$ (rather than mere homeomorphisms).

Example 1.3 The simplest Riemann surfaces are those defined by a single chart. Any connected open subset U in the plane is obviously a Riemann surface with the atlas consisting simply of the chart (U, Id). Particularly interesting cases are the whole *complex plane* $\mathbb{C}$, the *unit disc* $\mathbb{D} = \{z \in \mathbb{C} : |z| < 1\}$ and the *upper halfplane* $\mathbb{H} = \{z \in \mathbb{C} : \mathrm{Im}\, z > 0\}$.

Example 1.4 One chart is not enough to define a Riemann surface structure on the sphere

$$\mathbb{S}^2 = \left\{ (x, y, t) \in \mathbb{R}^3 : x^2 + y^2 + t^2 = 1 \right\}$$

since the whole sphere is not homeomorphic to an open subset of the plane. But one can consider the following two charts:

$$U_1 = \mathbb{S}^2 \smallsetminus \{(0,0,1)\}, \quad \varphi_1 (x, y, t) = \frac{x}{1-t} + i \frac{y}{1-t}$$

$$U_2 = \mathbb{S}^2 \smallsetminus \{(0,0,-1)\}, \quad \varphi_2 (x, y, t) = \frac{x}{1+t} - i \frac{y}{1+t}$$

From the identity $\dfrac{x - iy}{1 + t} = \dfrac{1 - t}{x + iy}$ it follows that the transition function is $\varphi_2 \circ \varphi_1^{-1}(z) = 1/z$, where z denotes the (complex) variable in $\varphi_1(U_1)$. Note that the domain where $\varphi_2 \circ \varphi_1^{-1}$ is defined is $\varphi_1(U_1 \cap U_2) = \mathbb{C} \smallsetminus \{0\}$.

Example 1.5 The name *Riemann sphere*, or *extended complex plane* is usually given to the Riemann surface $\widehat{\mathbb{C}}$ defined as follows. Add an aditional point to the complex plane $\mathbb{C}$, and denote it as ∞ (the notation indicates what the topology at this additional point is going to be: one gets close to ∞ by escaping from every point in the plane). A collection of fundamental neighbourhoods of ∞ is provided by the family of sets $D(\infty, R) = \{z \in \mathbb{C}, |z| > R\} \cup \{\infty\}$.

Denote $\widehat{\mathbb{C}} = \mathbb{C} \cup \{\infty\}$. The Riemann surface structure is determined by the following two charts:

$$U_1 = \mathbb{C}, \quad \psi_1(z) = z$$

$$U_2 = \widehat{\mathbb{C}} \setminus \{0\}, \quad \psi_2(z) = \begin{cases} 1/z \text{ if } z \neq \infty \\ \\ 0 \text{ if } z = \infty \end{cases}$$

Example 1.6 The complex projective line $\mathbb{P}^1 := \mathbb{P}^1(\mathbb{C})$ admits a Riemann surface structure with the following two charts:

$$U_0 = \{[z_0 : z_1] \text{ with } z_0 \neq 0\}, \ \phi_0([z_0 : z_1]) = \frac{z_1}{z_0}$$

$$U_1 = \{[z_0 : z_1] \text{ with } z_1 \neq 0\}, \ \phi_1([z_0 : z_1]) = \frac{z_0}{z_1}$$

The next examples are described in terms of algebraic (polynomial) equations. We shall refer to them as *curves*.

Example 1.7 (Hyperelliptic curves I) Consider first the algebraic equation $y^2 = \prod_{k=1}^{2g+1} (x - a_k)$, where $\{a_k\}_{k=1}^{2g+1}$ is a collection of $2g + 1$ distinct complex numbers, and let

$$\overset{\circ}{S} = \left\{ (x, y) \in \mathbb{C}^2 : y^2 = \prod_{k=1}^{2g+1} (x - a_k) \right\}$$

We shall now define a chart (U, φ) around each given point $P_0 = (x_0, y_0)$. As it is easier, we rather describe parametrizations:

- If $x_0 \neq a_i$ for all i (and so $y_0 \neq 0$), we take

$$\varphi^{-1}(z) = \left(z + x_0, \sqrt{\prod_{k=1}^{2g+1} (z + x_0 - a_k)} \right)$$

defined in the disc $\{|z| < \varepsilon\}$ with ε small enough for z not to reach any of the values a_i. The branch of the square root is chosen so that its value at $z = x_0$ equals y_0 (and not $-y_0$).
- For $P_0 = (a_j, 0)$, we take

$$\varphi_j^{-1}(z) = \left(z^2 + a_j, z\sqrt{\prod_{k \neq j}(z^2 + a_j - a_k)} \right), \ |z| < \varepsilon$$

again with ε small enough to guarantee that $z^2 + a_j$ does not reach a_k for every $k \neq j$. Note that if we had taken the first coordinate to be $z + a_j$ then the second one would not be a well-defined holomorphic function in $|z| < \varepsilon$. Also note that the choice of the branch of the square root is irrelevant. If we had defined a different parametrization $\widetilde{\varphi}_j^{-1}$ by using the opposite branch of the square root we would have parametrized the same subset of $\overset{\circ}{S}$ because of the identity $\widetilde{\varphi}_j^{-1}(z) = \varphi_j^{-1}(-z)$. A direct computation shows that $\varphi \circ \varphi_j^{-1}(z) = z^2 + a_j$.

One can give a simple argument to show that $\overset{\circ}{S}$ is connected. Whenever x describes a path joining x_0 to a_j, the map

$$
x \longmapsto \left(x, \sqrt{\prod_{i=1}^{2g+1} (x - a_k)} \right)
$$

where the root is determined by analytic continuation, describes a path in $\overset{\circ}{S}$ joining $\left(x_0, \sqrt{\prod_{k=1}^{2g+1} (x_0 - a_k)} \right)$ to $(a_j, 0)$. A precise definition of what we mean by the term *analytic continuation* is given in Section 2.9. It corresponds in this case to the simple idea of choosing the branch of the square root along the path in a way that makes the process continuous.

A compact surface S can be obtained out of $\overset{\circ}{S}$, in the same way we constructed the surface $\widehat{\mathbb{C}}$ by adding one abstract point to the complex plane $\mathbb{C}$, see Example 1.5. We also denote this additional point by ∞, and we define a coordinate neighbourhood as follows:

- A parametrization of a neighbourhood of $P_0 = \infty$ is given by

$$
\psi^{-1}(z) = \begin{cases} \left(\dfrac{1}{z^2}, \dfrac{1}{z^{2g+1}} \sqrt{\prod_{k=1}^{2g+1} (1 - a_k z^2)} \right) & \text{if } 0 < |z| < \varepsilon \\[4ex] \infty & \text{if } z = 0 \end{cases}
$$

This case is similar to that of φ_j in the sense that simply writing $1/z$ in the first coordinate would not work, and also because the choice of the branch of the square root is irrelevant. This way we have $\varphi \circ \psi^{-1}(z) = 1/z^2$, which is clearly a holomorphic function (its domain of definition does not contain $z = 0$). No computation is needed to check the compatibility of φ_j and ψ, since their domains of definition can be chosen to be disjoint sets.

Finally, we observe that the Riemann surface $S = \overset{\circ}{S} \cup \{\infty\}$ so constructed is compact, since we can decompose S as the union of two compact sets, namely

$$\left\{(x,y) \in \overset{\circ}{S} : \; |x| \leq 1/\varepsilon\right\} \cup \left(\left\{(x,y) \in \overset{\circ}{S} : \; |x| \geq 1/\varepsilon\right\} \cup \{\infty\}\right)$$

The first one is compact because it is a bounded closed subset of $\mathbb{C}^2$, and the second one because it agrees with $\psi^{-1}\left(\overline{\mathbb{D}(0, \sqrt{\varepsilon})}\right)$.

Example 1.8 (Hyperelliptic curves II) The compact surfaces constructed out of the algebraic curves $y^2 = \prod_{k=1}^{2g+2} (x - a_k)$ are also called hyperelliptic curves. The charts are defined as in the previous example, and only the compactification process is slightly different.

We can add a point ∞_1 with a parametrization

$$\psi_1^{-1}(z) = \begin{cases} \left(\dfrac{1}{z}, \dfrac{\sqrt{\prod_{k=1}^{2g+2}(1 - a_k z)}}{z^{g+1}}\right) & \text{if } 0 < |z| < \varepsilon \\[2em] \infty_1 & \text{if } z = 0 \end{cases}$$

The branch of the square root turns out to be relevant now. In fact, if in the expression above the symbol $\sqrt{w}$ denotes a given holomorphic choice of the square root in a neighbourhood of the point $w = 1$, then we can define a second mapping

$$\psi_2^{-1}(z) = \begin{cases} \left(\dfrac{1}{z}, \dfrac{-\sqrt{\prod_{k=1}^{2g+2}(1 - a_k z)}}{z^{g+1}}\right) & \text{if } 0 < |z| < \varepsilon \\[2em] \infty_2 & \text{if } z = 0 \end{cases}$$

to be the parametrization of a second abstract point which we have denoted by ∞_2. If we choose ε small enough in the definition of ψ_1 and ψ_2, both mappings parametrize disjoint sets, and this justifies that $\infty_2 \neq \infty_1$. In this respect the situation here is different from the previous example. There, the image of the punctured discs $\{0 < |z| < \varepsilon\}$ by the two possible parametrizations corresponding to the two choices of the square root would be the same open set U_ε, and so any two neighbourhoods $U_{\varepsilon_1} \cup \{\infty_1\}$ of ∞_1 and $U_{\varepsilon_2} \cup \{\infty_2\}$ of ∞_2 would have a non-empty intersection, hence the resulting space would not be Hausdorff, i.e. it would not even be a topological surface.

Remark 1.9 We anticipate that Examples 1.7 and 1.8 produce essentially the same collection of Riemann surfaces (see Example 1.83). These are called *hyperelliptic curves* when $g > 1$ and *elliptic curves* if $g = 1$.

Example 1.10 (Fermat curves) A similar technique produces a compact Riemann surface associated to the curve $x^d + y^d = 1$. Now we start with

$$\overset{\circ}{S} = \left\{ (x, y) \in \mathbb{C}^2 : x^d + y^d = 1 \right\}$$

and denote by ξ_d a chosen primitive root of unity of order d, as for example $\xi_d = e^{2\pi i/d}$. In what follows $\sqrt[d]{w}$ will stand for a determined holomorphic choice of the complex d-th root in a neighbourhood of the non-zero value in question in each case.

The Riemann surface structure in $\overset{\circ}{S}$ is given by the following charts:

- If $P_0 = \left(x_0, \xi_d^j \sqrt[d]{1 - x_0^d} \right)$ with $x_0 \neq \xi_d^k$ for $k = 1, \ldots, d$, we take

$$\varphi^{-1}(z) = \left(z, \xi_d^j \sqrt[d]{1 - z^d} \right)$$

which is defined in the disc $\{|z - x_0| < \varepsilon\}$.
- If $P_0 = \left(\xi_d^j, 0 \right)$, we take

$$\varphi_j^{-1}(z) = \left(z^d + \xi_d^j, z \sqrt[d]{- \prod \left(z^d + \xi_d^j - \xi_d^k \right)} \right)$$

where $|z| < \varepsilon$, and the product runs for $k \neq j$.

We can now compactify $\overset{\circ}{S}$, as in the previous examples, by adding d *points at infinity*, say $\infty_1, \ldots, \infty_d$:

- For $P_0 = \infty_j$, we take the parametrization

$$
\psi_j^{-1}(z) = \begin{cases} \left(\dfrac{1}{z}, \dfrac{\xi_d^j \sqrt[d]{(1 - z^d)}}{z} \right), & \text{if } |z| < \varepsilon \\[4mm] \infty_j, & \text{if } z = 0 \end{cases}
$$

Note again that if $|z| < 1$ then the mappings ψ_j^{-1} have disjoint images, hence the d points we have added to $\overset{\circ}{S}$ are in fact distinct.

Example 1.11 (p-gonal curves) Consider the algebraic curve

$$
y^p = (x - a_1)^{m_1} \cdots (x - a_r)^{m_r}
$$

where $1 \leq m_i < p$. Now the parametrizations that make this curve into a Riemann surface are

$$
\varphi^{-1}(z) = \left(z, \sqrt[p]{\prod_i (z - a_i)^{m_i}} \right)
$$

for a neighbourhood of a point (x, y) with $x \neq a_i$ and

$$
\varphi_i^{-1}(z) = \left(z^p + a_i, t^{m_i} \sqrt[p]{\prod_{j \neq i} (z^p + a_i - a_j)^{m_j}} \right)
$$

for a neighbourhood of $(a_i, 0)$.

Along the lines of the previous examples one can add a point at infinity if $\sum m_i$ is prime to p (or p points otherwise) to obtain a compact Riemann surface.

Throughout the text we will turn our attention several times (see Proposition 1.44 and Section 2.5.2) to a notorious Riemann surface of this type, which is the one associated to the algebraic equation $y^7 = x(x - 1)^2$. This is usually known as *Klein's curve of genus three*.

The next examples show how to construct some Riemann surfaces as quotient spaces.

Example 1.12 (The cylinder) $\mathbb{C}/\mathbb{Z}$ usually denotes the quotient set of the complex plane by the following equivalence relation. Two points in $\mathbb{C}$ are in the same class when they differ by an integer number. Thus, a point w is equivalent to all points $w + n$ with $n \in \mathbb{Z}$, i.e. to all its images by the translations $z \mapsto z + n$. The topology to be considered is, of course, the quotient topology.

If $U \subset \mathbb{C}$ is an open set such that no pair of its points belong to the same equivalence class, that is if the canonical projection $\pi : \mathbb{C} \to \mathbb{C}/\mathbb{Z}$ is injective when restricted to U, then we define a coordinate chart by $\varphi_U := (\pi_{|U})^{-1} : \pi(U) \to U$.

Suppose that two such coordinate neighbourhoods $\pi(U)$ and $\pi(V)$ have non-empty intersection. Let $P \in \pi(U) \cap \pi(V)$, that is $P = \pi(z_1) = \pi(z_2)$ with $z_1 \in U$ and $z_2 \in V$. Then we have $z_2 = z_1 + m$ for some $m \in \mathbb{Z}$. Therefore, in $\varphi_U(\pi(U) \cap \pi(V)) \subset U$, the transition function takes the form $\varphi_V \circ \varphi_U^{-1}(z) = z + m(z)$, with $m(z) \in \mathbb{Z}$. But $\varphi_V \circ \varphi_U^{-1}$ is a continuous function, hence $m(z)$ is locally constant. We thus see that the restriction of the transition functions to each connected component of their domain of definition is a translation by an integer m.

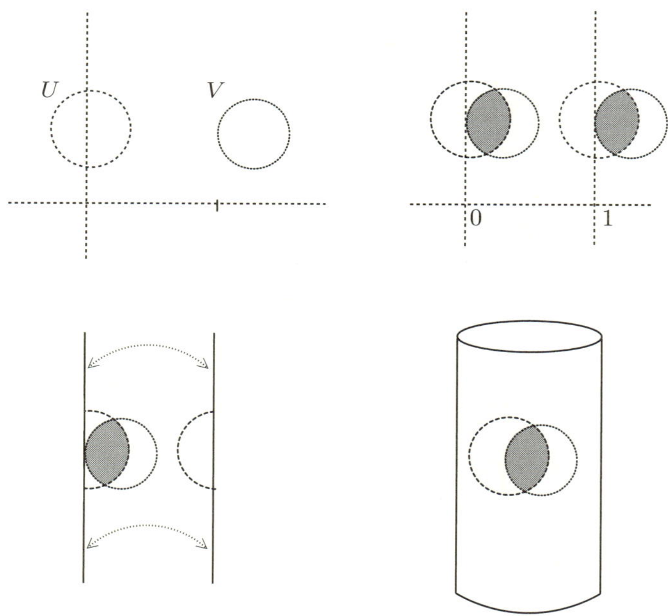

Fig. 1.2. $\mathbb{C}/\mathbb{Z}$ is topologically a cylinder.

The vertical strip $F = \{z \in \mathbb{C} : 0 \leq \operatorname{Re} z \leq 1\}$ is a *fundamental domain*, i.e. a subset of $\mathbb{C}$ containing at least one representative of every equivalence class and exactly one except in the boundary (cf. Definition 2.5). The surface $\mathbb{C}/\mathbb{Z}$ is obtained after glueing the two straight lines at the boundary of F. It is therefore a cylinder (see Figure 1.2).

Example 1.13 (A complex torus) As in the previous example, let us identify every $w \in \mathbb{C}$ with its images under all translations by Gaussian integers, that is complex numbers whose real and imaginary parts are both integer numbers. The classes for this equivalence relation are $[w] = \{w + n + mi, \ n, m \in \mathbb{Z}\}$. The corresponding quotient set

$$\mathbb{C}/\Lambda, \quad \text{where} \quad \Lambda = \mathbb{Z} \oplus \mathbb{Z}i$$

can be described as in the previous example. Arguing as above we see that the transition functions take in each connected component the form

$$\varphi_V \circ \varphi_U^{-1}(z) = z + \lambda, \ \lambda \in \Lambda$$

and the parallelogram in Figure 1.3 can be chosen as fundamental domain. Opposite sides are identified to form the quotient space, thus $\mathbb{C}/\Lambda$ is topologically a torus.

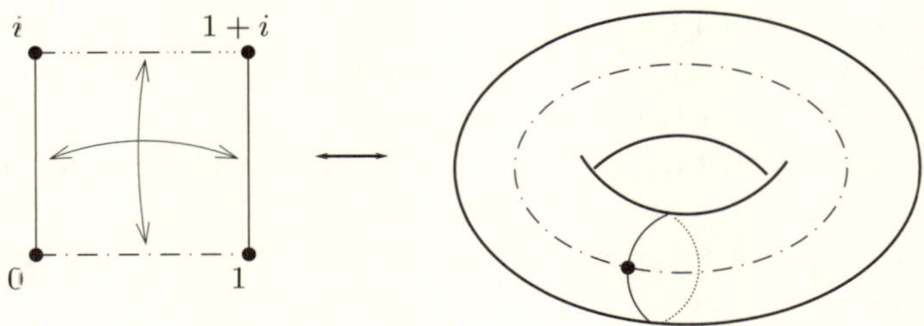

Fig. 1.3. $\mathbb{C}/\Lambda$ is topologically a torus.

Example 1.14 A Riemannian structure (i.e. a metric) on an orientable surface induces a Riemann surface structure whose charts (U_i, φ_i) are conformal bijections preserving orientation ([dC76]).

As the transition functions $\varphi_j \circ \varphi_i^{-1}$ must also be (orientation-preserving) conformal bijections between open subsets of $\mathbb{C}$, they are holomorphic ([Ahl78]).

1.1.2 Morphisms of Riemann surfaces

Complex analysis can be defined on a Riemann surface, the concept of holomorphy being the obvious one.

Definition 1.15 Let S be a Riemann surface and $f : S \longrightarrow \mathbb{C}$ a function. We say that f is *holomorphic* (resp. *meromorphic*) if, for any coordinate function φ, the function $f \circ \varphi^{-1}$ is holomorphic (resp. meromorphic) in the usual sense of complex analysis. The set of meromorphic functions on S is a field, which shall be denoted by $\mathcal{M}(S)$.

The same idea behind the previous definition, i.e. using local coordinates, can be applied to consider functions on a Riemann surface which take more general values than complex numbers. In fact, one can replace the target space $\mathbb{C}$ by any other Riemann surface.

Definition 1.16 A *morphism* between two Riemann surfaces S and S' is a continuous mapping $f : S \longrightarrow S'$ such that $\varphi' \circ f \circ \varphi^{-1}$ is a holomorphic function for every choice of coordinates φ in S and φ' in S' for which the composition makes sense. We will denote by $\mathrm{Mor}(S, S')$ the set of morphisms from S to S'.

Bijective morphisms are called *isomorphisms* and isomorphisms from a surface to itself are called *automorphisms*. The set of automorphisms of a given Riemann surface S forms a group. We shall denote it by $\mathrm{Aut}(S)$.

Remark 1.17 Let $f : S \longrightarrow S'$ be a non-constant morphism between connected compact Riemann surfaces. Since non-constant holomorphic maps are open maps, the image $f(S)$ must be simultaneously open and closed in S', and therefore equal to the whole S'. Thus, f is surjective.

Example 1.18 $\mathbb{H}$ and $\mathbb{D}$ are isomorphic Riemann surfaces via the map

$$\mathbb{H} \longrightarrow \mathbb{D}$$
$$z \longmapsto \frac{z - i}{z + i}$$

More generally, the Riemann mapping Theorem (see [Ahl78], [Con78]) states that any two simply connected open proper subsets of the plane are isomorphic Riemann surfaces.

Example 1.19 $\mathbb{C}$ and $\mathbb{D}$ are not isomorphic Riemann surfaces. In fact Liouville's theorem† ([Ahl78], [Con78]) shows that $\mathrm{Mor}(\mathbb{C}, \mathbb{D})$ consists only of constant maps, hence no isomorphism can exist between them. Nevertheless, $\mathbb{C}$ and $\mathbb{D}$ are diffeomorphic to each other by means of the diffeomorphism $\varphi(z) = \dfrac{z}{\sqrt{1 + \|z\|^2}}$, whose inverse is $\varphi^{-1}(w) = \dfrac{w}{\sqrt{1 - \|w\|^2}}$.

On the other hand $\mathbb{S}^2$ is compact, therefore it cannot be even homeomorphic to $\mathbb{C}$ or $\mathbb{D}$ (same comment applies to $\mathbb{P}^1$ and $\widehat{\mathbb{C}}$). In fact $\mathbb{S}^2, \mathbb{P}^1$ and $\widehat{\mathbb{C}}$ are all isomorphic to each other, the isomorphisms between them being

$$\mathbb{P}^1 \xrightarrow{\ f_1\ } \widehat{\mathbb{C}} \qquad\qquad \mathbb{S}^2 \xrightarrow{\ f_2\ } \widehat{\mathbb{C}}$$
$$[z_0 : z_1] \longmapsto \frac{z_1}{z_0} \qquad\qquad (x, y, t) \longmapsto \frac{x + iy}{1 - t}$$
$$[0 : 1] \longmapsto \infty \qquad\qquad (0, 0, 1) \longmapsto \infty$$

Straightforward computations show that the above mappings are indeed holomorphic. With the notation for the charts as in Examples 1.4, 1.5 and 1.6, we have for instance

$$\psi_1 \circ f_1 \circ \phi_1^{-1}(z) = \psi_1 \circ f_1 ([z : 1]) = \frac{1}{z}$$

or, if $z = \dfrac{x - iy}{1 + t}$,

$$\psi_1 \circ f_2 \circ \varphi_2^{-1}(z) = \psi_1 \circ f_2 (x, y, t) = \psi_1 \left(\frac{x + iy}{1 - t} \right) = \frac{1}{z}$$

---

† Liouville's theorem states that a *bounded* entire (i.e. holomorphic in the whole complex plane) function is necessarily constant.

Example 1.20 We now construct a morphism from a hyperelliptic curve to the sphere. Let S be the Riemann surface of the curve $y^2 = \prod_{k=1}^{2g+1} (x - a_k)$, as described in Example 1.7. Consider the map

$$S \xrightarrow{\ \mathbf{x}\ } \widehat{\mathbb{C}}$$
$$(x, y) \longmapsto x$$
$$\infty \longmapsto \infty$$

With the same notation as in Example 1.7, we can compute $\varphi_1 \circ \mathbf{x} \circ \varphi^{-1}(z) = z$. We also have $\varphi_1 \circ \mathbf{x} \circ \varphi_j^{-1}(z) = z^2 + a_j$ and $\varphi_2 \circ \mathbf{x} \circ \psi^{-1}(z) = z^2$. Therefore $\mathbf{x}$ is certainly holomorphic.

Example 1.21 Let now S be the Riemann surface of Fermat's curve $x^d + y^d = 1$, as described in Example 1.10. Consider the map

$$S \xrightarrow{\ \mathbf{x}\ } \widehat{\mathbb{C}}$$
$$(x, y) \longmapsto x$$
$$\infty_j \longmapsto \infty$$

Now the corresponding local expressions for $\mathbf{x}$ give functions of the form $z \longmapsto z$ or $z \longmapsto z^d + \zeta_d^j$. Thus $\mathbf{x}$ is holomorphic.

Example 1.22 (The complex conjugate Riemann surface) Given a Riemann surface S with atlas $\{(U_i, \varphi_i)\}$, we can construct a new Riemann surface which consists of the same underlying topological space endowed with the atlas $\{(U_i, \overline{\varphi}_i)\}$. In order to check that this is indeed a holomorphic atlas, note that $\overline{\varphi}_j \circ \overline{\varphi}_i^{-1}(z) = \overline{\varphi_j \circ \varphi_i^{-1}(\overline{z})}$. We will denote this new Riemann surface by $\overline{S}$ and refer to it as the *complex conjugate* of S. We make two observations:

(i) The identity map $\mathrm{Id} : S \longrightarrow \overline{S}$ is antiholomorphic, since clearly $\overline{\varphi}_i \circ \varphi_i^{-1}(z) = \overline{z}$. In general S and $\overline{S}$ will not be isomorphic.

(ii) If S and S' denote the Riemann surfaces of the hyperelliptic curves $y^2 = \prod(x - a_i)$ and $y^2 = \prod(x - \overline{a}_i)$ respectively, then

$S' \simeq \overline{S}$, an isomorphism being given by

$$S' \xrightarrow{\;\;F\;\;} \overline{S}$$
$$(x, y) \longmapsto (\overline{x}, \overline{y})$$

This is easily checked. For instance, near a point $P = (a, b) \in S'$ with $a \neq \overline{a}_i$ we have

$$\overline{\varphi}_i \circ F \circ \varphi_i^{-1}(x) \;=\; \overline{\varphi}_i \circ F\left(x + a, \sqrt{\prod(x + a - \overline{a}_i)}\right)$$

$$=\; \overline{\varphi}_i\left(\overline{x + a}, \overline{\sqrt{\prod(x + a - \overline{a}_i)}}\right)$$

$$=\; \overline{(x + a)} = x + a$$

Proposition 1.23 *Let us denote by $c_P \in \mathrm{Mor}(S, \widehat{\mathbb{C}})$ the constant morphism $c_P(x) = P$. Then*

$$\mathcal{M}(S) \equiv \mathrm{Mor}(S, \widehat{\mathbb{C}}) \smallsetminus \{c_\infty\}$$

Proof A morphism $f : S \to \widehat{\mathbb{C}}$ is clearly a holomorphic mapping outside the set $\{f^{-1}(\infty)\}$. At a point $P \in f^{-1}(\infty)$, choose a suitable coordinate function φ in S and let $\varphi_2(z) = 1/z$ be the chart around $\infty \in \widehat{\mathbb{C}}$. Then $\varphi_2 \circ f \circ \varphi^{-1}(z) = \dfrac{1}{f \circ \varphi^{-1}(z)} := h(z)$ for some holomorphic function h. Thus $f \circ \varphi^{-1}(z) = \dfrac{1}{h(z)}$, i.e. f is meromorphic.

Conversely, a meromorphic function f in S defines a continuous mapping $f : S \longrightarrow \widehat{\mathbb{C}}$ by simply sending the poles to $\infty \in \widehat{\mathbb{C}}$. By Riemann's removable singularity theorem† (see [Ahl78], [Con78]), the corresponding local expression of f is holomorphic in a small neighbourhood around each pole. Thus $f : S \longrightarrow \widehat{\mathbb{C}}$ is a morphism of Riemann surfaces. $\qquad\square$

Remark 1.24 The function $f : \mathbb{C} \smallsetminus \{0\} \to \mathbb{C}$ given by $f(z) = e^{1/z}$ does not extend to a morphism $f : \widehat{\mathbb{C}} \to \widehat{\mathbb{C}}$. The rule $0 \mapsto \infty$ would not produce a continuous function because there are values of z

---

† Riemann's removable singularity theorem states that a function that is holomorphic in a punctured disc $D \smallsetminus \{p\}$ and bounded in D extends to a holomorphic function in the whole D.

arbitrary close to zero (e.g. $z = -\dfrac{1}{n}$, $n \in \mathbb{N}$ and large) such that $f(z)$ stays away from ∞. This is nothing but a particular case of the Casorati–Weierstrass Theorem‡ (see [Con78], [Ahl78]).

Remark 1.25 Connected compact Riemann surfaces do not admit non-constant holomorphic funcions. Any given holomorphic function $f : S \longrightarrow \mathbb{C}$ is nothing more than a morphism from S to $\widehat{\mathbb{C}}$ that does not reach the value $\infty \in \widehat{\mathbb{C}}$ (Proposition 1.23). Now the statement follows from Remark 1.17.

Both the function field and the automorphism group are basic objects associated to a Riemann surface. It is not difficult to determine them in the case of the sphere $\mathbb{P}^1$.

Proposition 1.26 $\mathcal{M}(\mathbb{P}^1) = \mathbb{C}(z)$, *the field of rational functions in one variable.*

Proof Let f be a meromorphic function in $\mathbb{P}^1 = \widehat{\mathbb{C}}$ and suppose that $f(\infty) \neq \infty$ (if not, take $1/f$). Since the pole set of f is discrete and $\mathbb{P}^1$ is compact, there must be only finitely many poles, say $a_1, \ldots, a_n$. For each of them we can write locally

$$f(z) = \sum_{k=1}^{r_i} \frac{\lambda_k^i}{(z - a_i)^k} + h_i(z), \text{ with } h_i \text{ holomorphic at } a_i$$

Now, $f - \displaystyle\sum_{i=1}^{n} \sum_{k=1}^{r_i} \frac{\lambda_k^i}{(z - a_i)^k}$ is a holomorphic function in the whole $\widehat{\mathbb{C}}$, therefore it must be a constant function (Liouville's Theorem). It follows that f is a rational function. $\qquad\square$

Proposition 1.27 *The groups of automorphisms of* $\mathbb{P}^1$, $\mathbb{C}$, $\mathbb{D}$ *and*

---

‡ The weakest form of the Casorati–Weierstrass Theorem says that the image of any punctured neighbourhood of an essential singularity of a holomorphic function is dense in the whole complex plane $\mathbb{C}$.

$\mathbb{H}$ *are as follows:*

$$\mathrm{Aut}(\mathbb{P}^1) = \left\{ z \mapsto \frac{az+b}{cz+d}, \ a,b,c,d \in \mathbb{C}, \ ad-bc \neq 0 \right\}$$

$$= \mathbb{PSL}(2,\mathbb{C}), \ \textit{the group of Möbius transformations}$$

$$\mathrm{Aut}(\mathbb{C}) = \{ z \mapsto az+b, \ a,b \in \mathbb{C} \}$$

$$\mathrm{Aut}(\mathbb{D}) = \left\{ z \mapsto e^{i\theta} \frac{z-\alpha}{1-\overline{\alpha}z}, \ \alpha \in \mathbb{C}, |\alpha| < 1, \theta \in \mathbb{R} \right\}$$

$$= \left\{ z \mapsto \frac{\overline{a}z+\overline{b}}{bz+a}, \ a,b \in \mathbb{C}, |a|^2 - |b|^2 = 1 \right\}$$

$$\mathrm{Aut}(\mathbb{H}) = \left\{ z \mapsto \frac{az+b}{cz+d}, \ a,b,c,d \in \mathbb{R}, \ ad-bc = 1 \right\}$$

$$= \mathbb{PSL}(2,\mathbb{R})$$

Proof Let f be an automorphism of $\mathbb{P}^1 = \widehat{\mathbb{C}}$. By Proposition 1.23, f is a meromorphic function defined in $\mathbb{P}^1$, hence a rational function (Proposition 1.26). Let it be described as an irreducible fraction of the form

$$f(z) = \lambda \frac{(z-b_1) \cdots (z-b_d)}{(z-a_1) \cdots (z-a_n)}$$

For this fraction to define a bijective mapping the numerator and denominator cannot have a degree larger than 1. Thus f must have the form

$$f = \frac{az+b}{cz+d} =: M(z), \text{ with } M = \begin{pmatrix} a & b \\ c & d \end{pmatrix}$$

Note also that M must be invertible, since otherwise f would be constant. But then the matrix $\dfrac{1}{\sqrt{\det(M)}} M$ defines the same rational function as M, hence we can assume $\det(M) = 1$, i.e. $M \in \mathrm{SL}(2,\mathbb{C})$. Similarly, it is clear that $-M$ defines the same function as M, thus f is in fact determined by the class of M in $\mathbb{PSL}(2,\mathbb{C}) := \mathrm{SL}(2,\mathbb{C})/\langle \pm \mathrm{Id} \rangle$.

Now, if an automorphism f of $\mathbb{C}$ extends to (i.e. it is the restriction of) an automorphism of $\widehat{\mathbb{C}}$, then it is necessarily an affine mapping. But every automorphism of $\mathbb{C}$ does in fact extend, since the only possible obstacle would be ∞ being an essential singularity of f (instead of a removable singularity or a pole), but the Casorati–Weierstrass Theorem shows that this cannot happen. Indeed as f is bijective $f(\{|z| < 1\})$ and $f(\{|z| > 1\})$ would be disjoint open sets, hence the latter could not be a dense subset of $\mathbb{C}$.

Let now f be an automorphism of the disc such that $f(0) = \lambda$. Taking a Möbius transformation M that preserves $\mathbb{D}$ and satisfies $M(\lambda) = 0$ such as $M(z) = \dfrac{z - \lambda}{1 - \bar{\lambda}z}$ and applying Schwarz's Lemma† to the functions $h = M \circ f$ and h^{-1} we deduce first that $|h(z)| = |z|$ for all $z \in \mathbb{D}$ and second that $M \circ f(z) = e^{i\theta}z$ for some real number θ. It is enough now to compose with M^{-1} to obtain

$$f(z) = \frac{e^{i\theta}z + \lambda}{1 + e^{i\theta}\bar{\lambda}z} = e^{i\theta}\frac{z + e^{-i\theta}\lambda}{1 + e^{i\theta}\bar{\lambda}z}$$

which is the expression we were looking for if we set $\alpha = -e^{-i\theta}\lambda$. In order to obtain the second description of $\mathrm{Aut}(\mathbb{D})$, one simply has to write $e^{i\theta} = \dfrac{e^{i\theta/2}}{e^{-i\theta/2}}$ and then divide both numerator and denominator by the real number $\sqrt{1 - |\lambda|^2}$.

Finally, it is clear that the elements in $\mathbb{PSL}(2,\mathbb{R})$ are automorphisms of $\mathbb{H}$, since they are the restriction of automorphisms of $\widehat{\mathbb{C}}$ preserving $\mathbb{H}$.

Conversely, suppose $f \in \mathrm{Aut}(\mathbb{H})$, and let T be the isomorphism from $\mathbb{H}$ to $\mathbb{D}$ given by $z \mapsto \dfrac{z - i}{z + i}$. Since $T \circ f \circ T^{-1}$ is an automorphism of $\mathbb{D}$, it follows that f is a Möbius transformation. Now, the real line must be preserved by f, hence $f \in \mathbb{PSL}(2,\mathbb{R})$. $\qquad\square$

Remark 1.28 Clearly, given two triples of points $\{z_1, z_2, z_3\}$ and $\{w_1, w_2, w_3\}$ one can always find coefficients a, b, c, d such that the Möbius transformation $M(z) = \frac{az+b}{cz+d}$ sends z_i to w_i. In other words, $\mathbb{PSL}(2,\mathbb{C})$ acts transitively on triples of points of $\mathbb{P}^1$.

---

† Schwarz's Lemma says that a function f analytic in the unit disc $\mathbb{D}$ such that $f(0) = 0$ and $|f(z)| \leq 1$ for all $z \in \mathbb{D}$ verifies $|f'(0)| \leq 1$ and $|f(z)| \leq |z|$ for all $z \in \mathbb{D}$. Moreover, if $|f'(0)| = 1$ or if $|f(w)| = |w|$ for some $w \neq 0$, then there is a constant c with $|c| = 1$ such that $f(z) = c \cdot z$ for all $z \in \mathbb{D}$.

Remark 1.29 The Riemann surfaces in Examples 1.12 and 1.13 were defined by identifying all the possible images of the points in $\mathbb{C}$ by subgroups of its full group of automorphisms. Later on we will consider surfaces defined as quotients of $\mathbb{D}$ or $\mathbb{H}$.

Definition 1.30 Let f be a meromorphic function in a Riemann surface S and φ a chart around P such that $\varphi(P) = 0$. Let

$$f \circ \varphi^{-1}(z) = a_n z^n + a_{n+1} z^{n+1} + \cdots \ , \text{ with } a_n \neq 0$$

be the Laurent expansion of $f \circ \varphi^{-1}$ near $z = 0$. The integer n is called the *order* of f at P and denoted $\mathrm{ord}_P(f)$.

Clearly this definition would not make sense in the case where it depended on the choice of a chart. Let $\widetilde{\varphi}$ be another chart centred at P so that

$$f \circ \widetilde{\varphi}^{-1}(z) = b_m z^m + b_{m+1} z^{m+1} + \cdots \ , \text{ with } b_m \neq 0.$$

Since the transition functions are biholomorphic, we can write $\varphi \circ \widetilde{\varphi}^{-1}(z) = cz + \cdots$ with $c \neq 0$. Hence the identity

$$f \circ \widetilde{\varphi}^{-1} = \left(f \circ \varphi^{-1} \right) \circ \left(\varphi \circ \widetilde{\varphi}^{-1} \right)$$

can be written as

$$b_m z^m + b_{m+1} z^{m+1} + \cdots = a_n \left(cz + \cdots \right)^n + a_{n+1} \left(cz + \cdots \right)^{n+1} + \cdots$$

and it follows that $m = n$. The choice of the chart is therefore irrelevant.

Notice that $\mathrm{ord}_P(f) > 0$ (resp. $\mathrm{ord}_P(f) < 0$) means that P is a zero of f (resp. a pole).

Definition 1.31 Let $f : S_1 \longrightarrow S_2$ be a morphism of Riemann surfaces, $P \in S_1$ and $Q = f(P)$. Let ψ be a chart centred at $Q = f(P)$. The positive integer

$$m_P(f) := \mathrm{ord}_P(\psi \circ f)$$

is called the *multiplicity* of f at P. Equivalently,

$$m_P(f) = 1 + \mathrm{ord}_P(\psi \circ f)'$$

whether ψ is centred at Q or not.

When $m_P(f) \geq 2$ we say that $P \in S_1$ is a *branch point* (or *ramification point*) with branching order $m_P(f)$. We will distinguish this term from the term *branch value*, which we reserve for the image $Q = f(P) \in S_2$ of a branch point. Accordingly, morphisms with a non-empty set of branch values are called *ramified*.

The multiplicity of a morphism at a point is independent of the choice of charts. The proof of this fact goes exactly as the proof of the same statement for orders of functions (see above). We observe that the sets of branch points and branch values are discrete, therefore finite for compact Riemann surfaces. This is a consequence of the fact that zeros of holomorphic functions are isolated, since $m_P(f) > 1$ is equivalent to saying that the derivative of $(\psi \circ f \circ \varphi)$ vanishes at $\varphi(P)$.

The same arguments show also that given a point $P \in S$ where the meromorphic function $f : S \longrightarrow \widehat{\mathbb{C}}$ does not ramify, f (or $1/f$ in case P is a simple pole) can be chosen as a coordinate chart around P.

Example 1.32 Consider again the morphism

$$S \longrightarrow \widehat{\mathbb{C}}$$
$$(x, y) \longmapsto x$$
$$\infty \longmapsto \infty$$

which we studied in Example 1.20, where S is the Riemann surface of the curve $y^2 = \prod_{k=1}^{2g+1} (x - a_k)$ (Example 1.7). This is often called the *coordinate function* $\mathbf{x}$. Accordingly, there is an obviously defined coordinate function $\mathbf{y}$.

From the local expressions

$$\begin{aligned}
\varphi_1 \circ \mathbf{x} \circ \varphi^{-1}(z) &= z \\
\varphi_1 \circ \mathbf{x} \circ \varphi_j^{-1}(z) &= z^2 + a_j \\
\varphi_2 \circ \mathbf{x} \circ \psi^{-1}(z) &= z^2
\end{aligned}$$

it follows that $\mathbf{x}$ has branch points at $(a_1, 0), \ldots, (a_{2g+1}, 0), \infty$. The branching order at these points equals 2, and the corresponding branch values are $a_1, \ldots, a_{2g+1}, \infty$.

As a meromorphic function, $\mathbf{x}$ has a single pole of multiplicity 2 at ∞. Its zeros are located at the points $Q^\pm = \left(0, \pm\sqrt{\prod(-a_i)}\right)$, and both are simple.

As for the function $\mathbf{y}$, similar computations show that all the zeros are simple and located at the $2g+1$ points $(a_1, 0), \ldots, (a_{2g+1}, 0)$

whereas ∞ is the unique pole (of multiplicity $2g+1$). The order of the zeros and poles follow from the computations

$$\mathbf{y} \circ \varphi_i^{-1}(z) = \mathbf{y}\left(z^2 + a_i, z\sqrt{\prod_{k \neq i}\left(z^2 + a_i - a_k\right)}\right)$$

$$= z\sqrt{\prod_{k \neq i}\left(z^2 + a_i - a_k\right)}$$

and

$$\mathbf{y} \circ \psi^{-1}(z) = \mathbf{y}\left(\frac{1}{z^2}, \frac{1}{z^{2g+1}}\sqrt{\prod_{k=1}^{2g+1}\left(1 - a_k z^2\right)}\right)$$

$$= \frac{1}{z^{2g+1}}\sqrt{\prod_{k=1}^{2g+1}\left(1 - a_k z^2\right)}$$

respectively.

Example 1.33 Let S_1 an S_2 be the compact Riemann surfaces associated to the hyperelliptic curves of equation $y^2 = x^8 - 1$ and $y^2 = x^5 - x$ respectively. The map

$$S_1 \xrightarrow{\ f\ } S_2$$
$$(x, y) \longmapsto (x^2, xy)$$
$$\infty_1 \longmapsto \infty$$
$$\infty_2 \longmapsto \infty$$

is an *unramified* morphism, as can be shown by computation of the different local expressions of f with respect to the coordinates given in Examples 1.7 and 1.8. These computations are similar to those of the previous example, and we leave them as an exercise. We will come back to this morphism in Section 1.2.6, where the absence of ramification points will be shown in a different way.

1.1.3 Differentials

Let $\{(U_i, \varphi_i)\}_{i \in I}$ be the collection of charts (the atlas) that provides the Riemann surface structure in S. For $P \in U_i \cap U_j \subset S$, we write

$$\frac{dz_i}{dz_j}(P) := \left(\varphi_i \circ \varphi_j^{-1}\right)' (\varphi_j(P))$$

Definition 1.34 A *meromorphic differential* (or simply, a *differential*) ω is a collection of meromorphic functions $\{f_i : U_i \to \mathbb{C}\}$ such that $f_j = f_i \dfrac{dz_i}{dz_j}$ in the intersection $U_i \cap U_j$. This is usually written as $\omega_{|U_i} = f_i dz_i$, and accordingly the compatibility condition verified by the functions $\{f_i\}$ is expressed in the form $f_i dz_i = f_j dz_j$. If the functions f_i are holomorphic, ω is called *holomorphic*.

Note that the sum of two differentials and the product by a number is again a differential. In fact, the product $f\omega$ of a meromorphic function f by a differential ω is also a differential.

Definition 1.35 Let f be a meromorphic function. We define the meromorphic differential df as the collection

$$df = \left\{ \frac{\partial f}{\partial z_i}(P) := \left(f \circ \varphi_i^{-1}\right)' (\varphi_i(P)) \right\}$$

Proposition 1.36 *Given a meromorphic function f and a meromorphic differential ω, there exists another meromorphic function h such that $\omega = h \cdot df$.*

Proof We observe first that the quotient of two meromorphic differentials ω_1 and ω_2 defines a meromorphic function h simply by setting $h_{|U_i} = \dfrac{f_i^1}{f_i^2}$, where $\omega_1|_{U_i} = f_i^1 dz_i$ and $\omega_2|_{U_i} = f_i^2 dz_i$. If (U_j, φ_j) is another chart and $P \in U_i \cap U_j$ we have

$$h_{|U_j}(P) = \frac{f_j^1(P)}{f_j^2(P)} = \frac{f_i^1(P)\frac{dz_i}{dz_j}(P)}{f_i^2(p)\frac{dz_i}{dz_j}(P)} = \frac{f_i^1(P)}{f_i^2(P)} = h_{|U_i}(P)$$

and therefore h is well defined. The statement follows by taking $\omega_2 = df$ in the above argument. $\qquad\square$

Definition 1.37 Let ω be a meromorphic differential in S. We say that $P \in S$ is a *zero* or a *pole* of ω if it is a zero or a pole of f_i, where $P \in U_i$ and $\omega|_{U_i} = f_i dz_i$. In this case, we define the *order* of ω at P as $\mathrm{ord}_P \omega := \mathrm{ord}_P f_i$.

By Definition 1.34 these concepts are independent of the choice of charts.

Example 1.38 ($\mathbb{P}^1$ has no holomorphic differentials) Write $\mathbb{P}^1 \equiv \widehat{\mathbb{C}}$ and consider the identity function x and the corresponding differential $\omega = dx$. Clearly ω has neither zeros nor poles at points different from ∞. At ∞ a local computation gives

$$\frac{d(x \circ \psi_2^{-1}(z))}{dz} = \frac{d(1/z)}{dz} = -\frac{1}{z^2}$$

Thus, ω has a double pole at $P = \infty$. However, by Proposition 1.36 and Proposition 1.26 any meromorphic differential can be written as $\eta = R(x)dx$ for some rational function $R(x)$. Now, in order for η to be holomorphic, $R(x)$ cannot have poles, hence must be constant and, moreover, has to have a zero of order ≥ 2 at ∞. Impossible.

Example 1.39 (Holomorphic differentials on a torus) Let $S = \dfrac{\mathbb{C}}{\mathbb{Z} \oplus \mathbb{Z}i}$ be the Riemann surface introduced in Example 1.13.

We construct a differential on S simply by declaring

$$\omega_{|U_i} = f_i dz_i = 1 \cdot dz_i$$

To check that this assignment produces a differential, one simply has to recall that $\varphi_j \circ \varphi_i^{-1}(z) = z + m + ni$ with $m, n \in \mathbb{Z}$. Hence

$$f_j = 1 = 1 \cdot \frac{dz_i}{dz_j} = f_i \frac{dz_i}{dz_j}$$

Observe that this differential has neither zeros nor poles. It is usually denoted $\omega = dz$, but not in the sense of Definition 1.35 since $f([z]) = z$ does not define a function in S.

Example 1.40 (Holomorphic differentials on hyperelliptic curves) Let S be the Riemann surface associated to the curve

$$y^2 = (x - a_1) \cdots (x - a_{2g+1})$$

whose atlas was described in Example 1.7.

Consider first the differential $d\mathbf{x}$. Since the coordinate function $\mathbf{x}$ (Example 1.32) can be used as local coordinate at all points $P \neq (a_i, 0)$, $i = 1, \ldots, 2g+1$ and $P \neq \infty$ we see that $d\mathbf{x}$ has neither zeros nor poles at these points.

For $P = (a_i, 0)$ we have $d(\mathbf{x} \circ \varphi_i^{-1}(z)) = d(a_i + z^2) = 2z\,dz$, and for $P = \infty$ we have $d(\mathbf{x} \circ \psi^{-1}(z)) = d\left(\dfrac{1}{z^2}\right) = -\dfrac{2}{z^3}dz$. We conclude that $d\mathbf{x}$ has simple zeros at the points $P_i = (a_i, 0)$ and a pole of multiplicity 3 at ∞.

Now, to construct holomorphic differentials we also make use of the coordinate function $\mathbf{y}$ (see Example 1.32). We see that the only zero of the differential $\omega_1 = \dfrac{d\mathbf{x}}{\mathbf{y}}$ is ∞, and its order equals $2g+1-3 = 2g-2$. In turn, this implies that $\omega_2 = \dfrac{\mathbf{x} \cdot d\mathbf{x}}{\mathbf{y}}$ has two simple zeros at the points $Q^{\pm} = \left(0, \pm\sqrt{\prod(-a_i)}\right)$ and one zero of order $2g-2-2 = 2g-4$ at ∞. Proceeding in this way we see that

$$\omega_i = \frac{\mathbf{x}^{i-1} \cdot d\mathbf{x}}{\mathbf{y}} \quad (i = 1, \ldots, g)$$

is a holomorphic differential with two zeros of order $i-1$ at the points $Q^{\pm}$ and one zero of order $2(g-i)$ at the point ∞.

Example 1.41 (Holomorphic differentials on the Fermat curves) Let S be the Riemann surface associated to the curve $x^d + y^d = 1$ (see Example 1.10).

Arguing as in Example 1.40 one easily checks that in this case the differentials

$$\omega_{ij} = \frac{\mathbf{x}^i \cdot d\mathbf{x}}{\mathbf{y}^j}; \quad 2 \le j \le d-1,\ 0 \le i \le j-2$$

are holomorphic. To do that one first notes that now the function $\mathbf{x}$ (resp. $\mathbf{y}$) has d simple zeros at the points $P_k = (0, \xi_d^k)$ (resp. $Q_k = (\xi_d^k, 0)$) and d simple poles at the points ∞_k. Accordingly $d\mathbf{x}$ has zeros of order $d-1$ at the points Q_k and double poles at the points ∞_k.

Example 1.42 (Holomorphic differentials on Klein's curve) Let S be the Riemann surface associated to the algebraic curve $y^7 = x(x-1)^2$ considered in Example 1.11.

Clearly the differential $d\mathbf{x}$ (resp. the function $\mathbf{y}$) has a zero of order 6 (resp. a simple zero) at the point $P = (0,0)$, another zero of order 6 (resp. order 2) at the point $Q = (1,0)$ and a pole of multiplicity 8 (resp. multiplicity 3) at the point ∞. It follows that the differentials

$$\omega_1 = \frac{d\mathbf{x}}{\mathbf{y}^3}, \quad \omega_2 = \frac{(\mathbf{x}-1)d\mathbf{x}}{\mathbf{y}^5}, \quad \omega_3 = \frac{(1-\mathbf{x})d\mathbf{x}}{\mathbf{y}^6}$$

are holomorphic. Moreover, adding on the information that $(\mathbf{x}-1)$ has a zero at Q and a pole at ∞, both of order 7, we see that the zeros of these differentials are as follows:

- ω_1 vanishes at P with multiplicity 3 and ∞ with multiplicity 1.
- ω_2 vanishes at P with multiplicity 1 and Q with multiplicity 3.
- ω_3 vanishes at Q with multiplicity 1 and ∞ with multiplicity 3.

Remark 1.43 The reader should notice that for the differentials we have constructed in this section the following facts hold:

(i) For all differentials on the same Riemann surface S the number of poles minus the number of zeros, counting multiplicities, is a constant integer. This number agrees with the Euler–Poincaré characteristic $\chi = \chi(S)$ of the surface, to be introduced in the next section.

(ii) The holomorphic differentials constructed in each of Examples 1.38–1.42 are $\mathbb{C}$-linearly independent and its number equals $g = g(S) = \dfrac{2 - \chi(S)}{2}$, the genus of the surface, which is also introduced in the next section.

A deep result in the theory of Riemann surfaces, the Riemann–Roch Theorem, implies that (i) always occurs, that the dimension of the vector space of holomorphic differentials always equals $g(S)$ and that the g elements of any basis cannot have common zeros. This theorem will not be needed in this book. However, the last statement leads to the concepts of *canonical map* and *canonical curve*, which we next describe in the particular case of the Riemann surface associated to the curve $S = \{y^7 = x(x-1)^2\}$.

For $i = 1, 2, 3$, let $\omega_i = f_i dz_i$ represent the local expressions of the three differentials of Example 1.42 in a local chart (U_i, φ_i).

Then we can define a map

$$\Phi_i : U_i \longrightarrow \mathbb{C}^3$$

by the rule $\Phi_i(P) = (f_1(P), f_2(P), f_3(P))$.

Suppose that $P \in S$ lies in the intersection of two charts U_i and U_j. Then, by the very definition of a differential, we have

$$\Phi_j(P) = \frac{dz_i}{dz_j}(P)\Phi_i(P)$$

Now, since the three differentials have no common zeros and $\frac{dz_i}{dz_j}(P) \neq 0$, the local maps Φ_i glue together to define a global holomorphic map into the complex projective plane

$$\Phi : S \longrightarrow \mathbb{P}^2$$

By Remark 1.43 this is a general fact. The map

$$\Phi : S \longrightarrow \mathbb{P}^{g-1}$$
$$P \longmapsto (\omega_1(P) : \cdots : \omega_g(P))$$

is called the *canonical map* and its image the *canonical curve*.

The canonical curve of the Riemann surface associated to the curve $y^7 = x(x-1)^2$ is particularly beautiful. It is the projective plane curve defined by the homogeneus equation

$$X_0^3 X_1 + X_1^3 X_2 + X_2^3 X_0 = 0$$

This is a very much studied curve known as *Klein's quartic*, see [Kle78], [Lev99].

Proposition 1.44 *The canonical curve of the Riemann surface of $y^7 = x(x-1)^2$ is Klein's quartic.*

Proof Let $\Phi = (\omega_1, \omega_2, \omega_3)$ be the canonical map, where $\omega_1, \omega_2, \omega_3$ are the differentials defined in Example 1.42.

A simple calculation, using the relation $y^7 = x(x-1)^2$, shows that the expression

$$\left(\frac{1}{y^3}\right)^3 \left(\frac{x-1}{y^5}\right) + \left(\frac{x-1}{y^5}\right)^3 \left(\frac{1-x}{y^6}\right) + \left(\frac{1-x}{y^6}\right)^3 \frac{1}{y^3}$$

reduces to

$$\frac{1}{x^2(x-1)^3} - \frac{1}{x^3(x-1)^2} - \frac{1}{x^3(x-1)^3} = 0$$

This shows that $\mathrm{Im}(\Phi)$ is contained in Klein's quartic. Note that if $P = (0,0)$ and $Q = (1,0)$ the information provided in Example 1.42 yields $\Phi(P) = (0 : 0 : 1)$, $\Phi(Q) = (1 : 0 : 0)$ and $\Phi(\infty) = (0 : 1 : 0)$. To conclude that $\mathrm{Im}(\Phi)$ agrees with Klein's quartic we observe that in fact Φ is a bijection with inverse

$$\Phi^{-1}(x_0 : x_1 : x_2) = \left(\frac{x_1^3}{x_2^2 x_0} + 1, -\frac{x_1}{x_2} \right)$$

$\square$

Remark 1.45 Actually Proposition 1.44 shows in practice that in this case the canonical map is an isomorphism of Riemann surfaces. To make the proof complete we only need to specify what the Riemann surface structure on Klein's quartic is. This is done in the next example.

Example 1.46 (Non-singular plane curves) If $F(X_0, X_1, X_2)$ is a homogeneus polynomial of degree d such that the gradient $(F_{X_0}, F_{X_1}, F_{X_2})$ is non-zero at all points of the (projective) curve

$$S = \{P \in \mathbb{P}^2 \mid F(P) = 0\}$$

then S can be endowed with a Riemann surface structure as follows.

As usual, let us denote by $U_i = \{(x_0 : x_1 : x_2) \mid x_i \neq 0\}$ $(i = 0, 1, 2)$ the three standard open sets covering $\mathbb{P}^2$.

Let $P = (1 : x : y)$ be an arbitrary point in $S \cap U_0$ so that (x, y) is a zero of the polynomial $F_*(X_1, X_2) = F(1, X_1, X_2)$. Now, if

$$F_{X_2}(1, x, y) = (F_*)_{X_2}(x, y) \neq 0$$

(resp. if $F_{X_1}(1, x, y) = (F_*)_{X_1}(x, y) \neq 0$) we can use the Inverse Function Theorem to construct a parametrization near P of the form $x \longmapsto (1 : x : X_2(x))$ (resp. $y \longmapsto (1 : X_1(y) : y)$). The point is that $F_{X_1}(P) = 0$ and $F_{X_2}(P) = 0$ cannot occur simultaneously for a point $P \in S$, since in that case the Euler formula

$$X_0 F_{X_0} + X_1 F_{X_1} + X_2 F_{X_2} = d \cdot F$$

would imply that also $F_{X_0}(P) = 0$, which is a contradiction.

Of course, a construction of a parametrization around a point P in $S \cap U_1$ or $S \cap U_2$ can be achieved in a similar way.

1.2 Topology of Riemann surfaces

1.2.1 The topological surface underlying a compact Riemann surface

An orientation on a (differentiable) surface is given by an atlas such that the jacobian of the transition functions is positive everywhere.

Let γ be a piecewise differentiable Jordan curve encircling a point p in a surface X, that is $\gamma = \partial\Omega$, where Ω is diffeomorphic to a disc around p and $\partial\Omega$ stands for its boundary. Then an orientation on X induces an orientation (the positive orientation) on γ defined as follows. Let $q \in \partial\Omega$ and (U, φ) represent a small chart centreed at q (i.e. $\varphi(U) = \mathbb{D}$ and $\varphi(q) = 0$) such that $\varphi(U \cap \overline{\Omega}) = \overline{\mathbb{H}} \cap \mathbb{D}$. The positive orientation is the one that travels through $U \cap \partial\Omega$ from $\varphi^{-1}(-\epsilon)$ to $\varphi^{-1}(\epsilon)$. This notion does not depend on the choice of chart within the oriented atlas, as shown in Lemma 1.47 (see [BT82]). Thus an orientation on X determines a notion of positive rotation around any point p. Actually, when X is not assumed to be differentiable but merely a topological surface, the existence of a globally defined notion of positive rotation is usually taken as a definition of orientability.

Lemma 1.47 *Let $G : \mathbb{D} \to \mathbb{D}$ be a diffeomorphism whose jacobian is positive everywhere and such that $G(\mathbb{D} \cap \overline{\mathbb{H}}) = \mathbb{D} \cap \overline{\mathbb{H}}$. Then $G = (G_1, G_2)$ induces by restriction a diffeomorphism of $(-1, 1)$ $g(x) = G_1(x, 0)$ with a positive derivative at every point.*

Proof Write $G = (G_1, G_2)$. The function $x \longmapsto G_2(x, 0)$ vanishes identically by hypothesis, therefore $\partial_x G_2(x, 0) = 0$.

Now $y > 0$ implies $G_2(x, y) > 0$, thus $y \longmapsto G_2(x, y)$ is an increasing function near 0, therefore $\partial_y G_2(x, 0) > 0$. Hence

$$0 < \det DG(x, 0) = \begin{vmatrix} \partial_x G_1(x, 0) & \partial_y G_1(x, 0) \\ 0 & \partial_y G_2(x, 0) \end{vmatrix}$$

and therefore $g'(x) = \partial_x G_1(x, 0) > 0$. $\qquad\square$

Riemann surfaces come naturally equipped with an orientation. The reason is that if we write the transition functions in the form $\psi \circ \varphi^{-1}(x,y) = (u(x,y), v(x,y))$, the Cauchy–Riemann equations imply that

$$\left| \frac{\partial \left(\varphi_i \circ \varphi_j^{-1} \right)}{\partial (x,y)} \right| = \begin{vmatrix} u_x & u_y \\ v_x & v_y \end{vmatrix} = \begin{vmatrix} u_x & -v_x \\ v_x & u_x \end{vmatrix} = u_x^2 + v_x^2 > 0$$

A well-known theorem by Radó (see [Jos02] for a modern proof) states that all compact surfaces can be triangulated, that is X equals the union of subsets homeomorphic to triangles (or, more generally, polygons with an arbitrary number of sides) such that the intersection of two such triangles is either empty, one vertex or one edge. Since the proof of this result will not be included here, from now on by the term compact surface we shall mean a *triangulable* compact surface. At any rate, it will be a consequence of the existence of non-constant meromorphic functions (see Section 2.2) that compact Riemann surfaces can be triangulated.

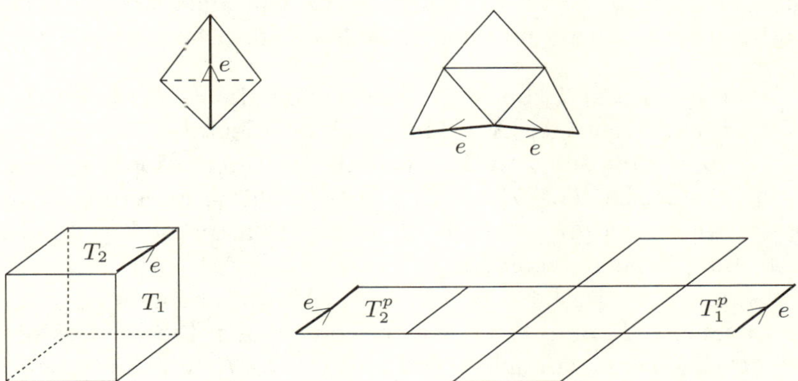

Fig. 1.4. A tetrahedron and a cube represented as a planar polygon with identified sides.

Using a triangulation, the topological space underlying a surface X can be described as a planar polygon R with a suitable equivalence relation on the boundary. This polygon is constructed by mapping homeomorphicaly a given triangle T to a convex planar triangle T^p and then extending this map to the neighbouring faces. The resulting polygon R gives back the surface X if we identify two edges whenever they come from the same edge of the triangulation in X. This is similar to the familiar process of cutting a polyhedron along certain edges to produce a planar figure. Prescribing the edge identifications (glueing instructions) on such a planar figure is needed to build the polyhedron back in three dimensional space (see Figure 1.4, where the two labels e indicate that the corresponding edges are identified).

Remark 1.48 The polygonal representation of a compact surface makes it clear that all compact surfaces can be endowed with a differentiable structure. If x lies in the boundary of R and it is not a vertex, one can take as chart an open set consisting of two small half discs around x, one for each of the two edges of R in which x lies (see Figure 1.6). If x is a vertex, the chart is made up of several circle sectors centred at the points equivalent to x. Conversely, if X is a compact differentiable surface, the same cut and paste process applied to a differentiable triangulation will lead us to a normalized polygon as above which represents a surface diffeomorphic to S. So from now on we will make no distinction between topological and differentiable compact surfaces.

Let now T_1 and T_2 be two adjacent triangles of a triangulation of an oriented surface X. The positive orientation in ∂T_1 and ∂T_2 induce opposite travel directions (or orientations) on their intersection edge $T_1 \cap T_2$. This will turn out to be an important observation regarding the glueing instructions needed to build X from the planar polygon R.

Assume that we have cut X open along a common edge e of two triangles T_1 and T_2. Let us denote by u, v the vertices of e, by T_i^p the planar triangle corresponding to T_i, and by e_i (resp. u_i, v_i) the edge of T_i^p corresponding to e (resp. the vertices of T_i^p corresponding to u_i, v_i). The orientation of the triangle T_i^p induces a preferred travel direction on e_i. By the observation above, if when travelling counterclockwise along ∂R we pass e_1 in the direction $u_1 \rightarrow v_1$, say, then we travel e_2 in the opposite direction $v_2 \rightarrow u_2$. This is why the edges e_1 and e_2 of ∂R will be denoted e and e^{-1}.

The $2n$ (pairwise identified) directed edges of a planar polygon R constructed from a triangulated orientable surface will be given labels a_i, b_i or a_i^{-1}, b_i^{-1} as described above. The above comment means that if we write down the sequence of labels we encounter when travelling, say, counterclockwise along the whole boundary ∂R, both symbols a_i, a_i^{-1} (resp. b_i, b_i^{-1}) must occur in the resulting *word* w. In other words, we never find a pair of (counterclockwise) directed edges labelled both with the same label, as it would happen in a surface containing a *Möbius strip* (Figure 1.5), the simplest example of a non-orientable surface.

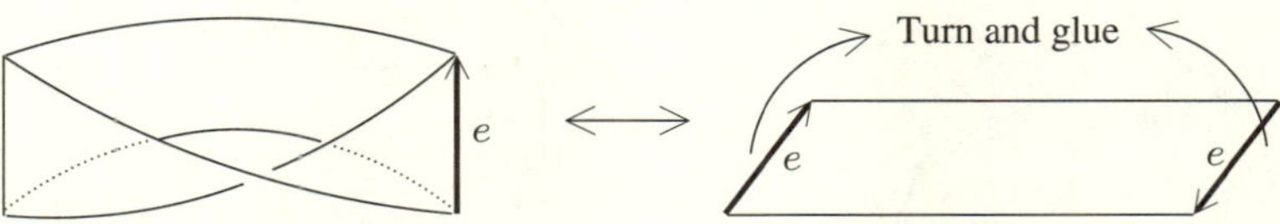

Fig. 1.5. A Möbius strip does not admit a Riemann surface structure.

Example 1.49 (Charts on the cylinder and on the Möbius strip) Consider the cylinder and the Möbius strip defined as a rectangle $[0,1] \times (-a, a)$ with the usual identification of sides (Figure 1.6). Then $\varphi = \mathrm{Id}$ serves as a chart around any interior point of the rectangle in both cases.

However, if D_1 and D_2 are the semidiscs indicated in Figure 1.6, an obvious chart around a point P lying in the rectangle's edge is

$$\varphi_1(x, y) = \begin{cases} (x, y) & \text{in } D_1 \\ (x - 1, y) & \text{in } D_2 \end{cases}$$

in the case of the cylinder, and

$$\psi_1(x, y) = \begin{cases} (x, y) & \text{in } D_1 \\ (x - 1, -y) & \text{in } D_2 \end{cases}$$

in the case of the Möbius strip. Note that in the second case we have $|\mathrm{Jac}(\psi_1 \circ \varphi^{-1})| > 0$ in D_1 and $|\mathrm{Jac}(\psi_1 \circ \varphi^{-1})| < 0$ in D_2, reflecting the non-orientability of the Möbius strip.

A fundamental fact is that a given polygon R representing a compact orientable surface can always be transformed by means of cut and paste operations to a new polygon R' that represents the same topological surface but is described now by a word w in

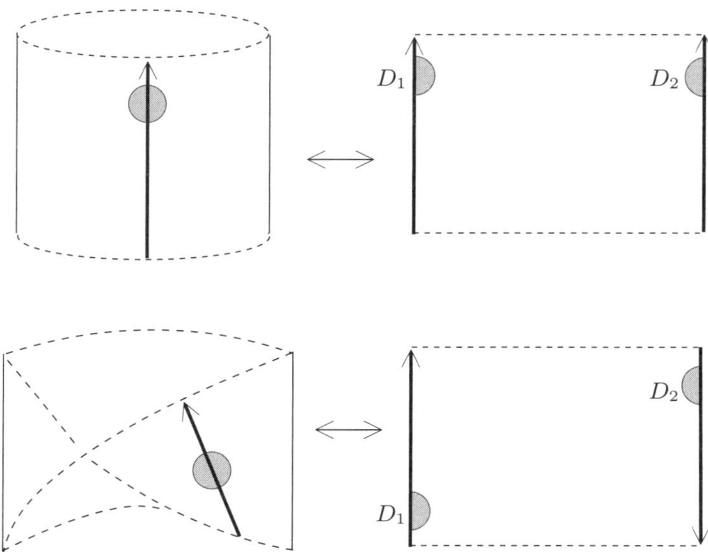

Fig. 1.6. Charts on the cylinder and the Möbius strip.

normalized form. By this we mean that w is one of the following words

$$\begin{cases} w_0 = aa^{-1} \\ \\ w_g = a_1 b_1 a_1^{-1} b_1^{-1} a_2 b_2 a_2^{-1} b_2^{-1} \cdots a_g b_g a_g^{-1} b_g^{-1} \end{cases} \tag{1.1}$$

where g is a positive integer called the *genus* of the surface.

It is quite obvious that the topological spaces determined by w_0, w_1, w_2, ... are homeomorphic to a sphere, a torus, a surface with two handles, etc. (see Figures 1.3 and 1.7).

We now describe the algorithm leading to the normalized form of our polygon R, following the account given in [FK92], see also [Mas91]. Only three operations are needed:

Operation one (erasing trivial data): a word of the form $w = aa^{-1}x \cdots$ can be transformed into $x \cdots$ (Figure 1.8).

Operation two (identification of vertices): assume that the two vertices of an edge of R represent distinct points P, Q of X.

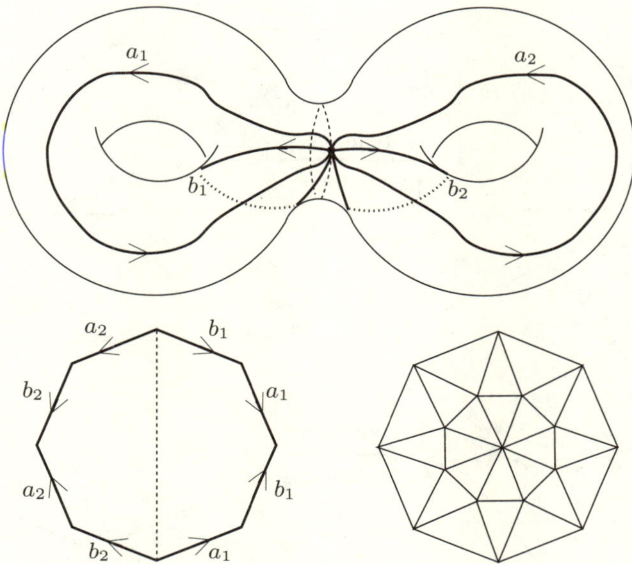

Fig. 1.7. A surface S of genus $g = 2$: the connected sum of two tori. A planar polygonal model and a triangulation of S are displayed below.

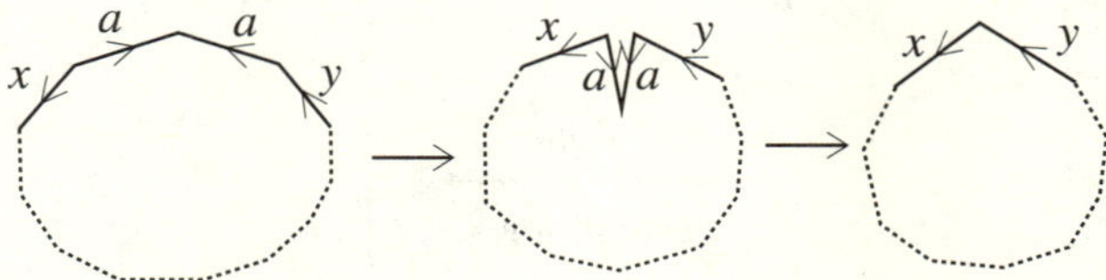

Fig. 1.8. Operation one: erasing trivial data.

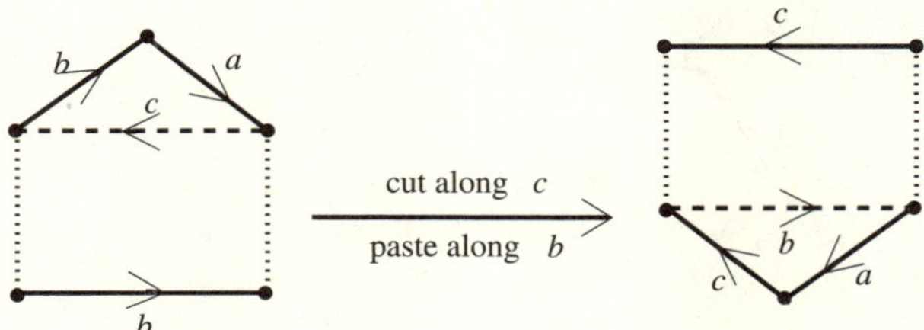

Fig. 1.9. Operation two: towards the identification of all vertices.

Compact Riemann surfaces and algebraic curves

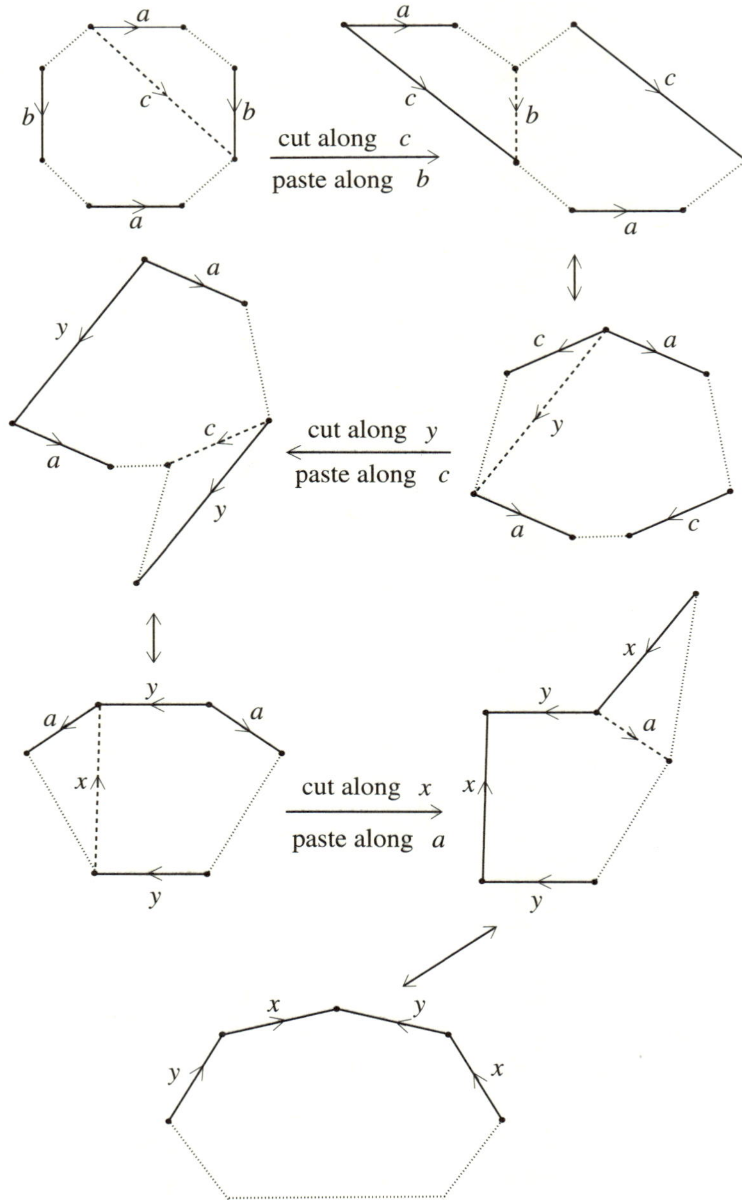

Fig. 1.10. Operation three: joining linked letters.

If no other vertex of R also corresponds to Q, two consecutive edges are identified and we can apply operation one unless R has precisely two edges, which in turn means that R is already in the normalized form described by w_0. In the other case, we can proceed as in Figure 1.9, cutting R along a suitable line to split it into two pieces that are then pasted together along another line to form a polygon R' representing the same topological surface.

The cutting line lies in the interior of R but in the border of R', and correspondingly the glueing line lies in the border of R but in the interior of R'. The effect of this operation is that the total number of vertices remains constant, while the number of vertices representing the point P increases by one. A suitable combination of operations one and two yields either the normal form w_0 or a polygon in which all vertices are identified.

Operation three (joining linked letters): a pair of letters a, b in w is said to be *linked* if the sequence $a \cdots b \cdots a^{-1} \cdots b^{-1}$ occurs in w. By the cutting and pasting procedure a linked pair can always be replaced by another linked pair of the form $\cdots xyx^{-1}y^{-1} \cdots$ (see Figure 1.10). Note that performing this operation in a polygon with all the vertices identified to each other yields a new polygon which still has all the vertices identified.

Example 1.50 Finding the normalized form aa^{-1} of the cube in Figure 1.4 requires only a repeated application of operation one.

Example 1.51 We show now a more complicated example of how the above method applies. Our goal is to compute the genus of the topological surface in Figure 1.11.

Direct inspection shows that there are two classes of vertices of the polygon (the white and the black points in Figure 1.11). In Figure 1.12 we show how to repeatedly apply operation two followed by a single application of operation one to obtain a polygon with only one class of vertices. Then we join linked letters (operation three), as shown in Figure 1.13. We find that the genus of the surface we started with is $g = 3$.

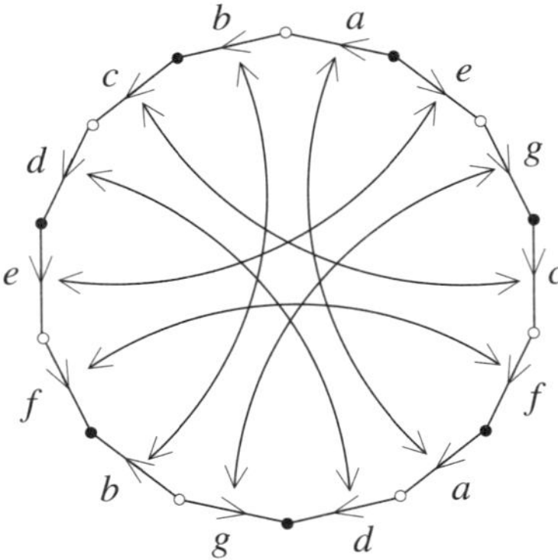

Fig. 1.11. The lines indicate the side pairings needed to build the topological surface corresponding to the word $w = abcdefb^{-1}gd^{-1}a^{-1}f^{-1}c^{-1}g^{-1}e^{-1}$.

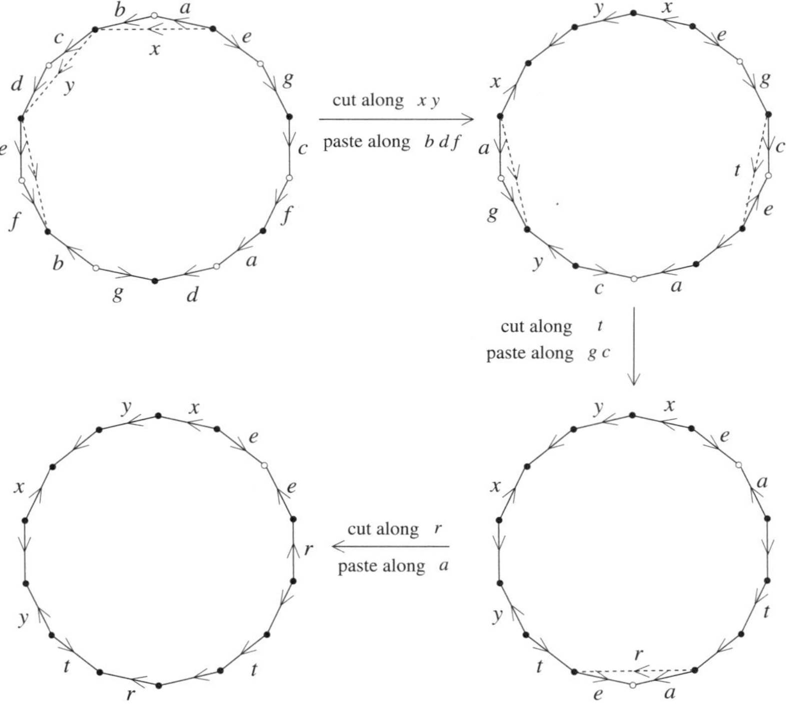

Fig. 1.12. Identification of all vertices.

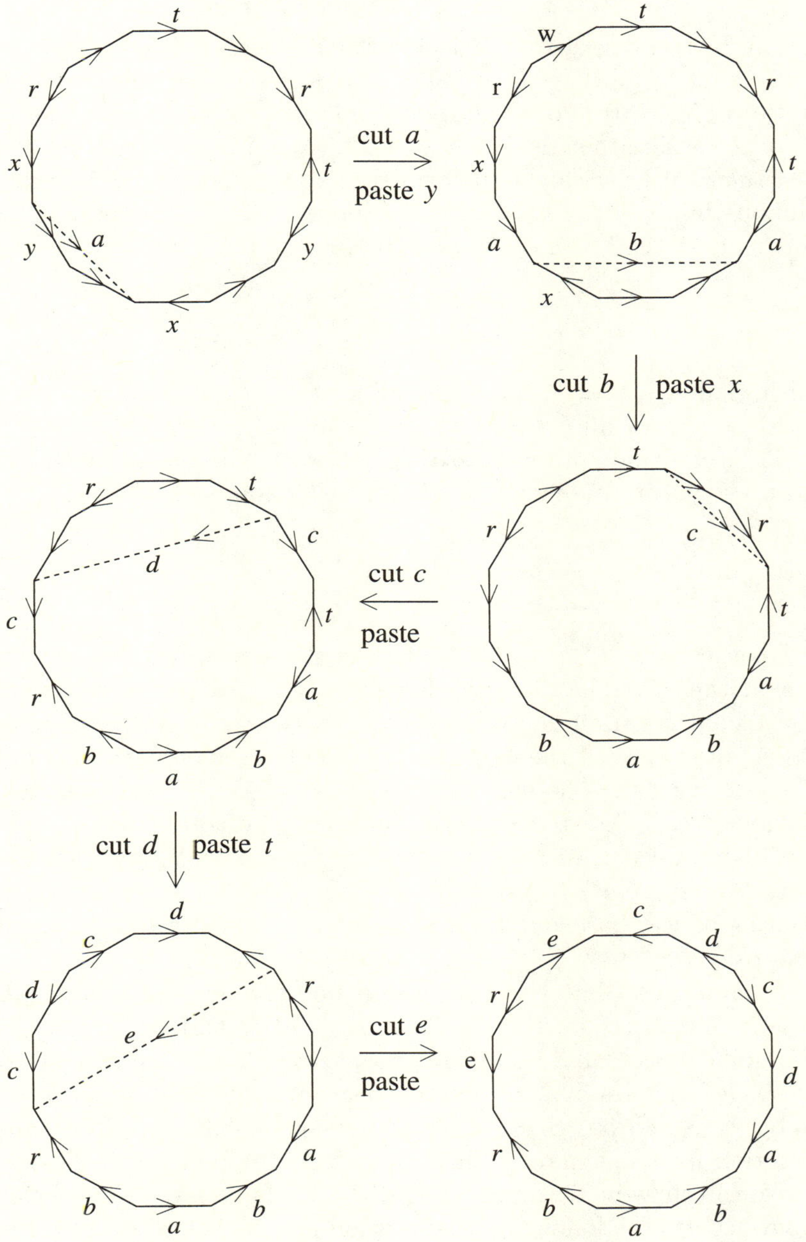

Fig. 1.13. Joining linked pairs.

1.2.2 The fundamental group

Let X be a compact topological surface. Recall that a continuous path $\gamma : I = [0,1] \to X$ is called a *loop with base point P* if its initial point $\gamma(0)$ and its endpoint $\gamma(1)$ agree with P. Two loops $\alpha, \beta : I \to X$ with same base point P are said to be *homotopically equivalent* if they can be deformed to each other through a continuous family $\{\gamma_s : I \to X\}_{s \in I}$ of loops with base point P, this means that there is a continuous map

$$\gamma : I \times I \longrightarrow X$$
$$(t,s) \longmapsto \gamma(t,s) =: \gamma_s(t)$$

such that $\gamma_s(0) = \gamma_s(1) = P$ for all $s \in I$ and $\gamma_0(t) = \alpha(t)$, $\gamma_1(t) = \beta(t)$ for all $t \in I$.

The set of homotopy classes of loops can be endowed with a group structure by means of the following composition law

$$[\alpha] \circ [\beta] = [\alpha\beta], \quad \text{where} \quad \alpha\beta(t) = \begin{cases} \alpha(2t) & \text{if } 0 \leq t \leq 1/2 \\ \beta(2t-1) & \text{if } 1/2 \leq t \leq 1 \end{cases}$$

In other words, $[\alpha] \circ [\beta]$ is the class of the loop which first follows α and then β. The identity element of the group is the class of the trivial constant loop ($\gamma(t) = P$ for all t), while the inverse $[\alpha]^{-1}$ is the class of the loop $t \to \alpha(1-t)$ (the same path as α but run backwards). Therefore, only the associativity is not entirely obvious (the interested reader may consult [Mas91]).

This group is called the *fundamental group* of X, and it is denoted by $\pi_1(X, P)$ or, very often, simply by $\pi_1(X)$ as different choices of the base point give rise to two essentially equivalent groups.

Assume now that the normalized polygon R of X has genus $g > 0$. Let a_i, b_i be the directed edges of R and let $P \in S$ be the point represented by any of the vertices. The directed edge a_i (resp. b_i) induces in X a loop α_i (resp. β_i) with base point at P. In other words, $\{\alpha_i, \beta_i\} \subset \pi_1(X, P)$, where we have used the same notation for loops and for their corresponding homotopy classes.

We will show in Theorem 2.6 that the fundamental group of such a surface has the following presentation in terms of generators and relations

$$\langle \alpha_1, \beta_1, \ldots, \alpha_g, \beta_g; \; \alpha_1\beta_1\alpha_1^{-1}\beta_1^{-1} \cdots \alpha_g\beta_g\alpha_g^{-1}\beta_g^{-1} = 1 \rangle \qquad (1.2)$$

In other words, $\alpha_1, \beta_1, \ldots, \alpha_g, \beta_g$ generate $\pi_1(X, P)$ and $w_g = 1$ is *essentially the unique relation in the group*, where w_g agrees with the word describing a normalized polygon model of X (see (1.1)). What 'essentially unique' means is that any other word on these generators representing the trivial loop is a product of conjugates of w_g and w_g^{-1}. A more precise explanation of this last sentence can be given as follows.

Let us denote by $\mathcal{F}_{2g}$ the set of words $w = w(a_i, b_i)$ on the symbols a_i, b_i and their inverses a_i^{-1}, b_i^{-1}, where $i = 1, \ldots, g$. This set can be made into a group called the *free group on* the $2g$ *generators* $\{a_i, b_i\}$ (see [Mas91]). The group operation is simply concatenation of words. The identity element 1 corresponds to the empty word and the existence of inverses is guaranteed by the cancellation law

$$a_i a_i^{-1} = a_i^{-1} a_i = b_i b_i^{-1} = b_i^{-1} b_i = 1$$

The rule that associates a_i to α_i and b_i to β_i defines a group epimorphism

$$\rho : \mathcal{F}_{2g} \longrightarrow \pi_1(X, P) \tag{1.3}$$
$$w(a_i, b_i) \longmapsto w(\alpha_i, \beta_i)$$

hence $\pi_1(X, P)$ is isomorphic to $\mathcal{F}_{2g}/\ker(\rho)$.

The word $w_g = a_1 b_1 a_1^{-1} b_1^{-1} \cdots a_g b_g a_g^{-1} b_g^{-1}$ lies in $\ker(\rho)$ because the loop $\rho(w_g) = \alpha_1 \beta_1 \alpha_1^{-1} \beta_1^{-1} \cdots \alpha_g \beta_g \alpha_g^{-1} \beta_g^{-1}$ can be seen as a counterclockwise walk along the boundary of the normalized polygon, and so it can be contracted to the point P through the interior of the polygon. Now, since the kernel is always a normal subgroup, $\ker(\rho)$ must contain the conjugates of w_g and w_g^{-1} and all their possible products. The statement above amounts to saying that these elements fill up the whole group $\ker(\rho)$.

Note that when $g = 1$ the presentation (1.2) implies that the two generators α_1, β_1 commute, and so the fundamental group is isomorphic to $\mathbb{Z}^2$. On the contrary, we see that the fundamental groups of the surfaces of genus $g > 1$ are not abelian groups. In fact, for each genus g the *abelianization* of the fundamental group is the group $\mathbb{Z}^{2g}$.

We recall that the abelianization $\overline{G}$ of a group G is the largest abelian quotient of G, more precisely

$$\overline{G} = \frac{G}{[G,G]}$$

where $[G,G]$, the commutator subgroup of G, is the group generated by all commutators, that is all elements of the form

$$[x,y] = xyx^{-1}y^{-1}, \quad x, y \in G$$

Note that $[G,G] \lhd G$ because the conjugate of a commutator is itself a commutator, and also that $\overline{G}$ is abelian, for clearly in the quotient we have $\overline{x}\,\overline{y}\,(\overline{x})^{-1}(\overline{y})^{-1} = 1$, that is $\overline{x}\,\overline{y} = \overline{y}\,\overline{x}$. It is also clear that $[G,G]$ is the smallest normal subgroup of G whose quotient is abelian.

When $G = \mathcal{F}_{2g}$, we have an epimorphism

$$\rho_1 : \mathcal{F}_{2g} \longmapsto \mathbb{Z}^g \oplus \mathbb{Z}^g = \mathbb{Z}^{2g}$$

defined by

$$a_i \longmapsto (e_i, 0) \tag{1.4}$$
$$b_i \longmapsto (0, e_i)$$

where e_i stands for the i-th vector of the canonical base of $\mathbb{Z}^g$. Since the target group is abelian, the kernel contains the subgroup $[\mathcal{F}_{2g}, \mathcal{F}_{2g}]$. But in fact $\ker(\rho_1) = [\mathcal{F}_{2g}, \mathcal{F}_{2g}]$ because ρ_1 induces an isomorphism

$$\frac{\mathcal{F}_{2g}}{[\mathcal{F}_{2g}, \mathcal{F}_{2g}]} \longmapsto \mathbb{Z}^{2g}$$

whose inverse is defined by inverting the arrows in (1.4) and then taking classes modulo $[\mathcal{F}_{2g}, \mathcal{F}_{2g}]$ (note that this inverse is well-defined because the target group is abelian).

On the other hand, we observe the following two facts relative to the epimorphism ρ in (1.3). First $\ker(\rho) \subset [\mathcal{F}_{2g}, \mathcal{F}_{2g}]$ and second $\rho([\mathcal{F}_{2g}, \mathcal{F}_{2g}]) = [\pi_1(X), \pi_1(X)]$.

Therefore, we can write

$$\mathbb{Z}^{2g} \simeq \frac{\mathcal{F}_{2g}}{[\mathcal{F}_{2g}, \mathcal{F}_{2g}]} \simeq \frac{\mathcal{F}_{2g}/\ker(\rho)}{[\mathcal{F}_{2g}, \mathcal{F}_{2g}]/\ker(\rho)} \simeq \frac{\pi_1(X)}{[\pi_1(X), \pi_1(X)]}$$

In other words, $\mathbb{Z}^{2g}$ is not only the abelianization of $\mathcal{F}_{2g}$ but also of $\pi_1(X)$. This yields the following:

Corollary 1.52 *If $g \neq g'$ the orientable compact surfaces defined by the words w_g and w'_g are not homeomorphic to each other.*

Proof If they were homeomorphic their fundamental groups would be isomorphic and so would be their abelianizations. But the theorem of classification of abelian groups shows that $\mathbb{Z}^{2g}$ is not isomorphic to $\mathbb{Z}^{2g'}$ if $g \neq g'$. $\square$

Remark 1.53 In particular, Corollary 1.52 shows that the genus is independent of the triangulation of the surface used to obtain a normalized polygon. Thus we are allowed to speak properly of the genus of a surface without reference to a particular triangulation.

1.2.3 The Euler–Poincaré characteristic

From a different point of view, the genus g of a compact orientable surface X may be defined in the following way.

Proposition 1.54 *Let X be a compact orientable topological surface of genus g.*

(i) *Let v, e and f be the number of vertices, edges and faces of a given triangulation of X. Then the integer*

$$\chi(X) := v - e + f$$

called the Euler–Poincaré characteristic *of X, is independent of the triangulation.*

(ii) *The genus and the Euler–Poincaré characteristic are related by*

$$\chi(X) = 2 - 2g$$

Proof (i) The proof will be a consequence of the following facts, each of which is easy to check:

- The corresponding statement for a closed topological disc D (instead of X) holds. In fact $\chi(D) = 1$.
  To see this one can argue by induction on the number n of triangles. If $n = 1$ then D itself is the only triangle of the triangulation and $\chi(D) = 3 - 3 + 1 = 1$. If D has been subdivided into n triangles we remove from D a triangle having one (or two) edges in the boundary of D to obtain

a new topological disc D' consisting of $n - 1$ triangles. To pass from D' back to D we have to add 1 edge and 1 face (or 1 vertex, 2 edges and 1 face), thus $\chi(D) = \chi(D') = 1$.

- If in the process of getting the normalized polygon out of a given triangulation of X we impose the condition that in operations two and three the cuts are performed along edges of the triangulation, then the triangulation is preserved throughout the whole process. Note that since these cutting lines may consist of several edges, some of the $4g$ sides (2 in the case $g = 0$) of our normal polygon may split into several edges of the triangulation.

- We can compute $\chi(X)$ by using the fact that the triangulation of X is induced by the triangulation of the normalized polygon R (which is a topological disc), after suitable identification of vertices and edges in the boundary. We note the following:

(a) The number of triangles (faces) is the same in R as in X.

(b) The e_i interior edges of the triangulation of R correspond to e_i different edges of the triangulation of X. However, the e_b boundary edges give rise to only $e_b/2$ edges in X, as they come identified in pairs.

(c) Likewise the v_i interior vertices of the triangulation of R give rise to v_i different vertices of the triangulation of X. But the situation is different for the $v_b = e_b$ boundary vertices. If we write $v_b = v + 4g$ (or $v_b = v + 2$ if $g = 0$) to count the $4g$ vertices (2 if $g = 0$) of the polygon separately from the rest of the vertices of the triangulation, we see that these v_b vertices correspond to only $v/2 + 1$ vertices in X ($v/2 + 2$ if $g = 0$).

Finally, we can write

$$
\begin{aligned}
\chi(X) &= \chi(R) - \left(\frac{v}{2} + 4g - 1\right) + \left(\frac{v + 4g}{2}\right) \\
&= 1 - \frac{v}{2} - 4g + 1 + \frac{v}{2} + 2g \\
&= 2 - 2g
\end{aligned}
$$

or

$$
\chi(X) = 1 - \frac{v}{2} + \frac{v + 2}{2} = 2
$$

if $g = 0$.

(ii) Has been proved already. $\qquad\qquad\square$

For example, the triangulation in Figure 1.7 has $V = 10$, $E = 36$ and $F = 24$, thus $\chi = 10 - 36 + 24 = -2 = 2 - 2 \cdot 2$, as expected.

1.2.4 The Riemann–Hurwitz formula for morphisms to the sphere

We already know that topologically $\mathbb{P}^1$ is the sphere and $\mathbb{C}/\mathbb{Z} \oplus \mathbb{Z}i$ is the torus, hence their genera are 0 and 1. But, how to determine the genus of surfaces such as the hyperelliptic and the Fermat curves? The tool for performing such computations will be the Riemann–Hurwitz formula.

Before stating the formula, let us show how it works in a particular case. Let S be the Riemann surface associated to the algebraic curve $y^2 = \prod_{k=1}^{2g+1} (x - a_k)$.

Remember that we have a morphism $\mathbf{x} : S \longrightarrow \widehat{\mathbb{C}}$, given by $\mathbf{x}(x, y) = x$ (see Example 1.20). We observe that every point of the sphere has two preimages except for the values $a_1, \ldots, a_{2g+1}, \infty$, whose preimage $\mathbf{x}^{-1}(a_k)$ is only the point $(a_k, 0)$, of multiplicity 2 (we will say that $\mathbf{x}$ has degree two). We now take a triangulation $\widehat{\mathbb{C}} = \bigcup_k T_k$ satisfying the following two conditions:

(i) All points $a_1, \ldots, a_{2g+1}, a_{2g+2} = \infty$ are vertices of some triangle.

(ii) The triangles have a diameter ε small enough for the balls $B(a_i, \varepsilon)$ not to contain a_j with $j \neq i$.

Under these conditions, the inverse image of $T_k \subset \widehat{\mathbb{C}}$ is the reunion of two triangles $\mathbf{x}^{-1}(T_k) = T_k^1 \cup T_k^2 \subset S$. The collection of all the so obtained triangles $\{T_k^1, T_k^2\}_k$ gives a triangulation of S.

Now, if $\chi(\widehat{\mathbb{C}}) = v - a + c = 2$ is Euler's formula for the triangulation $\{T_k\}_k$ of $\widehat{\mathbb{C}}$, the corresponding triangulation $\{T_k^1, T_k^2\}_k$ would give

$$\chi(S) = 2v - (2g + 2) - 2a + 2c = 2(v - a + c) - 2g - 2 = 2 - 2g$$

thus the genus of S equals precisely g.

We note in passing that, since g is an arbitrary integer, this example shows that any compact orientable surface can be endowed with a Riemann surface structure. In fact this can be done in infinitely many non-equivalent ways (see Corollary 2.49 and Section 2.6).

These computations can be generalized to arbitrary Riemann surfaces and arbitrary (non-constant) functions. Doing that requires a precise definition of the degree of a meromorphic function. For that we need to say something about the integration of differential forms.

If $c : I = [0,1] \to S$ is a (piecewise differentiable) path and ω is a meromorphic differential without poles in $c(I)$, we can compute the integral $\int_c \omega$. The result does not depend on the choice of coordinates, since if $\gamma = c_{|[a,b]}$ is a small piece of c lying inside $U_i \cap U_j$ the two possible ways to compute $\int_\gamma \omega$, namely

$$\int_\gamma \omega = \int_{\varphi_i \circ \gamma} f_i \circ \varphi_i^{-1}$$

and

$$\int_\gamma \omega = \int_{\varphi_j \circ \gamma} f_j \circ \varphi_j^{-1}$$

give the same result.

Indeed

$$\int_{\varphi_i \circ \gamma} f_i \circ \varphi_i^{-1} = \int_a^b \left(f_i \circ \varphi_i^{-1} \right) |_{\varphi_i \circ \gamma(t)} \, (\varphi_i \circ \gamma)' \, |_t dt$$

$$= \int_a^b (f_i \circ \gamma) \, |_t \, (\varphi_i \circ \gamma)' \, |_t dt$$

$$= \int_a^b (f_i \circ \gamma) \, |_t \left(\left(\varphi_i \circ \varphi_j^{-1} \right) \circ (\varphi_j \circ \gamma) \right)' \, |_t dt$$

$$= \int_a^b (f_i \circ \gamma) \, |_t \left(\varphi_i \circ \varphi_j^{-1} \right)' \, |_{\varphi_j \circ \gamma(t)} \, (\varphi_j \circ \gamma)' \, |_t dt$$

$$= \int_a^b (f_j \circ \gamma) \, |_t \, (\varphi_j \circ \gamma)' \, |_t dt$$

$$= \int_{\varphi_j \circ \gamma} f_j \circ \varphi_j^{-1}$$

Definition 1.55 Let P be a a point lying in a chart (U_j, φ_j). The *residue* of ω at P, denoted $\mathrm{res}_P(\omega)$, is defined as the a_{-1} coefficient in the Laurent expansion

$$f_j \circ \varphi_j^{-1}(z) = \frac{a_{-n}}{(z - \varphi_j(P))^n} + \cdots + \frac{a_{-1}}{(z - \varphi_j(P))} + a_0 + \cdots$$

By the one complex variable residue theorem we can write

$$\mathrm{res}_P(\omega) \;=\; \mathrm{res}_{\varphi_j(P)}(f_j \circ \varphi_j^{-1})$$

$$=\; \frac{1}{2\pi i} \int_{\partial B(\varphi_j(P),\varepsilon)} f_j \circ \varphi_j^{-1}$$

$$=\; \frac{1}{2\pi i} \int_{\partial B(P,\varepsilon)} \omega$$

where $B(P,\varepsilon)$ stands for a small neighbourhood of P diffeomorphic to the disc $B(\varphi_j(P),\varepsilon)$.

It follows that the residue of a differential at a point is independent of the choice of coordinates, as it can be computed by integration along any small loop around P.

Theorem 1.56 (Residue Theorem) *The sum of all the residues of a meromorphic differential in a compact Riemann surface equals zero. That is*

$$\sum_{P \in S} \mathrm{res}_P(\omega) = 0$$

for any meromorphic differential ω in the compact Riemann surface S.

Proof Let $P_1, \ldots, P_r$ be the poles of ω. Consider a triangulation $\bigcup_j T_j$ of $S \setminus \bigsqcup_k B(P_k, \varepsilon)$ where each triangle T_j is contained in a coordinate neighbourhood U_j (see Figure 1.14).

Then

$$\sum_{P \in S} \mathrm{res}_P(\omega) = \frac{1}{2\pi i} \sum_k \int_{\partial B(P_k,\varepsilon)} \omega = \frac{1}{2\pi i} \sum_j \int_{\partial T_j} \omega_{|U_j}$$

since the edges of the triangulation that do not lie in $\partial B(P_k, \varepsilon)$ are integrated twice in the last sum, but in opposite directions.

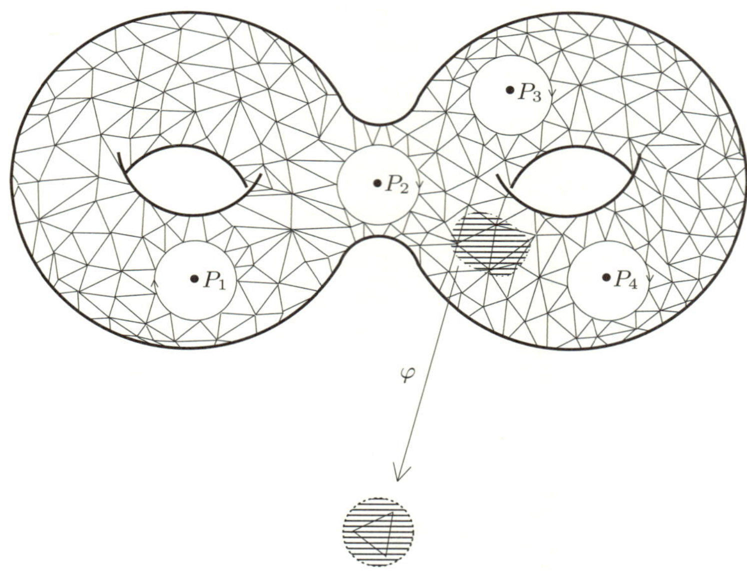

Fig. 1.14. Integration along $\cup_i \partial B(P_i, \varepsilon)$ equals integration along $\cup_j \partial T_j$.

Now let $\omega_{|U_j} = f(z)dz$ where $f \circ \varphi_j^{-1} = u + iv$ is holomorphic. By Green's formula we have

$$\int_{\partial T_j} \omega = \int_{\partial(\varphi_j(T_j))} (u + iv)dx + (-v + iu)dy$$

$$= \int_{\varphi_j(T_j)} [(-v_x - u_y) + i(u_x - v_y)] \, dx \wedge dy$$

$$= 0$$

where the last equality is a consequence of the Cauchy–Riemann equations. $\qquad\square$

Proposition 1.57 (Degree of a meromorphic function) *Let* $f : S \longrightarrow \widehat{\mathbb{C}}$ *be a non-constant morphism. The non-negative integer*

$$\deg(f) := \sum_{f(P)=c} m_P(f)$$

called the degree *of f, is independent of the choice of $c \in \widehat{\mathbb{C}}$.*

Proof Let $c \neq \infty$ and $P \in f^{-1}(c)$ a point of branching order $m_P(f) = n$. Then the Taylor expansion of $f - c$ around P is

$$(f - c) \circ \varphi_j^{-1}(z) = a_n z^n + a_{n+1} z^{n+1} + \cdots$$

hence

$$\frac{\left((f - c) \circ \varphi_j^{-1}\right)'(z)}{\left((f - c) \circ \varphi_j^{-1}\right)(z)} = \frac{n}{z} + \cdots$$

In particular, P is a pole of the differential $\omega = \dfrac{d(f - c)}{(f - c)}$ with residue n. Similarly, when P is a pole of f, the residue of ω at P equals $-m_P(f)$.

The Residue Theorem applied to ω yields

$$0 = \sum_{P \in S} \mathrm{res}_P(\omega) = \sum_{f(P)=c} m_P(f) - \sum_{f(P)=\infty} m_P(f)$$

$$\square$$

Thus, the degree of f is simply the number of points in every fibre $f^{-1}(c)$ such that c is not a branch value (see Definition 1.31). If we take multiplicities into account this number remains constant at all values $c \in \mathbb{P}^1$.

Example 1.58 According to Proposition 1.27, every morphism $f : \widehat{\mathbb{C}} \longrightarrow \widehat{\mathbb{C}}$ is a rational function, that is $f(z) = p(z)/q(z)$ for certain polynomials $p, q \in \mathbb{C}[z]$. The degree of f is in this case the maximum of the degrees of p and q.

Example 1.59 The considerations made at the beginning of the section for computing the genus of the Riemann surface of the curve $y^2 = \prod_{k=1}^{2g+1} (x - a_k)$ show formally that $\deg(\mathbf{x}) = 2$. In the same way, one sees that the degree of the second coordinate function $\mathbf{y} : S \longrightarrow \widehat{\mathbb{C}}$ equals $2g + 1$. Note that the degree of the function $\mathbf{x}$ agrees with the degree of the variable y in the defining equation, and conversely. Similarly, both coordinate functions of the Fermat curve $x^d + y^d = 1$ have degree d.

We now compute the genus of a compact Riemann surface endowed with an arbitrary morphism to $\mathbb{P}^1$.

Theorem 1.60 (Riemann–Hurwitz for morphisms to $\mathbb{P}^1$)
Let S be a compact Riemann surface, and $f : S \longrightarrow \mathbb{P}^1$ a morphism of degree $d \geq 1$, then the genus g of S is given by the formula

$$2g - 2 = -2d + \sum_{P \in S} (m_P(f) - 1)$$

Proof Compare the Euler characteristic of S and $\mathbb{P}^1$ in the same way we did in the previous example. Since f is a local homeomorphism outside the branch points $P_1, \ldots, P_n$, a triangulation $\mathbb{P}^1 = \bigcup_k T_k$ whose triangles T_k are small enough and such that $f(P_1), \ldots, f(P_n)$ belong to the set of vertices, induces a triangulation $S = \bigcup_k f^{-1}(T_k)$, where $f^{-1}(T_k) = T_k^1 \cup \cdots \cup T_k^d$.

Now, if $2 = v - a + c$ is the Euler formula for the triangulation $\{T_k\}_k$ of the sphere $\mathbb{P}^1$, the formula corresponding to the triangulation $\{T_k^1, \ldots, T_k^d\}_k$ of S is $2 - 2g = v' - a' + c'$, where clearly $a' = da$ and $c' = dc$. As for v', we do not have $v' = dv$ since, as we already observed in the previous example, those vertices which are branch values will have less than d preimages.

Around a branch point P, the function f can be locally written as $(\varphi' \circ f \circ \varphi^{-1})(z) = c_m z^m + c_{m+1} z^{m+1} + \cdots = \psi(z)^m$, where

$$\psi(z) = z \sqrt[m]{c_m + c_{m+1}z + \cdots}$$

is near the origin a bijective function. Therefore, in a small neighbourhood of P the function f takes once the value $f(P)$ and m times the rest of the values. In other words, the inverse image of a triangle T_k with a vertex at $f(P)$ is then the reunion of m triangles that have P as a common vertex. In case we had set $v' = dv$, this single vertex P would have been counted m times. It is clear then that the right computation is $v' = dv - \sum_{P \in S} (m_P(f) - 1)$, and the formula follows. $\qquad\square$

Remark 1.61 A straightforward consequence of the formula of Riemann–Hurwitz is that every morphism $f : S \to \mathbb{P}^1$ which is not bijective *is necessarily ramified.*

Example 1.62 (The genus of Fermat curves) Let S be the compact Riemann surface associated to the curve

$$x^d + y^d = 1$$

Consider the morphism $\mathbf{x} : S \longrightarrow \mathbb{P}^1$ described in Example 1.21, given as $\mathbf{x}(x, y) = x$. The ramification points are $Q_j = \left(\xi_d^j, 0\right)$ with $j = 1, \ldots, d$, all of them with branching order equal to d (since the local expression of $\mathbf{x}$ has the form $s \longmapsto s^d + \xi_d^j$). The Riemann–Hurwitz formula gives

$$2g - 2 = -2d + \sum_{Q_j} (d - 1) = -2d + d\,(d - 1)$$

It follows that $g = \dfrac{(d - 1)\,(d - 2)}{2}$.

Example 1.63 Consider the p-gonal algebraic curve

$$y^p = (x - a_1)^{m_1} \cdots (x - a_r)^{m_r}$$

of Example 1.11. Consider the degree p morphism $\mathbf{x} : S \longrightarrow \mathbb{P}^1$ given by $\mathbf{x}(x, y) = x$. The branching values are $a_1, a_2, \ldots, a_r, \infty$ or $a_1, a_2, \ldots, a_r$, depending on whether $\sum m_i$ is prime to p or not. Since the branching order is always p, the genus g of S is given by

$$g = \begin{cases} \dfrac{(p - 1)(r - 1)}{2} & \text{if } \sum m_i \text{ is prime to } p \\[2em] \dfrac{(p - 1)(r - 2)}{2} & \text{otherwise} \end{cases}$$

Note that in the case of Klein's Riemann surface $y^7 = x(x - 1)^2$ we have $p = 7$, $r = 2$, $m_1 = 1$ and $m_2 = 2$, hence $g = 3$.

1.2.5 Coverings

We now summarize some basic facts about covering space theory.

Definition 1.64 A continuous mapping $p : E \longrightarrow X$ between topological surfaces E and X is a *covering map* (or more simply a *covering*) if for every $x \in X$ there is a neighbourhood V such that $p^{-1}(V) = \bigcup U_i$, where the sets U_i are pairwise disjoint and the restriction $p|_{U_i} : U_i \longrightarrow V$ is a homeomorphism (we say then that V is *well covered* by p).

We shall often write $p^{-1}(V) = \bigsqcup U_i$ to indicate that the sets U_i are pairwise disjoint.

In case X has a holomorphic structure, E inherits a unique Riemann surface structure such that p is holomorphic. This structure is given by charts of the form $(U_i, \varphi_j \circ p)$, where (V_j, φ_j) is a chart in X and $p(U_i) = V_j$ is a well-covered neighbourhood. The transition functions are then $(\varphi_k \circ p) \circ (\varphi_j \circ p)^{-1} = \varphi_k \circ \varphi_j^{-1}$ and the local expression of $p : E \longrightarrow X$ in these charts is $\varphi_j \circ p \circ (\varphi_j \circ p)^{-1} = \mathrm{Id}$, thus p is certainly holomorphic.

If p is also holomorphic with respect to another chart (U, ψ), we have, by definition, that $(\varphi_j \circ p) \circ \psi^{-1}$ is holomorphic. But then (U, ψ) is compatible with all the charts $(U_i, \varphi_j \circ p)$ described above. This shows the uniqueness of the holomorphic structure on X.

Covering space theory applies naturally in more general settings than the theory of surfaces. The simplest non-trivial example can be given in dimension one.

Example 1.65 (The universal cover of the circle) The circle $X = \mathbb{S}^1$ admits the helix

$$\widetilde{X} = \left\{ \left(e^{2\pi i t}, t\right) \mid t \in \mathbb{R} \right\} \subset \mathbb{R}^3$$

as covering space, where the covering map is $p\left(e^{2\pi i t}, t\right) = e^{2\pi i t}$ (see Figure 1.15).

Note that the helix $\widetilde{X}$ is topologically equivalent to the real line.

Given a covering $p : E \longrightarrow X$, every path $\gamma : I = [0,1] \longrightarrow X$ can be lifted to E, meaning that there is a path $\widetilde{\gamma}$ in E such that $p \circ \widetilde{\gamma} = \gamma$. The lift $\widetilde{\gamma}$ is uniquely determined once its initial point in E is chosen. This is clear in the previous example, and it is easy to understand in general. Suppose that $p^{-1}(V) = \bigsqcup U_i$ with $p : U_i \simeq V$, and that V contains a piece of γ of the form $\gamma([0, \varepsilon])$. Once we choose the initial point $e_0 \in p^{-1}(\gamma(0)) \subset E$, which is equivalent to choosing one of the open sets U_i, the lift of the piece $\gamma([0, \varepsilon]) \subset V$ can only be $p|_{U_i}^{-1} \circ \gamma$ (see Figure 1.16). Then we do the same with the second piece $\gamma([\varepsilon, \varepsilon']) \subset V'$ and continue this process until the whole γ *is covered*. Clearly by the continuity assumption this can be done in a unique way.

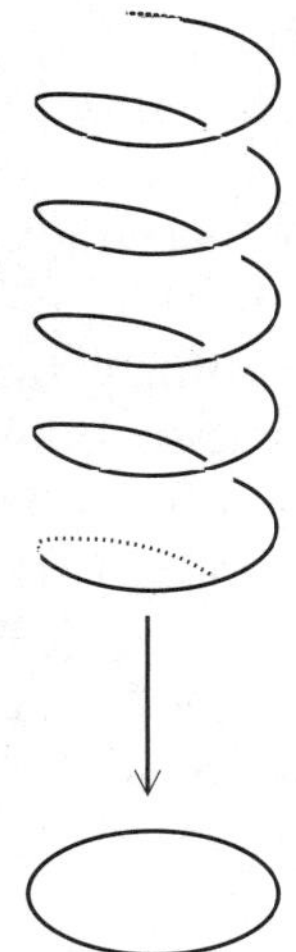

Fig. 1.15. The universal covering of the circle.

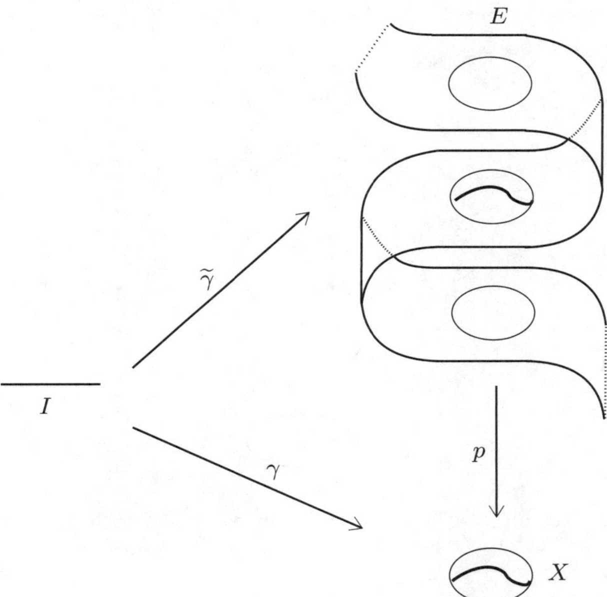

Fig. 1.16. Lifting a path.

In fact a covering is nothing but a local homeomorphism with the property that paths can be lifted (this is called the *path lifting property*).

Note that the fibres of a covering are always discrete sets. Another important property is that if X is connected (as we are always tacitly assuming), two arbitrary fibres $p^{-1}(x)$ and $p^{-1}(x')$ have the same cardinality, the so-called *degree* of the covering. To show this, take a path γ joining x to x'. For every $e \in p^{-1}(x)$, let us denote by $\widetilde{\gamma}_e$ the lift of γ with initial point at $\widetilde{\gamma}_e(0) = e$. Then the correspondence $e \to \widetilde{\gamma}_e(1)$ determines a bijection between $p^{-1}(x)$ and $p^{-1}(x')$, since $\widetilde{\gamma}_{e_1}(1) = \widetilde{\gamma}_{e_2}(1) = e'$ would mean that the path γ^{-1} would have two different lifts with e' as initial point.

Example 1.66 The truncated helix $Y = \left\{ \left(e^{2\pi i t}, t \right) \mid t > 0 \right\} \subset \mathbb{R}^3$ is not a covering of $\mathbb{S}^1$, despite the fact that $p : Y \longrightarrow \mathbb{S}^1$ is still a local homeomorphism. What is missing here is the path lifting property. See Figure 1.17.

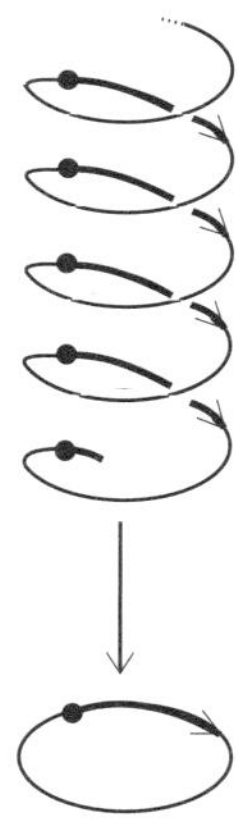

Fig. 1.17. Not all local homeomorphisms enjoy the path lifting property.

In fact, in covering space theory not only paths can be lifted. Any continuous mapping $f : Y \longrightarrow X$ can be lifted to a continuous mapping $\widetilde{f} : Y \longrightarrow E$, provided Y is *simply connected* (that is, with trivial fundamental group). Of course we say that $\widetilde{f}$ is a lift of f if $p \circ \widetilde{f} = f$.

In order to define $\widetilde{f}$ we start by choosing a point $y_0 \in Y$ and its image $\widetilde{f}(y_0) = e_0$ in the set $p^{-1}(f(y_0))$, the p-fibre of $f(y_0)$. This choice will determine $\widetilde{f}$ completely. For an arbitrary y, connect y_0 to y by a path $\gamma_0 : I = [0,1] \to Y$, and lift the path $f \circ \gamma_0$, with initial point at e_0. One can define then $\widetilde{f}(y) = \widetilde{f \circ \gamma_0}(1)$.

Remark 1.67 In this construction it is crucial to observe that the definition of $\widetilde{f}$ does not depend on the choice of γ_0 but only on that of e_0. This is because if γ_1 is another path joining y_0 to y and $\{\gamma_t : I \to Y\}_{t \in [0,1]}$ is a homotopy between them (which must exist, since Y is simply connected), then $t \to \widetilde{f \circ \gamma_t}(1) \in p^{-1}(y)$ is a continuous map that takes its values in a discrete set, and therefore it is a constant map.

An *isomorphism* of two coverings of X, $p_i : E_i \longrightarrow X$ ($i = 1, 2$) is a homeomorphism $f : E_1 \longrightarrow E_2$ such that the following diagram commutes:

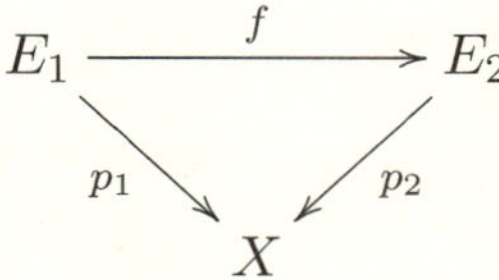

The group of automorphisms of a covering $p : E \longrightarrow X$, to be denoted $\mathrm{Aut}(E, p)$, is usually referred to as the *group of covering transformations* or *covering group* of p. If $f : E_1 \longrightarrow E_2$ is an isomorphism of coverings, then $\mathrm{Aut}(E, p_2) = f \circ \mathrm{Aut}(E, p_1) \circ f^{-1}$.

Before describing the basic properties of coverings, we introduce some terminology to refer to the nature of the action of the covering group $\mathrm{Aut}(E, p)$ on E.

Definition 1.68 Let E be a topological surface and G a subgroup of homeomorphisms of E. Then:

(i) The action of G on E is called *free* if the non-trivial elements of G do not fix any point of E.

(ii) The action of G on E is called *properly discontinuous* if for each $e \in E$ there are at most finitely many transformations $g_1 = \mathrm{Id}, \dots, g_r \in G$ fixing e, and there exists a neighbourhood U_e such that $g(U_e) \cap U_e = \emptyset$ for all $g \in G \setminus \{g_1, \dots, g_r\}$.

Clearly, when the action of G is both free and properly discontinuous, the neighbourhood U_e in condition (ii) can be chosen so that $U_e \cap \gamma(U_e) = \emptyset$ for all $\gamma \neq \mathrm{Id}$.

For an arbitrary subgroup G of homeomorphisms of E the rule

$$e \sim e' \Leftrightarrow e' = g(e) \quad \text{for some } g \in G$$

clearly defines an equivalence relation on E. Let us denote the corresponding quotient space by $X = E/G$, the equivalence class of $e \in E$ by $[e] = [e]_G$, and by

$$\begin{aligned} \pi_G : E &\longrightarrow X \\ e &\longmapsto [e] \end{aligned}$$

the natural quotient map.

When the action of G on E is free and properly discontinuous, the map $\pi : E \longrightarrow X$ is a covering map with covering group G. The image $\pi(U)$ of the open set $U = U_e$ in Definition 1.68 is a neighbourhood of $\pi(e)$ well covered by π, and we have $\pi^{-1}(\pi(U)) = \bigsqcup_{g \in G} g(U)$.

For a subgroup $H \leq G$ let $Y = E/H$. Then

$$\begin{aligned} f : Y &\longrightarrow X \\ [e]_H &\longmapsto [e]_G \end{aligned}$$

is an intermediate covering map and the diagram

$$\begin{array}{ccc} & E & \\ {\scriptstyle \pi_H}\swarrow & & \searrow{\scriptstyle \pi_G} \\ Y & \xrightarrow{\quad f \quad} & X \end{array}$$

commutes. Now

$$f^{-1}\left(\pi_G(U)\right) = \pi_H\left(\bigsqcup_{g \in G} g(U)\right) = \pi_H\left(\bigsqcup_{g \in J} g(U)\right),$$

where J is a set of representatives of the right cosets of G modulo H.

For an example of this construction, take $E = \mathbb{C}$, $G = \mathbb{Z} \oplus \mathbb{Z}i$ and $H = \mathbb{Z}$ to get a commutative diagram

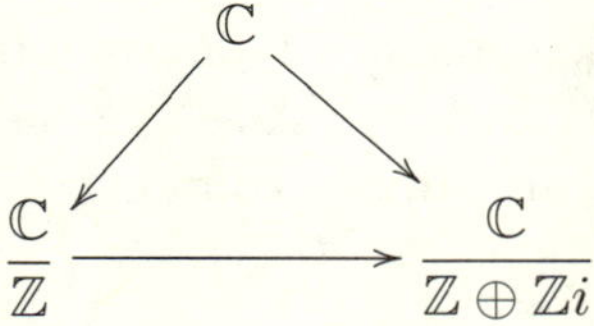

The bottom map $f : \dfrac{\mathbb{C}}{\mathbb{Z}} \longrightarrow \dfrac{\mathbb{C}}{\mathbb{Z} \oplus \mathbb{Z}i}$ is a covering map (even a morphism of Riemann surfaces) from the cylinder to the torus (cf. Examples 1.12 and 1.13).

The following theorem summarizes the main properties of covering spaces.

Theorem 1.69 *Let X be a connected topological surface. Then:*

(i) *There is always a covering $\pi : \widetilde{X} \longrightarrow X$, with $\widetilde{X}$ connected and simply connected.*

(ii) *$\widetilde{X}$ is unique up to isomorphism. It is called the* universal *covering space of X.*

(iii) *The group $\mathrm{Aut}(\widetilde{X}, \pi)$ of covering transformations of the universal covering $\pi : \widetilde{X} \to X$ can be identified with the fundamental group $\pi_1(X)$.*

(iv) *The action of $\mathrm{Aut}(\widetilde{X}, \pi)$ on $\widetilde{X}$ is free and properly discontinuous, preserves every fibre and is transitive in each of them. In particular π induces a homeomorphism*

$$\widetilde{X}/\mathrm{Aut}(\widetilde{X}, \pi) \longrightarrow X$$

$$[\tilde{x}] \longmapsto \pi(\tilde{x})$$

(v) *Any covering of X is isomorphic to a covering of the form $\widetilde{X}/G \longrightarrow \widetilde{X}/\mathrm{Aut}(\widetilde{X}, \pi) \simeq X$, where G is a certain subgroup of $\mathrm{Aut}(\widetilde{X}, \pi)$.*

(vi) *If $\pi : \widetilde{X} \longrightarrow X$ is a holomorphic covering, then the covering group $\mathrm{Aut}(\widetilde{X}, \pi)$ is a group of holomorphic transformations, the map $\pi : \widetilde{X}/\mathrm{Aut}(\widetilde{X}, \pi) \simeq X$ is a biholomorphism, and any other holomorphic covering of X is isomorphic to a projection of the form $\widetilde{X}/G \longrightarrow \widetilde{X}/\mathrm{Aut}(\widetilde{X}, \pi) \simeq X$.*

Proof In order to illustrate the arguments it is convenient to keep in mind the case of the universal covering of $\mathbb{S}^1$ described in Example 1.65. We shall give a brief sketch of the proof. For full details the reader may consult [Ful95] or [Mas91].

(ii) If $\pi_1 : X_1 \to X$ and $\pi_2 : X_2 \to X$ are two coverings such that both X_1 and X_2 are simply connected, we have a diagram

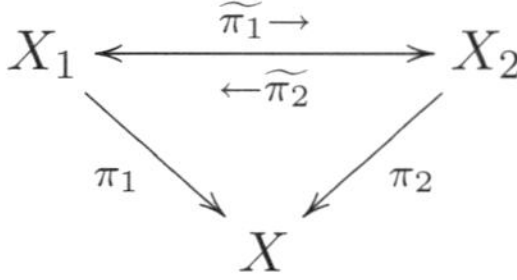

where the map $\widetilde{\pi_1}$ is any lift of π_1 to X_2. Its inverse is the unique lift $\widetilde{\pi_2}$ of π_2 to X_1 satisfying $\widetilde{\pi_2} \circ \widetilde{\pi_1}(x) = x$, at some point $x \in X_1$.

(iii) Let us choose points $x_0 \in X$ and $\widetilde{x}_0 \in \pi^{-1}(x_0)$ in $\widetilde{X}$. Given $f \in \mathrm{Aut}(\widetilde{X}, \pi)$, we can construct a loop in X by first choosing any path σ_f in $\widetilde{X}$ joining $\widetilde{x}_0$ to $f(\widetilde{x}_0)$ and then projecting it down to the path $\gamma_f := \pi \circ \sigma_f$ in X. Clearly γ_f is a loop with base point x_0.

Conversely, given $\gamma \in \pi_1(X, x_0)$ denote by $f_\gamma \in \mathrm{Aut}(\widetilde{X}, \pi)$ the unique lift of the covering map $\pi : \widetilde{X} \longrightarrow X$ which sends $\widetilde{x}_0$ to the endpoint of the lifted loop $\widetilde{\gamma}_{\widetilde{x}_0}$ (note that an argument similar to the one used in Remark 1.67 shows that this endpoint depends only on the homotopy class of γ). Then the isomorphism $\mathrm{Aut}(\widetilde{X}, \pi) \simeq \pi_1(X, x_0)$ is given by

$$
\begin{array}{ccc}
\mathrm{Aut}(\widetilde{X}, \pi) & \longleftrightarrow & \pi_1(X, x_0) \\
f & \longmapsto & \gamma_f \\
f_\gamma & \longmapsfrom & \gamma
\end{array}
$$

Note that the path $\sigma_g^f : t \to f \circ \sigma_g(t)$ connects $f(\widetilde{x}_0)$ to $f \circ g(\widetilde{x}_0)$ and projects via π to the same loop γ_g as σ_g. Therefore, the product loop $\gamma_f \gamma_g$ can be obtained by projection of the composed path $\sigma_f \sigma_g^f$. Since this is a path connecting $\widetilde{x}_0$ to $f \circ g(\widetilde{x}_0)$, we conclude that $\gamma_f \gamma_g = \gamma_{f \circ g}$. This shows that the rule $f \mapsto \gamma_f$ defines a group homomorphism.

(iv) If $f \in \mathrm{Aut}(\widetilde{X}, \pi)$, then f permutes the points inside each fibre $\pi^{-1}(x) = \{\widetilde{x}_i\}$ or, equivalently, the disjoint open sets U_i of which the inverse image $\pi^{-1}(V_x)$ of a well-covered neighbourhood

V_x of x consists of. Therefore, if $f(U_j) \cap U_k \neq \emptyset$ then $f(U_j) = U_k$ and $j = k$. In particular, $f(\tilde{x}_j) = \tilde{x}_j$ (indeed $f|_{U_j} = \mathrm{Id}$). This implies that the action of $\mathrm{Aut}(\tilde{X}, \pi)$ is properly discontinuous. It also implies that the set $\{x : f(x) = x\}$ is simultaneously open and closed, thus empty or the whole $\tilde{X}$. Therefore either $f = \mathrm{Id}$ or it does not fix points.

Since the elements of $\mathrm{Aut}(\tilde{X}, \pi)$ are nothing but the different lifts of the mapping $\pi : \tilde{X} \longrightarrow X$, they are in one-to-one correspondence with the different points lying in a given fibre (as we know, each of them determines a lift). Thus $\mathrm{Aut}(\tilde{X}, \pi)$ acts transitively in each fibre. Finally, the map

$$\tilde{X}/\mathrm{Aut}(\tilde{X}, \pi) \longrightarrow X$$
$$[\tilde{x}] \longmapsto \pi(\tilde{x})$$

is a homeomorphism since the diagram

$$\tilde{X}/\mathrm{Aut}(\tilde{X}, \pi) \longrightarrow X$$

commutes and π is a covering map.

(v) If $p : E \to X$ is an arbitrary covering space, we have the following commutative diagram:

The mapping $\tilde{\pi} : \tilde{X} \to E$ defines the universal covering of E. Thus by iv) we have $E \simeq \tilde{X}/\mathrm{Aut}(\tilde{X}, \tilde{\pi})$. Moreover $G = \mathrm{Aut}(\tilde{X}, \tilde{\pi})$ is a subgroup of $\mathrm{Aut}(\tilde{X}, \pi)$, since if $f : \tilde{X} \longrightarrow \tilde{X}$ verifies $\tilde{\pi} \circ f = \tilde{\pi}$ then $\pi \circ f = p \circ \tilde{\pi} \circ f = p \circ \tilde{\pi} = \pi$.

(vi) Given $f \in \mathrm{Aut}(\tilde{X}, \pi)$, the identity $\pi \circ f = \pi$ means that we can locally describe f as a composition of holomorphic mappings, namely $f_{|U_i} = \left(\pi_{|U_j} \right)^{-1} \circ \pi_{|U_i}$.

(i) Finally, let us say a word about the construction of the universal covering $\pi : \tilde{X} \longrightarrow X$. Choose a point $x_0 \in X$. Since $\tilde{X}$ is going to be a connected covering space of X, any given point, say

$\widetilde{x}$, is going to be the endpoint of a lift of a certain path α joining x_0 to $\pi(\widetilde{x}) \in X$. Moreover, according to the comments preceding Theorem 1.69, any other path α_1 homotopically equivalent to α would produce the same endpoint. This suggests defining $\pi : \widetilde{X} \to X$ as

$$\widetilde{X} := \{(x, \alpha) : x \in X, \alpha \in \Pi(x_0, x)\} \longrightarrow X$$
$$(x, \alpha) \longmapsto x$$

where $\Pi(x_0, x)$ denotes the set of homotopy classes of paths connecting x_0 to x. Since we want π to be a local homeomorphism, the topology in $\widetilde{X}$ is determined by declaring as a fundamental system of neighbourhoods of the point (x, α) the collection of sets of the form

$$\widetilde{U}_{(x,\alpha)} = \{(y, \beta) : y \in V_x, \beta = c \circ \alpha, \ c : I \to V_x \text{ connects } x \text{ to } y\}$$

where V_x is a neighbourhood of x homeomorphic to a disc (see Figure 1.18).

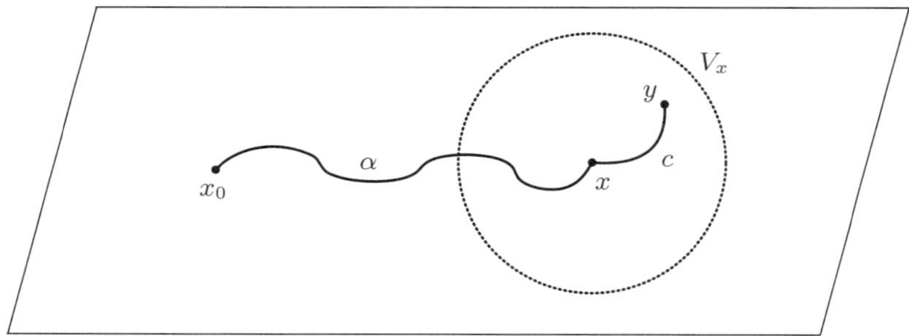

Fig. 1.18. Fundamental neighbourhoods of the universal covering.

$\square$

The following examples are meant to illustrate Theorem 1.69.

Example 1.70 The helix $\widetilde{X} = \left\{\left(e^{2\pi it}, t\right), \ t \in \mathbb{R}\right\}$ in Example 1.65 is the universal covering of $\mathbb{S}^1$. The covering transformations have the form $f\left(e^{2\pi it}, t\right) = \left(e^{2\pi it}, t + n\right)$, therefore we have $\mathrm{Aut}(\widetilde{X}, p) \simeq \mathbb{Z} = \pi_1\left(\mathbb{S}^1\right)$, as expected. Moreover, the fibre of

$e^{2\pi it} \in \mathbb{S}^1$ is $p^{-1}\left(e^{2\pi it}\right) = \left\{\left(e^{2\pi it}, t+n\right) : n \in \mathbb{Z}\right\}$, which can be obviously identified with $\mathbb{Z}$.

Example 1.71 The universal covering of $\mathbb{C}^* = \mathbb{C} \setminus \{0\}$ is

$$\begin{aligned} \pi : \mathbb{C} &\longrightarrow \mathbb{C}^* \\ z &\longmapsto \exp(z) = e^{2\pi iz} \end{aligned}$$

The fibre above $w = e^{2\pi iz}$ is $\pi^{-1}(w) = \{z + n : n \in \mathbb{Z}\}$. We see that $\mathbb{Z} \simeq \operatorname{Aut}(\mathbb{C}, \pi)$ via the correspondence

$$n \longleftrightarrow T_n, \quad \text{where } T_n(z) = z + n$$

We also find that $\mathbb{Z} \simeq \pi_1\left(\mathbb{C}^*, 1\right)$ via the identification

$$n \longleftrightarrow \gamma_n = \pi\left([0, n]\right) = \left\{e^{2\pi it} : t \in [0, n]\right\}$$

where we have taken $x_0 = 1 \in \mathbb{C}^*$ and $\widetilde{x_0} = 0 \in \mathbb{C}$. We see that γ_n corresponds to *turning n times around the origin*.

As the general theory predicts, the covering π identifies $\mathbb{C}/\mathbb{Z}$ with $\mathbb{C}^*$. More precisely, part (iv) of Theorem 1.69 states that the map

$$\begin{aligned} \mathbb{C}/\mathbb{Z} &\longrightarrow \mathbb{C}^* \\ [z]_{\mathbb{Z}} &\longmapsto e^{2\pi iz} \end{aligned}$$

is biholomorphic. Likewise part (v) states that any other covering $p : E \to \mathbb{C}^*$ is isomorphic to a covering of the form

$$\begin{aligned} \mathbb{C}/n\mathbb{Z} &\longrightarrow \mathbb{C}/\mathbb{Z} \simeq \mathbb{C}^* \\ [z]_{n\mathbb{Z}} &\longmapsto [z]_{\mathbb{Z}} \simeq e^{2\pi iz} \end{aligned}$$

which, in turn, is isomorphic to the covering map

$$\begin{aligned} \mathbb{C}^* &\longrightarrow \mathbb{C}^* \\ w &\longmapsto w^n \end{aligned}$$

for we have the following commutative diagram

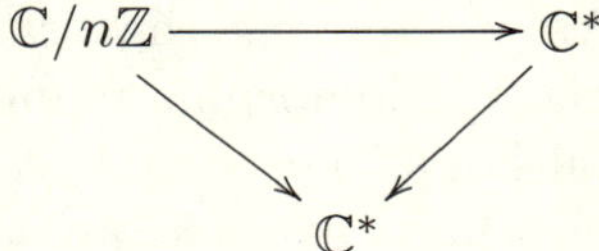

where the maps are given by

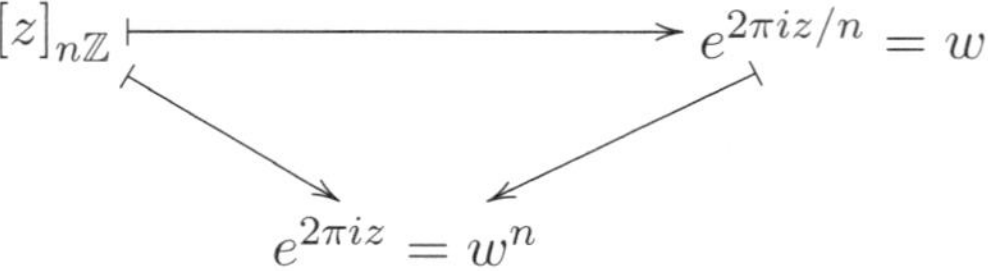

Example 1.72 The universal covering of the punctured disc

$$\mathbb{D}^* = \mathbb{D} \smallsetminus \{0\}$$

is

$$\mathbb{H} \xrightarrow{\ \pi\ } \mathbb{D}^*$$
$$z \longmapsto e^{2\pi i z}$$

Everything goes as in the previous example. In fact both examples are topologically equivalent. We shall remark for further use the fact that, as in the previous case, $\mathbb{D}^*$ admits one covering of each degree. The degree n covering being

$$
\begin{array}{ccc}
\mathbb{H}/n\mathbb{Z} & \longrightarrow & \mathbb{H}/\mathbb{Z} \\
\uparrow\downarrow & & \uparrow\downarrow \\
\mathbb{D}^* & \longrightarrow & \mathbb{D}^* \\
z & \longmapsto & z^n
\end{array}
$$

Example 1.73 Let $w_1, w_2 \in \mathbb{C}$ be linearly independent over $\mathbb{R}$ and consider the abelian group $\Lambda = \mathbb{Z}w_1 \oplus \mathbb{Z}w_2$. This group acts by translations on $\mathbb{C}$ giving rise to a complex torus $\mathbb{C}/\Lambda$. It is clear that the quotient map $\mathbb{C} \to \mathbb{C}/\Lambda$ is the universal covering map and so Λ is isomorphic to the fundamental group of the torus (see Figure 1.19). According to the constructions described in this section, an element $\gamma \in \pi_1(\mathbb{C}/\Lambda)$ is precisely the projection of a segment $[0, \lambda]$, with $\lambda \in \Lambda$.

1.2.6 Ramified coverings

Let $f : X \longrightarrow Y$ be a non-constant morphism between compact Riemann surfaces. After removing the branch values and their preimages in X, we get a holomorphic mapping $f^* : X^* \longrightarrow Y^*$, which clearly is a local homeomorphism. We claim that, in fact, it is a covering map. To show this, take an arbitrary point $y \in Y^*$

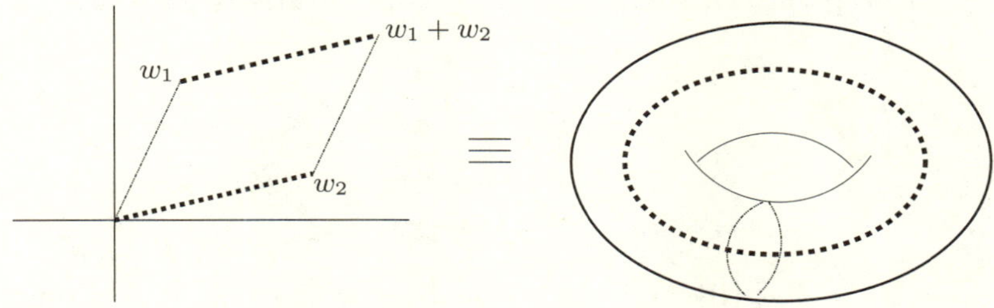

Fig. 1.19. $\Lambda \simeq \pi_1\,(\mathbb{C}/\Lambda)$.

and set $f^{-1}(y) = \{x_1, \ldots, x_d\}$. Let V be a neighbourhood of y and $U_1, \ldots, U_d$ neighbourhoods of $x_1, \ldots, x_d$ such that $f|_{U_i} : U_i \to V$ is a homeomorphism.

We claim that V can be taken small enough so that in fact $f^{-1}(V) = U_1 \sqcup \cdots \sqcup U_d$. If this were not true, we would have a sequence of points $y_n \in V$ approaching y and such that each fibre $f^{-1}(y_n)$ contains a point $x'_n \notin \bigcup U_j$. Let then $x \in X$ be a limit point of this sequence. Since f is continuous we must have $f(x) = y$, therefore x is one of the points $x_j \in f^{-1}(y)$. But then, for n large enough, we should have that $x'_n \in U_j$, which is a contradiction.

Let $y \in Y$ be any point and V any neighbourhood of y isomorphic to the unit disc. Suppose that all points in $V^* = V \smallsetminus \{y\}$ are regular values of f. If the decomposition of the inverse image of V as disjoint union of its connected components is

$$f^{-1}(V) = U_1 \cup \cdots \cup U_r$$

then the decomposition of the inverse image of V^* must be

$$f^{-1}(V^*) = U_1^* \cup \cdots \cup U_r^*$$

with $U_i^* = U_i \smallsetminus \{f^{-1}(y)\}$. Each restriction $f^* : U_i^* \longrightarrow V^* \simeq \mathbb{D}^*$ is again a covering map. By Example 1.72, this covering must be isomorphic to $\mathbb{D}^* \ni z \longmapsto z^{m_i} \in \mathbb{D}^*$ for certain natural number

m_i. More precisely, there must be a commutative diagram

$$
\begin{array}{ccc}
U_i^* & \xrightarrow{\ f^*\ } & V^* \\
\varphi_i \downarrow & & \downarrow \psi \\
\mathbb{D}^* & \longrightarrow & \mathbb{D}^* \\
z & \longmapsto & z^{m_i}
\end{array}
$$

where $\mathbb{D}^* = \mathbb{D} \setminus \{0\}$ and the vertical arrows are analytic isomorphisms.

It follows that $U_i^* = U_i \setminus \{x_i\}$ for certain $x_i \in f^{-1}(y)$. We now invoke Riemann's removable singularity theorem to deduce that φ_i (resp. ψ) extends to an isomorphism which sends x_i (resp. y) to the centre of $\mathbb{D}$. Therefore, near x_i the map f *is of the form* $z \longmapsto z^{m_i}$, i.e. there are charts (U_i, φ_i) of x_i and (V, ψ) of y such that the local expression of f is $z \longmapsto z^{m_i}$. This shows firstly that m_i is the order of f at x_i, and secondly that U_i contains exactly m_i of the d preimages of every unbranched value in V. In particular $\sum m_i = d$.

Summarizing, we have proved the following:

Theorem 1.74 (Local structure of morphisms of Riemann surfaces) *Let* $X \xrightarrow{\ f\ } Y$ *be a non-constant morphism of compact Riemann surfaces.*

(i) *Let* $\Sigma = \Sigma_f \subset Y$ *denote the set of branch values of* f. *Then the restriction*

$$
X^* = X \setminus f^{-1}(\Sigma) \xrightarrow{\ f\ } Y \setminus \Sigma = Y^*
$$

is a covering in the sense of Section 1.2.5.

(ii) *Let* $y \in Y$ *and set* $f^{-1}(y) = \{x_i\}$. *Let* V_y *be any neighbourhood of* y, *isomorphic to a disc, that contains no branch values of* f *apart from (possibly)* y. *Then* $f^{-1}(V_y) = \bigsqcup U_i$, *where each* U_i *is isomorphic to a disc,* $x_i \in U_i$, *and the restriction of* f *to* U_i *is of the form* $z \longmapsto z^{m_i}$, *where* $m_i = m_{x_i}(f)$.

If $V_y^* = V_y \setminus \{y\}$ *then* $f^{-1}(V_y^*) = \bigsqcup U_i^*$, *where* $U_i^* = U_i \setminus \{x_i\}$ *is isomorphic to the punctured disc* $\mathbb{D} \setminus \{0\}$.

(iii) *The number* $\sum_{\{x \,\mid\, f(x)=y\}} m_x(f)$ *does not depend on the choice of* $y \in Y$.

Definition 1.75 Let $f : X \to Y$ be a non-constant morphism of compact Riemann surfaces and let $y \in Y$ be any point. The number

$$\deg(f) = \sum_{\{x \ \mid \ f(x)=y\}} m_x(f)$$

s called the *degree* of f.

It is because of Theorem 1.74 that in the theory of Riemann surfaces the term *covering* is often used to refer to an arbitrary morphism between compact Riemann surfaces $f : X \longrightarrow Y$, whether it is unramified or ramified (in which case f would not be a covering map in the sense of Section 1.2.5). Accordingly, the term *covering transformation* is also used for the case of (possibly ramified) coverings $f : X \longrightarrow Y$. It refers to an automorphism σ of X such that $f \circ \sigma = f$.

If in some particular occasion we need to be more precise about the smoothness of a morphism $f : X \longrightarrow Y$ we shall speak of ramified (or branched) and unramified (or unbranched) coverings.

We can now state the most general version of the Riemann–Hurwitz formula.

Theorem 1.76 (The general Riemann–Hurwitz formula)
Let $f : X \to Y$ be a degree d morphism of compact Riemann surfaces of genera g and g' respectively. Then

$$2g - 2 = d\left(2g' - 2\right) + \sum_{x \in X} \left(m_x(f) - 1\right).$$

To prove this result one simply has to make the obvious adaptation of the proof of Proposition 1.60.

Corollary 1.77 *With the notation of the previous theorem, we have:*

    (i) *In any case, $g \geq g'$. In particular, if $g = 0$ then necessarily $g' = 0$.*

    (ii) *If $g' = 0$ and $g > 0$, then f is necessarily ramified.*

    (iii) *If $g' = 1$, f is unramified if and only if $g = 1$.*

    (iv) *If f is unramified and $g' > 1$, then either $g = g'$ and $d = 1$, or $g > g'$.*

The proof is an easy consequence of the Riemann–Hurwitz formula.

Example 1.78 Let S_1 and S_2 be the Riemann surfaces associated to the curves $x^{2n}+y^{2n}=1$ and $y^2=x^{2n}-1$ respectively. Consider the map

$$S_1 \xrightarrow{\ f\ } S_2$$
$$(x,y) \longmapsto (x, iy^n)$$

According to Examples 1.10 and 1.8, the genera of the two curves equal $(2n-1)(2n-2)/2$ and $n-1$ respectively. We know also that in both cases the x coordinate can be used as local parameter at all but finitely many points (and the points at infinity). Thus, the local expression of f near all but finitely many points is simply $x \mapsto x$. The results of the next section will show that this is enough to ensure that f is holomorphic everywhere (alternatively one can directly check analiticity at each of these points as we did in Examples 1.20 or 1.21).

As every point of the target curve whose y coordinate is non-zero has exactly n preimages by f, the degree of f equals n. Denote by ξ_{2n} a primitive $2n$-th root of unity. The point $(\xi_{2n}^j, 0) \in S_1$ is a branch point of f of multiplicity n, as it is the unique preimage of the point $(\xi_{2n}^j, 0) \in S_2$.

Recall that the curve $x^{2n} + y^{2n} = 1$ has $2n$ points at infinity, whereas the curve $y^2 = x^{2n} - 1$ has only two. Then, necessarily each of these two latter points has exactly n preimages and no further ramification occurs at infinity.

The Riemann–Hurwitz formula gives in this case

$$2\frac{(2n-1)(2n-2)}{2} - 2 = n(2(n-1) - 2) + 2n(n-1)$$

which is correct since both sides equal $4n^2 - 6n$.

Example 1.79 Let S_1 and S_2 be the Riemann surfaces associated to the curves $y^2 = x^8 - 1$ and $y^2 = x^5 - x$. Consider again the morphism defined in Example 1.33 by

$$S_1 \xrightarrow{\;f\;} S_2$$
$$(x, y) \longmapsto (x^2, xy)$$
$$\infty_1 \longmapsto \infty$$
$$\infty_2 \longmapsto \infty$$

Clearly $\deg(f) = 2$, as every point $(a, b) \in S_2$ with $a \neq 0$ has two preimages, namely $\pm(\sqrt{a}, b/\sqrt{a})$. Moreover, this statement holds true also for the points $(0, 0), \infty \in S_2$, whose two preimages are $(0, \pm i)$ and ∞_1, ∞_2 respectively. We thus see that f is an unramified morphism whose Riemann–Hurwitz formula reduces to the identity $2 \cdot 3 - 2 = 2 \cdot (2 \cdot 2 - 2) + 0$.

1.2.7 Auxiliary results about the compactification of Riemann surfaces and extension of maps

Let us start this section with an unramified holomorphic covering of finite degree $f^* : X^* \longrightarrow Y^*$, where Y^* is a *Riemann surface of finite type*, that is Y^* has been obtained from a compact Riemann surface Y by removing finitely many points.

Let y be a point in $Y \smallsetminus Y^*$ and (V_y, φ_y) a coordinate disc around it. Set $V_y^* = V_y \smallsetminus \{y\}$ and let $f^{-1}(V_y^*) = U_{y,1}^* \cup \cdots \cup U_{y,r}^*$ be the decomposition of $f^{-1}(V_y^*)$ as the union of its connected components. Each of them is clearly a holomorphic covering of V_y^*, hence, by Example 1.72, there must be a commutative diagram of the form

$$
\begin{array}{ccc}
U_{y,i}^* & \xrightarrow{\;f\;} & V_y^* \\
\varphi_i \downarrow & & \downarrow \varphi_y \\
\mathbb{D}^* & \longrightarrow & \mathbb{D}^* \\
z & \longmapsto & z^{m_i}
\end{array}
$$

where the vertical arrows are holomorphic isomorphisms.

This leads us to *create* an additional point P_y^i that will play the role of the centre of $U_{y,i}^*$. We extend our holomorphic isomorphism $\varphi_i : U_{i,y}^* \to \mathbb{D}^*$ to a bijection $\varphi_i : U_{y,i} = U_{y,i}^* \cup \{P_y^i\} \longrightarrow \mathbb{D}$ by setting $\varphi_i\left(P_y^i\right) = 0$, and declare φ_i to be a homeomorphism. This way we have defined a new topological space $X := X^* \cup \left\{P_y^i\right\}_{i,y}$. We can provide X with a Riemann surface structure by adding to the atlas of X^* the charts $(U_{y,i}, \varphi_i)$ we have just introduced. Whenever an old chart (U, ψ) intersects one of the new ones, the transition function $\varphi_i \circ \psi^{-1}$ is holomorphic in $U \cap U_{y,i} \subset U_{y,i}^*$, as this is precisely what saying that φ_i is holomorphic means. Moreover, setting $f(P_y^i) = y$ we extend the map $f^* : X^* \longrightarrow Y^*$ to a map $f : X \longrightarrow Y$. Locally we have $\varphi_y \circ f \circ \varphi_i^{-1}(z) = z^{m_i}$. In other words, f is a morphism with $\mathrm{m}_{P_y^i} f = m_i$.

One can also show that X is compact. For this, it is enough to check that $X \smallsetminus \{\bigcup U_{y,i}\}$ is compact (assuming that the nighborhoods V_y have been chosen sufficiently small so that the closures of the open sets $U_{y,i}$ are compact). What is clear is that its image $f(X \smallsetminus \{\bigcup U_{y,i}\}) = Y \smallsetminus \{\bigcup V_y\}$ is compact. Therefore, there is a finite collection of well-covered open sets $V_1, \ldots, V_n$ such that $Y \smallsetminus \{\bigcup V_y\} \subset V_1 \cup \cdots \cup V_n$. From this we deduce that the closed set $X \smallsetminus \{\bigcup U_{y,i}\} = f^{-1}(Y \smallsetminus \{\bigcup V_y\})$ is contained in the compact set $\overline{f^{-1}(V_1)} \cup \cdots \cup \overline{f^{-1}(V_n)}$, hence it is also compact.

Summarizing, we have proved the following result:

Lemma 1.80 (Extension of morphisms) *Let Y be a compact Riemann surface, $\Sigma \subset Y$ a finite subset, $Y^* = Y \smallsetminus \Sigma$. Assume that $f^* : X^* \longrightarrow Y^*$ is an unramified holomorphic covering of finite degree. Then there exists a unique compact Riemann surface $X \supset X^*$ such that f^* extends to a unique morphism $f : X \to Y$. Moreover, $X \smallsetminus X^*$ is a finite set.*

Proof Only the uniqueness of X requires still a proof. This is done in the following Proposition 1.81, where it is shown that the compactification X of X^* is even independent of f. $\square$

Proposition 1.81 *Let X_1 (resp. X_2) be a compact Riemann surface, and $\Sigma_1 \subset X_1$ (resp. $\Sigma_2 \subset X_2$) be a finite set. Assume that $X_1^* = X_1 \smallsetminus \Sigma_1$ and $X_2^* = X_2 \smallsetminus \Sigma_2$ are isomorphic. Then X_1 and X_2 must be isomorphic too.*

Proof Let $\varphi : X_1^* \to X_2^*$ be an isomorphism and let $V_1, \ldots, V_n$ be disjoint coordinate discs around the different points of Σ_2. Take a point $x \in \Sigma_1$ and let $U \subset X_1$ be a coordinate disc around it such that $U \cap X_1^* = U \smallsetminus \{x\}$. Let $\{x_n\} \subset U \cap X_1^*$ be an arbitrary sequence of points converging to x, and let $y \in X_2$ be a limit point of the sequence $\{\varphi(x_n)\}$ in X_2. Then $y \in \Sigma_2$, since $y \in X_2^*$ would imply that $\varphi^{-1}(y)$ is a limit point of the sequence $\{x_n\}$, hence we would have $x = \varphi^{-1}(y) \in X_1^*$, which is a contradiction. This means that if U has been chosen sufficiently small then $\varphi(U \smallsetminus \{x\})$ is contained in $V_1 \cup \cdots \cup V_n$, hence, by connectedness, in just one of them. Now the removable singularity theorem allows us to extend the map $\varphi : U \smallsetminus \{x\} \to X_2$ to the whole U by the rule $\varphi(x) = y$.

After proceeding in the same way at all points, we get a holomorphic map $\widehat{\varphi} : X_1 \to X_2$ that has degree 1, i.e. it is an isomorphism. $\qquad\square$

Because of these extension properties we will use a notation such as $\left\{y^2 = \prod_{k=1}^{2g+1} (x - a_k)\right\}$ to refer not only to the finite points of the curve $y^2 = \prod_{k=1}^{2g+1} (x - a_k)$ but sometimes also to the compact Riemann surface associated to it.

Let us give now a couple of examples which illustrate the usefulness of Lemma 1.80.

Example 1.82 (The hyperelliptic involution) Consider the map J from the Riemann surface $\left\{y^2 = \prod_{k=1}^{2g+1} (x - a_k)\right\}$ to itself (see Example 1.7) defined by $J(x, y) = (x, -y)$. Away from the points $(a_k, 0)$ and ∞, this is clearly a bijective holomorphic map. Now by Lemma 1.80 we know that J defines an automorphism (usually known as the *hyperelliptic involution*) without having to check analyticity at the unpleasant points.

Note that $\langle J \rangle$ is the full group of covering transformations of the degree two ramified covering of the Riemann sphere given by the x coordinate (see Example 1.59).

Example 1.83 (Normalized model of hyperelliptic curves) Here we show that the curves introduced in Examples 1.7 and 1.8 produce equivalent families of Riemann surfaces. In fact we do a little more than that. We prove that their corresponding coordinate function $\mathbf{x}$ produce equivalent families of coverings. In more precise terms, we show that every covering of the form

$$\left\{y^2 = \prod_{i=1}^{n} (x - a_i)\right\} \longrightarrow \widehat{\mathbb{C}}$$

$$(x, y) \longmapsto x$$

where $n = 2g + 2$ is equivalent to one of the same form with $n = 2g + 1$ and viceversa.

In order to find values $b_1, \ldots, b_{2g+1}$ and the isomorphisms F and f that make commutative the diagram

$$
\left\{ y^2 = \prod_{i=1}^{2g+2} (x - a_i) \right\} \xrightarrow{\ F\ } \left\{ y^2 = \prod_{j=1}^{2g+1} (x - b_j) \right\}
$$
$$
\downarrow \mathbf{x} \qquad\qquad\qquad\qquad \downarrow \mathbf{x}
$$
$$
\widehat{\mathbb{C}} \xrightarrow{\qquad\ f\ \qquad} \widehat{\mathbb{C}}
$$

we first observe that f must be a Möbius transformation that sends the branch value set $\{a_1, \ldots, a_{2g+2}\}$ of the first covering to the branch value set $\{b_1, \ldots, b_{2g+1}, \infty\}$ of the second one. We can therefore choose f to be defined by $f(z) = \dfrac{1}{z - a_{2g+2}}$, so that $f(a_{2g+2}) = \infty$ and $b_i = f(a_i) = \dfrac{1}{a_i - a_{2g+2}}$. Accordingly, the isomorphism F will take the form $F(x, y) = \left(\dfrac{1}{x - a_{2g+2}}, y_1 \right)$, where $y_1 = y_1(x, y)$ must satisfy the equation

$$
y_1^2 = \prod_{i=1}^{2g+1} \left(\frac{1}{x - a_{2g+2}} - \frac{1}{a_i - a_{2g+2}} \right)
$$

To compute y_1 we proceed as follows:

$$
\left(\frac{y}{(x - a_{2g+2})^{g+1}} \right)^2 = \left(\frac{x - a_1}{x - a_{2g+2}} \right) \cdots \left(\frac{x - a_{2g+1}}{x - a_{2g+2}} \right)
$$
$$
= \left(1 - \frac{a_1 - a_{2g+2}}{x - a_{2g+2}} \right) \cdots \left(1 - \frac{a_{2g+1} - a_{2g+2}}{x - a_{2g+2}} \right)
$$
$$
= - \prod_{i=1}^{2g+1} (a_i - a_{2g+2}) \left(\frac{1}{x - a_{2g+2}} - \frac{1}{a_i - a_{2g+2}} \right)
$$

hence we can take

$$
y_1 = \frac{y}{(x - a_{2g+2})^{g+1} \sqrt{ - \prod_{i=1}^{2g+1} (a_i - a_{2g+2}) }}
$$

One can even do the following normalization. The identity

$$\frac{y^2}{(b_2 - b_1)^{2g+1}} = \frac{\prod_{i=1}^{2g+1}(x - b_i)}{(b_2 - b_1)^{2g+1}}$$

$$= \prod_{i=1}^{2g+1}\left(\frac{(x - b_1) - (b_i - b_1)}{b_2 - b_1}\right)$$

shows that the hyperelliptic covering

$$\{y^2 = (x - b_1)\cdots(x - b_{2g+1})\} \longrightarrow \widehat{\mathbb{C}}$$

$$(x, y) \longmapsto x$$

is isomorphic to the *normalized hyperelliptic covering*

$$\{y^2 = x(x - 1)(x - c_3)(x - c_4)\cdots(x - c_{2g+1})\} \longrightarrow \widehat{\mathbb{C}}$$

$$(x, y) \longmapsto x$$

where $c_i = \dfrac{b_i - b_1}{b_2 - b_1}$ and the isomorphism is given by

$$F(x, y) = \left(\frac{x - b_1}{b_2 - b_1}, \frac{y}{\sqrt{(b_2 - b_1)^{2g+1}}}\right)$$

1.3 Curves, function fields and Riemann surfaces

So far, we have worked with some concrete examples of Riemann surfaces obtained from certain algebraic curves. Now we prove the fact that every algebraic curve determines a compact Riemann surface.

We start with an auxiliary result of algebraic geometric nature.

Lemma 1.84 *Let K be an algebraically closed field (such as $\overline{\mathbb{Q}}$ and $\mathbb{C}$) and $F(X,Y), G(X,Y) \in K[X,Y]$ polynomials in two variables with coefficients in K. The following statements hold:*

> (i) *(Weak form of Bezout's Theorem) If F and G are relatively prime then the curves $F(x,y) = 0$ and $G(x,y) = 0$ intersect only at finitely many points. Moreover, these points have coordinates in K.*

(ii) *(Weak form of Nullstellensatz) If F is irreducible and G vanishes at all points of the curve $F(x,y) = 0$ then F divides G.*

Proof (i) Using the tautological isomorphism $K[X,Y] \equiv K[X][Y]$ we may regard F and G as elements of $K(X)[Y]$. By hypothesis F and G are coprime in $K[X][Y]$, hence they are still coprime in $K(X)[Y]$ (Gauss Lemma, see e.g. [Lan84]). Let

$$1 = AF + BG$$

be a Bezout identity in $K(X)[Y]$. Getting rid of the denominators, we obtain

$$q(X) = A_1 F + B_1 G$$

where $q(X) \in K[X]$ and $A_1 = qA, B_1 = qB$ lie in $K[X,Y]$.

Now, in case F and G had infinitely many common solutions $\{(x_n, y_n)\}_{n=1}^{\infty}$ then all values in the necessarily infinite sequence $\{x_n\}_{n=1}^{\infty}$ would be roots of the polynomial $q(X)$, which is a contradiction. Moreover, if a point $P = (x, y)$ belongs to the intersection of both curves then $q(x) = 0$ and, since $q(X) \in K[X]$ and K is algebraically closed, the first coordinate x must lie in K. But since F has coefficients in K the second coordinate y must lie in K too.

(ii) By (i) F and G cannot be coprime which, F being irreducible, can only mean that F divides G. $\qquad\square$

Remark 1.85 Lemma 1.84 contains weak versions of two celebrated theorems in algebraic geometry, namely Bezout's Theorem and Hilbert's Nullstellensatz. Bezout's Theorem states that if in (i) we set $\deg(F) = n$ and $\deg(G) = m$ then the number of common solutions is always $\leq nm$ with equality if the points are suitable counted. The Nullstellensatz for curves states that if in (ii) the polynomial F is not assumed to be irreducible then the correct conclusion is that F divides some power of G.

Theorem 1.86 *Let*

$$F(X,Y) = p_0(X)Y^n + p_1(X)Y^{n-1} + \cdots + p_n(X)$$

$$= q_0(Y)X^m + q_1(Y)X^{m-1} + \cdots + q_m(Y)$$

be an irreducible polynomial. If $n \geq 1$ define

$$S_F^X = \left\{ (x,y) \in \mathbb{C}^2 \mid F(x,y) = 0, F_Y(x,y) \neq 0, p_0(x) \neq 0 \right\}$$

and, similarly, if $m \geq 1$ set

$$S_F^Y = \left\{ (x,y) \in \mathbb{C}^2 \mid F(x,y) = 0, F_X(x,y) \neq 0, q_0(y) \neq 0 \right\}$$

Then:

(i) *S_F^X and S_F^Y are connected Riemann surfaces on which the coordinate functions $\mathbf{x}$ and $\mathbf{y}$ are holomorphic functions.*

(ii) *There exists a unique compact and connected Riemann surface $S = S_F$ that contains S_F^X and S_F^Y.*

(iii) *The coordinate functions $\mathbf{x}$ and $\mathbf{y}$ extend to meromorphic functions on S.*

(iv) *The branching points of $\mathbf{x}$ (resp. $\mathbf{y}$) lie in the finite set $S \smallsetminus S_F^X$ (resp. $S \smallsetminus S_F^Y$).*

Proof The holomorphic structure in S_F^X is completely clear. It arises from *solving* y in terms of x, something possible thanks to the implicit function theorem. It is obvious that $\mathbf{x}$ and $\mathbf{y}$ are holomorphic functions, and so is the fact that $\mathbf{x} : S_F^X \longrightarrow \mathbf{x}\left(S_F^X\right) \subset \widehat{\mathbb{C}}$ is a covering map (of degree $n = \deg_Y F$). Moreover, since the polynomials F and F_Y have only finitely many zeros in common (Lemma 1.84), we see that $\mathbf{x}\left(S_F^X\right)$ fills $\widehat{\mathbb{C}}$ except for finitely many values $\{a_1, \ldots, a_r, \infty\}$. Let now W be a connected component of S_F^X. Clearly, the restriction $\mathbf{x} : W \longrightarrow \widehat{\mathbb{C}} \smallsetminus \{a_1, \ldots, a_r, \infty\}$ is a covering map with degree $d \leq n$. By Lemma 1.80, there is a unique morphism of compact Riemann surfaces $\mathbf{x} : \widehat{W} \to \widehat{\mathbb{C}}$ extending the map $\mathbf{x}$.

We would like to see that $W = S_F^X$, i.e. that S_F^X is already connected. We consider the symmetric functions

$$s_1(x) = \sum y_i(x), \ \ s_2(x) = \sum y_i(x)y_j(x), \ \ \ldots, s_d(x) = \prod y_i(x)$$

where the points $(x, y_1(x)), \ldots, (x, y_d(x)) \in S_F^X$ are the preimages of $x \in \widehat{\mathbb{C}} \setminus \{a_1, \ldots, a_r, \infty\}$ via the function $\mathbf{x} : W \longrightarrow \mathbb{C}$. In particular, $y_1(x), \ldots, y_d(x)$ are roots of $F(x, Y)$ when considered as a polynomial in one variable. Each of them is a holomorphic function defined only in a certain open set of $\widehat{\mathbb{C}}$. However, the functions $s_i(x)$ are well-defined holomorphic functions in the whole $\widehat{\mathbb{C}} \setminus \{a_1, \ldots, a_r, \infty\}$. On the other hand, it can be easily seen that near a_k the roots $y_k(x)$ are bounded in terms of the coefficients of the polynomial $F(x, Y) \in \mathbb{C}[Y]$ (see Lemma 1.88 below). Similarly, the functions $1/y_k(x)$ remain bounded near ∞. Therefore, each function $s_i(x)$ extends to a meromorphic function defined in the whole $\widehat{\mathbb{C}}$, hence it can be identified to a rational function $s_i(x) \in \mathbb{C}(x)$.

We now consider the polynomial

$$G(X, Y) = s(X)(Y^d - s_1(X)Y^{d-1} + s_2(X)Y^{d-2} - \cdots \pm s_d(X))$$

where $s(X)$ is the least common multiple of the denominators of the rational functions $s_i(X)$. Any point $P \in W$ can be written as $P = (x, y_j(x))$ for some $j \in \{1, \ldots, d\}$. Therefore,

$$\begin{aligned}
G(P) &= s(x)(y_j^d(x) - s_1(x)y_j^{d-1}(x) + \cdots \pm s_d(x)) \\[2mm]
&= s(x) \textstyle\prod_{i=1}^d (y_j(x) - y_i(x)) \\[2mm]
&= 0
\end{aligned}$$

The conclusion is that $G(X, Y)$ vanishes identically in W (as does also the irreducible polynomial $F(X, Y)$). By Lemma 1.84 the polynomial G has to be a multiple of F, hence $\deg_Y(G) \geq \deg_Y(F)$. It follows that $d = n$. Moreover, as by construction the coefficients of G as a polynomial in $\mathbb{C}[X][Y]$ are coprime, one easily sees that $F = G$. In particular $W = S_F^X$.

Of course the proof of the statement relative to S_F^Y is the same. Since S_F^X and S_F^Y coincide apart from finitely many points, Proposition 1.81 implies that they have a common compactification S_F.
$\square$

Example 1.87 If $F(X, Y) = Y$ then $S_F^X = \{(x, 0) \mid x \in \mathbb{C}\} \equiv \mathbb{C}$ and so $S_F = \widehat{\mathbb{C}}$.

Lemma 1.88 *If $\alpha^n + c_1 \alpha^{n-1} + \cdots + c_n = 0$ then $|\alpha| < 2 \max |c_k|^{1/k}$.*

Proof Take $c = \max |c_k|^{1/k}$ and divide by c^n to obtain

$$\left|\frac{\alpha}{c}\right|^n \leq \left|\frac{\alpha}{c}\right|^{n-1} + \left|\frac{\alpha}{c}\right|^{n-2} + \cdots + \left|\frac{\alpha}{c}\right| + 1 = \frac{|\alpha/c|^n - 1}{|\alpha/c| - 1}$$

Now, if $|\alpha/c| \geq 2$ we would get $|\alpha/c|^n \leq |\alpha/c|^n - 1$, which is a contradiction. $\square$

Having proved that every irreducible curve is a compact Riemann surface, a natural question is whether the converse statement also holds, that is one would like to know if all compact Riemann surfaces are obtained in this way. The answer to this question is affirmative; proving it is the next section's main goal.

1.3.1 The function field of a Riemann surface

Recall that if S is a compact Riemann surface, we denote by $\mathcal{M}(S)$ the field of its (meromorphic) functions.

Proposition 1.89 *Suppose $f \in \mathcal{M}(S)$ has degree n. Then the field extension $\mathbb{C}(f) \subset \mathcal{M}(S)$ has degree $\leq n$. (In fact the degree equals exactly n, as we will soon see.)*

Proof It is enough to show that every $h \in \mathcal{M}(S)$ satisfies a polynomial of degree $\leq n$ with coefficients in $\mathbb{C}(f)$. Then the result will follow from the Primitive Element Theorem†, see [Lan84].

Let $y_1(x), \ldots, y_n(x) \in S$ be the preimages of x by f, counted according to their multiplicities. Consider the expressions

$$b_1(x) = \sum h\left(y_i(x)\right)$$

$$b_2(x) = \sum h\left(y_i(x)\right) h\left(y_j(x)\right)$$

$$\vdots$$

$$b_n(x) = \prod h(y_i(x))$$

and set

$$p(y) = \prod \left(h(y) - h\left(y_i(f(y))\right)\right) = \sum (-1)^k b_k(f(y)) h(y)^{n-k}$$

Arguing as in the proof of Theorem 1.86, it can be easily seen that the symmetric functions $b_i(x)$ define functions on the whole

---

† One possible version of the Primitive Element Theorem states that any finite extension $k \hookrightarrow K = k(\alpha_1, \ldots, \alpha_n)$ between fields of characteristic zero (such as ours) is generated by a single *primitive element* $\beta \in K$, which can be chosen to be of the form $\beta = k_1\alpha_1 + \cdots + k_n\alpha_n$.

$\mathbb{P}^1$; they are therefore rational functions which we denote $b_i(x)$. On the other hand, it is clear that $p(y)$ defines a local function that vanishes identically, since one of the preimages $y_i(f(y))$ must coincide with y.

Then the polynomial with coefficients in $\mathbb{C}(f)$ that has h as one of its roots is going to be

$$P(Y) = Y^n - b_1(f)Y^{n-1} + \cdots \pm b_n(f) = \sum (-1)^k b_k(f) Y^{n-k}$$

since the value of the function

$$P(h) = \sum (-1)^k b_k(f) h^{n-k}$$

at an arbitrary point $y \in S$ is $P(h)(y) = p(y) = 0$. $\qquad\square$

We need at this point a result that will be proved in the next chapter (Corollary 2.12, Proposition 2.16).

Theorem 1.90 *Given two points P and Q of a compact Riemann surface S there exists a meromorphic function $\varphi \in \mathcal{M}(S)$ such that $\varphi(P) = 0$ and $\varphi(Q) = \infty$.*

We often refer to this statement as the *separation property of the field of meromorphic functions*. We must stress that this is a highly non-trivial result. For instance, we already know that S does not admit non-constant holomorphic functions $\psi : S \longrightarrow \mathbb{C}$ (Remark 1.25).

Theorem 1.91 *Let $\mathcal{M}(S) = \mathbb{C}(f, h)$, and let $F(X, Y)$ be an irreducible polynomial such that $F(f, h) \equiv 0$. Then the rule*

$$S \xrightarrow{\;\;\Phi\;\;} S_F$$
$$P \longmapsto (f(P), h(P))$$

defines an isomorphism.

Proof We start by showing that Φ is a well-defined map. With the notation of Theorem 1.86, let $\mathbf{x}\left(S_F^X\right) = \widehat{\mathbb{C}} \smallsetminus \{a_1, \ldots, a_r, \infty\}$. Set $B = \{a_1, \ldots, a_r, \infty\}$, and $S^0 = S \smallsetminus f^{-1}(B)$. We have the

following commutative diagram

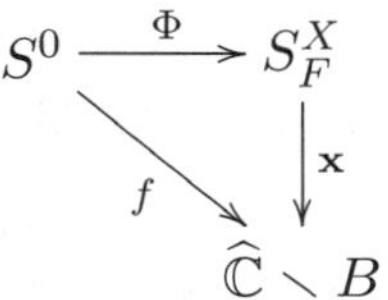

We note that if $f(P) = a \in \widehat{\mathbb{C}} \setminus \{a_1, \ldots, a_r, \infty\}$, then the value of h at P must be one of the n distinct roots of $F(a, Y)$, hence $\Phi(P)$ is a well-defined point of S_F^X for every $P \in S^0$. Now, by the results in Section 1.2.7, in order to be able to extend Φ to the whole S we only need to show that $\Phi : S^0 \longrightarrow S_F^X$ is a covering map. This is a consequence of the fact that $\mathbf{x}$ and f are so. Indeed if, with the notation of Theorem 1.74, $f^{-1}(V_a) = \bigsqcup U_i$ and $\mathbf{x}^{-1}(V_a) = \bigsqcup W_j$, then $\Phi^{-1}(W_j)$ can only be a disjoint union of a number of the open sets U_i.

It remains to be shown that Φ has degree 1. Suppose not; then the fibres of all but finitely many points $Q = (a, b) \in S_F^X$ would contain at least two points Q_1, Q_2. Let now φ be an arbitrary meromorphic function. As $\mathcal{M}(S)$ is generated by f and h, it follows that φ can be expressed as a rational function in f and h, say

$$\varphi = \frac{\sum a_{ij} f^i h^j}{\sum b_{ij} f^i h^j}$$

hence

$$\varphi(Q_1) = \frac{\sum a_{ij} a^i b^j}{\sum b_{ij} a^i b^j} = \varphi(Q_2)$$

This means that for all these pairs of points any meromorphic function takes the same value at Q_1 and Q_2. Thus, no meromorphic function can have a zero at Q_1 and a pole at Q_2, contradicting Theorem 1.90. $\qquad\square$

Remark 1.92 The fact that f and h generate the function field is not used in the first part of the previous proof. Therefore, the same argument shows that

$$S \xrightarrow{\;\;\Phi\;\;} S_F$$
$$P \longmapsto (f(P), h(P))$$

defines a morphism for every pair of functions f and h such that $F(f, h) \simeq 0$.

Corollary 1.93 *Let (F) denote the ideal of $\mathbb{C}[X, Y]$ generated by F. Then:*

(i) *The correspondence determined by $X \to f$, $Y \to h$ defines a $\mathbb{C}$-isomorphism from the quotient field of $\mathbb{C}[X, Y] / (F)$ to $\mathcal{M}(S)$.*

(ii) *The correspondence determined by $X \to \mathbf{x}$, $Y \to \mathbf{y}$ defines a $\mathbb{C}$-isomorphism from the quotient field of $\mathbb{C}[X, Y] / (F)$ to $\mathcal{M}(S_F)$. In particular $\mathcal{M}(S_F) = \mathbb{C}(\mathbf{x}, \mathbf{y})$.*

(iii) $F(\mathbf{x}, Y) \in \mathbb{C}(\mathbf{x})[Y]$ *(resp. $F(f, Y) \in \mathbb{C}(f)[Y]$) is the minimal polynomial of $\mathbf{y}$ over $\mathbb{C}(\mathbf{x})$ (resp. h over $\mathbb{C}(f)$).*

(iv) $\deg(f) = [\mathcal{M}(S) : \mathbb{C}(f)]$.

Proof (i) As $F(f, h) = 0 \in \mathcal{M}(S)$ the assignment $X \to f$ and $Y \to h$ defines a homomorphism of $\mathbb{C}$-algebras

$$\rho : \mathbb{C}[X, Y] / (F) \to \mathcal{M}(S)$$

It remains to show that its kernel is precisely the ideal (F). Now, saying that $G(X, Y)$ lies in $\ker(\rho)$ means that $G(f, h) = 0 \in \mathcal{M}(S)$, which by Theorem 1.78 is equivalent to saying that $G(X, Y)$ vanishes identically on the curve $F(x, y) = 0$, which by part (ii) of Lemma 1.84 means, in turn, that $G \in (F)$.

(ii) is a equivalent to (i), once Theorem 1.91 has been established.

(iii) Obvious.

(iv) $[\mathcal{M}(S) : \mathbb{C}(f)]$ is the degree of the minimal polynomial of h over $\mathbb{C}(f)$, namely $F(f, Y) \in \mathbb{C}(f)[Y]$. This degree is of course $\deg_Y(F)$, which clearly is also the degree of the function $\mathbf{x}$, which, by Theorem 1.91, is the same as the degree of f. $\qquad\square$

Remark 1.94 We have shown the equivalence between the following classes of objects:

(1) Compact Riemann surfaces S.

(2) *Function fields in one variable* (i.e. finite extensions of $\mathbb{C}(X)$).

(3) Irreducible algebraic curves $C : F(X, Y) = 0$.

We pass from (1) to (2) by considering the field $\mathcal{M}(S)$. The link from (2) to (3) is performed by choosing a pair of generators f, h and an irreducible algebraic relation $F(f, h) = 0$ between them. Finally, the bridge from (3) to (1) connects an irreducible algebraic curve $F(X, Y)$ to the Riemann surface S_F. We observe that the correspondence $(1) \Rightarrow (2)$ is more natural than the others, in the sense that it does not depend on any choices.

In fact this is a *functorial correspondence*, which means that the rule that associates to each Riemann surface S its function field $\mathcal{M}(S)$ and to each morphism of Riemann surfaces $f : S_1 \longrightarrow S_2$ the $\mathbb{C}$−algebra homomorphism $f^* : \mathcal{M}(S_2) \longrightarrow \mathcal{M}(S_1)$ defined by $f^*(\varphi) = \varphi \circ f$ is a (contravariant) *functor* from the category of compact Riemann surfaces to the category of function fields. Furthermore, this functor is actually an equivalence of categories.

Proposition 1.95 *The functor described above establishes an isomorphism between the categories of Riemann surfaces and function fields in one variable.*

Proof We have to prove that the following two statements hold:

(i) If $f, h \in \mathrm{Mor}(S_1, S_2)$ are such that $f^* = h^*$ then $f = h$.

(ii) If $\varphi : \mathcal{M}_2 \longrightarrow \mathcal{M}_1$ is an arbitrary $\mathbb{C}$-algebra homomorphism (i.e. homomorphism over $\mathbb{C}$) between two fuction fields $\mathcal{M}_2$ and $\mathcal{M}_1$ then there are Riemann surfaces S_1, S_2 and $f \in \mathrm{Mor}(S_1, S_2)$ such that the following diagram commutes:

$$
\begin{array}{ccc}
\mathcal{M}(S_2) & \xrightarrow{\ f^*\ } & \mathcal{M}(S_1) \\
\uparrow\downarrow & & \uparrow\downarrow \\
\mathcal{M}_2 & \xrightarrow{\ \varphi\ } & \mathcal{M}_1
\end{array}
$$

where the vertical arrows are field isomorphisms over $\mathbb{C}$.

(i) Let $f, h \in \mathrm{Mor}(S_1, S_2)$ and $x \in S_1$ such that $f(x) = P$ and $h(x) = Q$ with $P \neq Q$. By Theorem 1.90 there exists $\varphi \in \mathcal{M}(S_2)$ such that $\varphi(P) \neq \varphi(Q)$. Therefore, $(f^*\varphi)(x) = \varphi(f(x))$ does not agree with $(h^*\varphi)(x) = \varphi(h(x))$. This proves the first statement.

(ii) Let $\varphi : \mathcal{M}_2 \longrightarrow \mathcal{M}_1$ a $\mathbb{C}$-algebra homomorphism of fields (hence necessarilly an injection).

For $i = 1, 2$ let x_i, y_i be generators of $\mathcal{M}_i$ such that y_i is algebraic over $\mathbb{C}(x_i)$. As above, choose an irreducible polynomial $F_i(X, Y)$ satisfying $F_i(x_i, y_i) = 0$.

Consider the following commutative diagram:

$$\begin{array}{ccc} \mathcal{M}(S_{F_2}) & \xrightarrow{\;\widetilde{\varphi}\;} & \mathcal{M}(S_{F_1}) \\ \Big\downarrow{\alpha_2} & & \Big\downarrow{\alpha_1} \\ \mathcal{M}_2 & \xrightarrow{\;\varphi\;} & \mathcal{M}_1 \end{array}$$

where each α_i is an isomorphism defined by sending the coordinate functions $\mathbf{x}, \mathbf{y}$ of the Riemann surface S_{F_i} to the generators x_i, y_i of $\mathcal{M}_i$ (Corollary 1.93), and $\widetilde{\varphi} = \alpha_i^{-1}\varphi\alpha_2$.

Put $\alpha_1^{-1}\varphi(x_2) = R_1(\mathbf{x}, \mathbf{y}) \in \mathcal{M}(S_{F_1})$ and $\alpha_1^{-1}\varphi(y_2) = R_2(\mathbf{x}, \mathbf{y}) \in \mathcal{M}(S_{F_1})$ where R_1 and R_2 are certain rational fuctions in two variables. We can now write

$$0 = F_2(x_2, y_2) \in \mathcal{M}(S_{F_2})$$

and therefore

$$\begin{aligned} 0 &= \alpha_1^{-1}\varphi(F_2(x_2, y_2)) \\[1em] &= F_2(\alpha_1^{-1}\varphi(x_2), \alpha_1^{-1}\varphi(y_2)) \\[1em] &= F_2(R_1(\mathbf{x}, \mathbf{y}), R_2(\mathbf{x}, \mathbf{y})) \in \mathcal{M}(S_{F_1}) \end{aligned}$$

Now Remark 1.92 tells us that the rule

$$f(x, y) = (R_1(x, y), R_2(x, y))$$

defines a morphism $f : S_{F_1} \longrightarrow S_{F_2}$.

Moreover, we claim that $f^* = \widetilde{\varphi}$ as required. This is because f^* and $\widetilde{\varphi}$ coincide on the generators $\mathbf{x}, \mathbf{y}$ of $\mathcal{M}(S_2)$, e.g.

$$f^*(\mathbf{x}) = R_1(\mathbf{x}, \mathbf{y}) = \alpha_1^{-1}\varphi(x_2) = \alpha_1^{-1}\varphi(\alpha_2(\mathbf{x})) = \widetilde{\varphi}(\mathbf{x})$$

$\square$

1.3.2 *Manipulating generators of a function field*

We devote this section to describe the kind of manipulations that are often performed when dealing with function fields. The point is that one can choose the generators of a given function field so as to obtain convenient associated curves.

To illustrate this idea we first work out the case of the Riemann surfaces of genus 3 associated to curves of the form

$$F(X,Y) = Y^7 - P(X) \in \mathbb{C}[X,Y]$$

We will find that among them there are only two isomorphism classes, namely those corresponding to $F_1(X,Y) = Y^7 - X(X-1)$ and $F_2(X,Y) = Y^7 - X(X-1)^2$.

According to the preceding results this statement is equivalent to saying that among the finite field extensions $\mathcal{K} = \mathbb{C}(x,y)$ of the field of rational functions $\mathbb{C}(x)$ satisfying a relation of the form $y^7 = P(x)$ whose corresponding Riemann surfaces have genus three there are only two different isomorphism classes.

So, let us assume that $\mathcal{K} = \mathbb{C}(x,y)$ with

$$y^7 = (x - a_1)^{m_1}(x - a_2)^{m_2} \cdots (x - a_k)^{m_k} \tag{1.5}$$

We proceed in several steps:

Step 1: suppose that $m_1 = 7m + n$ with $0 \leq n < 7$. Then, if we set $y_1 = \dfrac{y}{(x - a_1)^m}$ we see that $\mathcal{K}$ agrees with $\mathbb{C}(x,y_1)$ with $y_1^7 = (x - a_1)^n(x - a_2)^{m_2} \cdots (x - a_r)^{m_r}$. Proceeding in the same way with the rest of the factors we conclude that $\mathcal{K}$ has generators, still denoted x and y, satisfying an equation of the form

$$y^7 = (x - a_1)^{m_1}(x - a_2)^{m_2} \cdots (x - a_r)^{m_r}; \quad 1 \leq m_i < 7 \tag{1.6}$$

Step 2: now, by Example 1.11, in order for the equation 1.6 to define a Riemann surface of genus 3, one of the following two possibilities must occur:

(a) $r = 2$ and $m_1 + m_2$ is prime to 7.

(b) $r = 3$ and $m_1 + m_2 + m_3$ is multiple of 7.

Next we observe that case (b) can be transformed into case (a). To see this we start with an equation of the form

$$y^7 = (x - a_1)^{m_1}(x - a_2)^{m_2}(x - a_3)^{m_3}$$

with $m_1 + m_2 + m_3 = 7m$.

Therefore

$$\left(\frac{y}{(x-a_3)^m}\right)^7 = \left(\frac{x-a_1}{x-a_3}\right)^{m_1}\left(\frac{x-a_2}{x-a_3}\right)^{m_2}$$

$$= \left(1 - \frac{a_1-a_3}{x-a_3}\right)^{m_1}\left(1 - \frac{a_2-a_3}{x-a_3}\right)^{m_2}$$

$$= c\left(\frac{1}{a_1-a_3} - \frac{1}{x-a_3}\right)^{m_1}\left(\frac{1}{a_2-a_3} - \frac{1}{x-a_3}\right)^{m_2}$$

for a suitable constant c.

This shows that the generators of $\mathcal{K}$ given by $\dfrac{1}{x-a_3}, \dfrac{c^{-1/7}y}{(x-a_3)^m}$ are related by an equation of the form

$$y^7 = (x-b_1)^{m_1}(x-b_2)^{m_2}, \quad 1 \le m_i < 7 \tag{1.7}$$

Step 3: we now rewrite the identity (1.7) as

$$y^7 = ((x-b_1) - 0)^{m_1}((x-b_1) - (b_2-b_1))^{m_2}, \quad 1 \le m_i < 7$$

which, chosing $(x-b_1)$ and y as new generators, takes the form $y^7 = x^{m_1}(x-c)^{m_2}$, a relation which in turn can be written as $(py)^7 = (qx)^{m_1}(qx-1)^{m_2}$ for suitable non-zero constants c, p and q. Thus, in terms of the generators qx and py our algebraic curve becomes

$$y^7 = x^{m_1}(x-1)^{m_2}, \quad 1 \le m_i < 7 \tag{1.8}$$

Step 4: let us now choose integers d and $n > 0$ satisfying a Bezout identity $nm_1 + 7d = 1$. Then our relation (1.8) implies that

$$y^{7n} = x^{nm_1}(x-1)^{nm_2} = \frac{x}{x^{7d}}(x-1)^{nm_2}, \quad 1 \le m_i < 7$$

Now replacing y with yx^d and arguing as in the first step if $nm_2 \ge 7$ we conclude that $\mathcal{K}$ can be generated by two elements, still denoted x, y satisfying an equation of the form

$$y^7 = x(x-1)^m, \quad 1 \le m < 6 \tag{1.9}$$

Now we claim that:

(1) The values $m = 1, 3, 5$ produce isomorphic fields.
(2) The values $m = 2, 4$ produce also isomorphic fields.

To check statement (1) we start making $m = 3$ and observe that $y^7 = x(x-1)^3 \Rightarrow y^{7\cdot5} = x^5(x-1)^{3\cdot5}$. From here we see that the generators $x_1 = x, y_1 = \dfrac{y^5}{(x-1)^2}$ are related by the equation $y_1^7 = x_1^5(x_1 - 1)$, which can be turned into $y^7 = x(x-1)^5$, arguing as in the third step.

In turn,

$$y^7 = x^5(x-1) \quad \Rightarrow \quad (y/x)^7 = \left(\frac{1}{x}\right)\left(1 - \frac{1}{x}\right)$$

$$\Rightarrow \quad \left(\frac{\xi_{14}y}{x}\right)^7 = \left(\frac{1}{x}\right)\left(\frac{1}{x} - 1\right)$$

Hence, if we take $\dfrac{\xi_{14}y}{x}, \dfrac{1}{x}$ as new generating pair of functions we arrive at the desired equation $y^7 = x(x-1)$. This completes the proof of claim (1) except that in this case it is not entirely obvious that the functions $x_1 = x, y_1 = \dfrac{y^5}{(x-1)^2}$ do indeed generate $\mathcal{K}$. In order to leave no doubt of that we observe that

$$y^7 = x^5(x-1) \Rightarrow y = \frac{(y^5)^3}{(y^7)^2} = \frac{(y^5)^3}{(x^5(x-1))^2}$$

Turning now to statement (2), we see that

$$y^7 = x(x-1)^4 \Rightarrow y^{7\cdot2} = x^2(x-1)^{2\cdot4} \Rightarrow \left(\frac{y^2}{x-1}\right)^7 = x^2(x-1)$$

Thus the elements $x, \dfrac{y^2}{x-1}$ satisfy the equation $y^7 = x^2(x-1)$. Note that x and $\dfrac{y^2}{x-1}$ are generators because of the identity

$$y = \frac{y^7}{(y^2)^3} = \frac{x(x-1)^4}{(y^2)^3}$$

Finally, one argues as in the third step to turn $y^7 = x^2(x-1)$ into $y^7 = x(x-1)^2$.

We have therefore proved the following:

Proposition 1.96 *There are, up to isomorphism, at most two 7-gonal Riemann surfaces of genus 3. More precisely, if the compact*

Riemann surface S_F associated to a polynomial F of the form $F(X,Y) = Y^7 - P(X)$ has genus 3, then $S_F \simeq \{y^7 = x(x-1)\}$ or $S_F \simeq \{y^7 = x(x-1)^2\}$.

As another application of these techniques we prove the following result:

Proposition 1.97 *Let S be a compact Riemann surface such that the field $\mathcal{M}(S)$ is generated by two functions x and y of degree 2. Then the genus of S is at most 1.*

Proof By the results in Section 1.3.1 we know that S is isomorphic to S_F where $F = F(X,Y)$ is an irreducible polynomial such that $\deg_Y(F) = \deg_X(F) = \deg(x) = \deg(y) = 2$.

Thus f and h satisfy an algebraic relation of the form

$$A(x) + B(x)y + C(x)y^2 = 0 \tag{1.10}$$

where A, B and C are polynomials in one variable of degree at most 2.

Multiplying identity (1.10) by $C(x)$ and setting $y_1 = C(x)y$ we get

$$0 = A(x)C(x) + B(x)y_1 + y_1^2 = A(x)C(x) - \frac{B^2(x)}{4} + \left(\frac{B(x)}{2} + y_1\right)^2$$

so, if we take x and $y_2 = \dfrac{B(x)}{2} + y_1 = \dfrac{B(x)}{2} + C(x)y$ as generators of $\mathcal{M}(S)$, we see that $S \simeq \{y^2 = p(x)\}$, where $p(x)$ is given by $p(x) = \dfrac{B^2(x)}{4} - A(x)C(x)$.

Since $\deg(p(x)) \leq 4$, the result follows from Example 1.83 and the fact that $\{y^2 = (x - a_1)(x - a_2)(x - a_3)\}$ has genus 1, see the beginning of Section 1.2.4. $\qquad\square$

This last result will be used later to show that the two Riemann surfaces $S_1 = \{y^7 = x(x-1)\}$ and $S_2 = \{y^7 = x(x-1)^2\}$ in Proposition 1.96 are in fact not isomorphic to each other. More precisely, we will prove that S_2 is not isomorphic to any hyperelliptic curve (see Example 2.50), whereas the obvious change of generators $x_1 = y$, $y_1 = x - \frac{1}{2}$ transforms the relation $y^7 = x(x-1)$ into $y_1^2 = x_1^7 + \frac{1}{4}$ (hence S_1 is the Riemann surface associated to a hyperelliptic curve).

2

Riemann surfaces and discrete groups

2.1 Uniformization

We begin this chapter with one of the most important results in the theory of Riemann surfaces:

Uniformization Theorem *Every simply connected Riemann surface is isomorphic to* $\mathbb{D}, \mathbb{C}$ *or* $\widehat{\mathbb{C}}$.

This is a very deep result which we shall assume throughout this book. For a proof the reader may consult [Bea84], [FK92].

As we saw in Section 1.2.5, covering space theory shows that every Riemann surface S can be represented as $S = \widetilde{S}/G$, where the natural projection $\pi : \widetilde{S} \longrightarrow \widetilde{S}/G = S$ is the universal covering and $G = \mathrm{Aut}(\widetilde{S}, \pi) \subset \mathrm{Aut}(\widetilde{S})$ (see Theorem 1.69). Recall that the action of G on $\widetilde{S}$ is free and properly discontinuous and that G is isomorphic to $\pi_1(S)$. Now, the Uniformization Theorem provides the three possible candidates for the universal covering space $\widetilde{S}$. On the other hand, the automorphism groups of these three surfaces are very well known (Proposition 1.27).

We are interested in the case when S is a compact Riemann surface. If the genus g of S equals 1, then $\pi_1(S)$ is isomorphic to $\mathbb{Z} \oplus \mathbb{Z}$ (see Example 1.73), therefore it is an abelian group. On the other hand, the fundamental group of S is not abelian if $g > 1$ (this was announced in Section 1.2.2 and will be proved in Section 2.1.2).

These are the ingredients required to obtain the following description of compact Riemann surfaces in terms of their universal coverings:

Theorem 2.1 (Uniformization of compact Riemann surfaces) *According to their universal coverings, compact Riemann surfaces can be classified as follows:*

- *$\widehat{\mathbb{C}}$ is the only compact Riemann surface of genus 0.*
- *Every compact Riemann surface of genus 1 can be described in the form $\mathbb{C}/\Lambda$, where Λ is a lattice, that is $\Lambda = w_1\mathbb{Z} \oplus w_2\mathbb{Z}$ for two complex numbers w_1, w_2 such that $w_1/w_2 \notin \mathbb{R}$ acting on $\mathbb{C}$ as a group of translations.*
- *Every compact Riemann surface of genus greater than 1 is isomorphic to a quotient $\mathbb{H}/K$, where $K \subset \mathrm{PSL}(2,\mathbb{R})$ acts freely and properly discontinuously.*

Proof We know from the Uniformization Theorem that $\widehat{\mathbb{C}}$ is the only simply connected compact Riemann surface. It is in fact the only surface that has $\widetilde{S} = \widehat{\mathbb{C}}$ as universal covering, since every transformation $M \in \mathrm{Aut}(\widehat{\mathbb{C}}) = \mathrm{PSL}(2,\mathbb{C})$ has one or two fixed points (alternatively use Corollary 1.77). Thus, Riemann surfaces of genus ≥ 1 must have $\mathbb{C}$ or $\mathbb{H}$ as universal covering space.

If S has genus 1 then $\widetilde{S} = \mathbb{C}$ and $S = \mathbb{C}/G$. One can rule out the possibility $\widetilde{S} = \mathbb{H}$ observing that $\mathrm{Aut}(\mathbb{H}) = \mathrm{PSL}(2,\mathbb{R})$ does not contain any subgroup G isomorphic to $\pi_1(S) \simeq \mathbb{Z} \oplus \mathbb{Z}$ acting freely and properly discontinuously on $\mathbb{H}$ (see Lemma 2.2). Moreover, since the transformations of the form $z \mapsto az + b$ with $a \neq 1$ have fixed points, we deduce that G is an abelian group with two generators $T_{w_1} : z \mapsto z + w_1$ and $T_{w_2} : z \mapsto z + w_2$, where w_1, w_2 are linearly independent over $\mathbb{R}$ as otherwise the quotient would not be compact.

Consider now the case of genus $g \geq 2$. Now $\widetilde{S} \neq \mathbb{C}$ since (as we have just mentioned) in order for a group G to act freely on $\mathbb{C}$ it has to consist entirely of translations, hence it must be abelian. But the fundamental group of a surface of genus $g \geq 2$ is not abelian (see Section 1.2.2). $\qquad\square$

Lemma 2.2 *No subgroup of $\mathrm{PSL}(2,\mathbb{R})$ isomorphic to $\mathbb{Z} \oplus \mathbb{Z}$ acts freely and properly discontinuously on $\mathbb{H}$.*

Proof The proof of the fact that such a group $\Gamma = \langle \alpha, \beta \rangle$ cannot exist consists of several steps, each of which is easy to check. See also [Sie88], Volume I, Section 2.10.

(i) Since Γ acts freely on $\mathbb{H}$, the fixed points of α and β in $\widehat{\mathbb{C}}$ must lie in $\mathbb{R} \cup \{\infty\}$.

(ii) There exists a transformation $\gamma \in \mathbb{PSL}(2,\mathbb{R})$ that maps the fixed point set of β to $\{\infty\}$ (in case there is only one fixed point) or to $\{0, \infty\}$ (in case there are two).

(iii) Conjugation by γ gives a new group with the same properties as Γ, namely

$$\Gamma' = \gamma \circ \Gamma \circ \gamma^{-1} = \langle \alpha_1 = \gamma \circ \alpha \circ \gamma^{-1}, \beta_1 = \gamma \circ \beta \circ \gamma^{-1} \rangle \subset \mathbb{PSL}(2,\mathbb{R})$$

where now $\beta_1(z) = z + l$ (in case there is only one fixed point) or $\beta_1(z) = kz$ (in case there are two).

(iv) The relation $\alpha_1 \circ \beta_1 = \beta_1 \circ \alpha_1$ implies $\alpha_1(z) = z + l_1$ (if $\beta_1(z) = z + l$) or $\alpha_1(z) = k_1 z$ (if $\beta_1(z) = kz$).

(v) We can assume $l_1/l \notin \mathbb{Z}$ (resp. $\log k_1 / \log k \notin \mathbb{Z}$), for otherwise Γ' would be a cyclic group.

(vi) Given $\varepsilon > 0$, there exist a couple of integers $m, n \in \mathbb{Z}$ such that $|ml_1 + nl| < \varepsilon$ or $|m \log k_1 + n \log k| < \varepsilon$ respectively. Hint: assume $|l_1| < |l|$ and choose $m_1 \in \mathbb{Z}$ such that $|l/l_1 - m_1| < 1$, then proceed in the same manner with $l_2 := l - m_1 l_1$ to obtain a sequence of real numbers $l_k = a_k l_1 + b_k l$ (with $a_k, b_k \in \mathbb{Z}$) such that $|l_{k+1}| < |l_k|$. This implies that the sequence $l_{k+1} - l_k$ has a subsequence converging to zero.

(vii) Considering $\alpha_1^m \circ \beta_1^n(z)$ we deduce that the action of the group cannot be properly discontinuous. $\qquad\square$

Remark 2.3 Point (iv) in the previous Lemma implies that for every $\alpha \in \mathbb{PSL}(2,\mathbb{R})$, the subgroup $\mathrm{Comm}(\alpha) = \{\beta \mid \alpha\beta = \beta\alpha\}$, known as the *commutator* of α, is abelian.

We are already used to quotients of the form $\mathbb{C}/\Lambda$, that produce Riemann surfaces of genus 1. We now would like to construct some examples of higher genus. For this purpose it will be convenient, or even necessary, to recall the basics of the hyperbolic metric.

2.1.1 $\mathbb{PSL}(2,\mathbb{R})$ *as the group of isometries of hyperbolic space*

The *hyperbolic metric* on the upper half plane $\mathbb{H}$ is defined by

$$\frac{|dz|^2}{(\mathrm{Im}\, z)^2} := \frac{(dx)^2 + (dy)^2}{y^2}$$

where $|dz|^2$ stands for the Euclidean metric $(dx)^2 + (dy)^2$. This notation is commonly used in complex variable theory because the result of transforming $|dz|^2$ by a holomorphic mapping f is simply $|f'(z)|^2|dz|^2$. Indeed if $f = u + iv$ the Cauchy–Riemann equations allow us to write

$$
\begin{aligned}
(du)^2 + (dv)^2 &= (u_x dx + u_y dy)^2 + (v_x dx + v_y dy)^2 \\[2mm]
&= (u_x^2 + v_x^2)(dx)^2 + (u_y^2 + v_y^2)(dy)^2 \\[2mm]
&\quad + 2(u_x u_y + v_x v_y)(dx)(dy) \\[2mm]
&= |f'(z)|^2 (dx)^2 + |f'(z)|^2 (dy)^2 + 0 \cdot (dx)(dy)
\end{aligned}
$$

The *length of a curve* $\gamma(t) = (x(t), y(t))$ and the *area of a set* $E \subset \mathbb{H}$ are therefore given by the formulas

$$
l(\gamma) = \int \frac{\sqrt{x'(t)^2 + y'(t)^2}}{y(t)}\,dt, \qquad a(E) = \int \frac{dxdy}{y^2}
$$

It is obvious that hyperbolic lengths and areas become much bigger than their Euclidean counterparts as one moves towards the boundary of $\mathbb{H}$. On the other hand, since the hyperbolic metric is conformal to the Euclidean metric $dx^2 + dy^2$, hyperbolic angles coincide with Euclidean angles.

Let $M \in \mathrm{Aut}(\mathbb{H}) \simeq \mathbb{PSL}(2, \mathbb{R})$ be given as $M(z) = \dfrac{az + b}{cz + d}$, where $ad - bc = 1$. A simple computation shows that

$$
M'(z) = \frac{1}{(cz + d)^2} \quad \text{and} \quad \mathrm{Im}\, M(z) = \frac{\mathrm{Im}\, z}{|cz + d|^2} \tag{2.1}
$$

This easily implies the invariance of the hyperbolic metric with respect to the action of the automorphisms of $\mathbb{H}$. In fact $\mathbb{PSL}(2, \mathbb{R})$ agrees with the group of all orientation-preserving isometries of $\mathbb{H}$, the reason being that since such an isometry f must preserve angles, it necessarily has to be holomorphic ([Ahl78]), in other words $f \in \mathrm{Aut}(\mathbb{H}) = \mathbb{PSL}(2, \mathbb{R})$. This explains why the hyperbolic metric on $\mathbb{H}$ is so natural from the point of view of complex analysis.

One can easily check that the isomorphism of Riemann surfaces in Example 1.18, namely

$$\mathbb{D} \xrightarrow{\quad L \quad} \mathbb{H}$$

$$z \longmapsto \frac{i(1+z)}{1-z}$$

transforms the hyperbolic metric on $\mathbb{H}$ into the metric

$$\frac{|L'(z)|^2|dz|^2}{(\text{Im } L(z))^2} = 4\frac{|dz|^2}{(1-|z|^2)^2} = 4\frac{(dx)^2+(dy)^2}{(1-x^2-y^2)^2}$$

which is thereby called the *hyperbolic metric on* the unit disc $\mathbb{D}$. One may, and we will certainly do in the sequel, work with any of the two models of the *hyperbolic plane* at convenience.

Since the elements of $\mathbb{P}\text{SL}(2,\mathbb{R})$ have determinant equal to 1, they are completely classified by their trace, up to conjugation. On the other hand, the trace $\text{Tr}(M)$ gives information about the fixed point set of the transformation $M(z) = \dfrac{az+b}{cz+d}$. This is because the equation $M(z) = z$ is equivalent to the quadratic equation $cz^2 + (d-a)z - b = 0$, whose discriminant is $\text{Tr}(M)^2 - 4$.

These two observations allow the following classification of isometries of $\mathbb{H}$:

- If $|\text{Tr}(M)| > 2$ then M acts fixed point freely on $\mathbb{H}$, as the two fixed points of M as a Möbius transformation acting on $\widehat{\mathbb{C}}$ lie on the real line $\mathbb{R} \cup \{\infty\} = \partial\mathbb{H}$. This kind of isometry, known as *hyperbolic isometry*, acts on $\mathbb{H}$ as a translation (the left picture in Figure 2.1), pushing the whole space from one of the fixed points (the repelling point) to the other one (the attracting point). The transformation is conjugate to $z \longmapsto \lambda z$ with $\lambda \in \mathbb{R}$.
- If $|\text{Tr}(M)| = 2$ the two fixed points of the previous case collapse to a single double point in $\partial\mathbb{H}$. The attracting and repelling points of the previous case coincide, and the action of M on $\mathbb{H}$ is often described as a rotation of the whole space around a point not belonging to the hyperbolic plane (the middle picture in Figure 2.1). This type of isometry is called *parabolic*, and is conjugate to $T(z) = z + 1$.
- If $|\text{Tr}(M)| < 2$ there is a point in $z_0 \in \mathbb{H}$ fixed by M. Then M, often called a *hyperbolic rotation*, is conjugate to a rotation $R_\theta : z \longmapsto e^{i\theta}z$ defined in the unit disc. Any isomorphism from

$\mathbb{H}$ to $\mathbb{D}$ that sends z_0 to $0 \in \mathbb{D}$, e.g. $z \mapsto \dfrac{z - z_0}{z - \overline{z_0}}$, will conjugate M to R_θ. The action of M on $\mathbb{H}$ is a rotation of the whole space around the fixed point (the right picture in Figure 2.1). This kind of transformation is called an *elliptic isometry*.

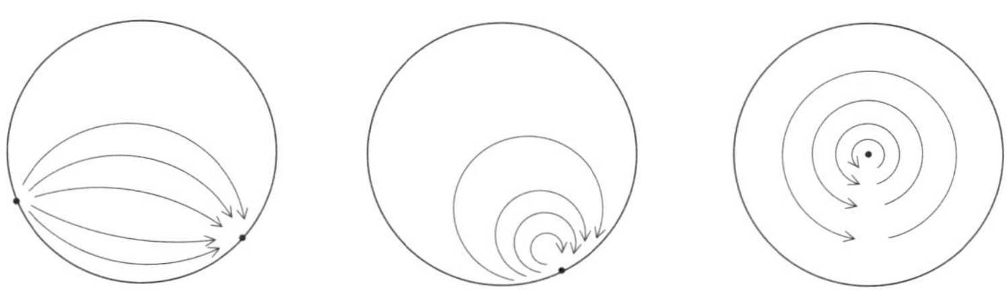

Fig. 2.1. The action of a hyperbolic, a parabolic and an elliptic isometry of the hyperbolic plane.

Since $\mathbb{PSL}(2, \mathbb{R}) \leq \mathbb{PSL}(2, \mathbb{C})$, the isometries of hyperbolic space possess all the properties enjoyed by general Möbius transformations. Firstly, they preserve generalized circles, where by a *generalized circle* we mean either a usual circle or a straight line (a circle of infinite radius). To see this, it is enough to check that the equation of a generalized circle can always be written in terms of the complex coordinate z in the form

$$Az\overline{z} + Bz + \overline{B}\overline{z} + C = 0$$

where B is a complex number and A and C are real numbers. The shape of such an equation is obviously preserved under Möbius transformations.

Another interesting object preserved by Möbius transformations is the *cross-ratio*. Recall that if (z_0, z_1, z_2, z_3) is a 4-tuple of distinct points in $\widehat{\mathbb{C}}$, the cross-ratio $\lambda = (z_0, z_1; z_2, z_3)$ is defined to be $T(z_0)$, where T is the unique element of $\mathbb{PSL}(2, \mathbb{C})$ satisfying $T(z_1) = 0$, $T(z_2) = 1$ and $T(z_3) = \infty$. Since such T is given by $T(z) = \dfrac{(z - z_1)(z_2 - z_3)}{(z_1 - z_2)(z_3 - z)}$ one can explicitly write

$$\lambda = (z_0, z_1; z_2, z_3) = \frac{(z_0 - z_1)(z_2 - z_3)}{(z_1 - z_2)(z_3 - z_0)}$$

The relevance of this quantity lies in the fact that $\mathbb{PSL}(2, \mathbb{C})$ acts transitively on the set of 4-tuples in $\widehat{\mathbb{C}}$ with given cross-ratio.

This can be seen as follows ([JS87]). Let $M \in \mathbb{PSL}(2, \mathbb{C})$ and denote $w_j = M(z_j)$ for $0 \le j \le 3$. Let U be the unique element of $\mathbb{PSL}(2, \mathbb{C})$ sending w_1, w_2 and w_3 to $0, 1$ and ∞ respectively, so that $(w_0, w_1; w_2, w_3) = U(w_0)$. Now UM is the unique element of $\mathbb{PSL}(2, \mathbb{C})$ sending z_1, z_2 and z_3 to $0, 1$ and ∞ respectively, hence

$$(z_0, z_1; z_2, z_3) = UM(z_0) = U(w_0) = (w_0, w_1; w_2, w_3)$$

Conversely, if $(z_0, z_1; z_2, z_3) = \lambda = (w_0, w_1; w_2, w_3)$, then there exist $U, T \in \mathbb{PSL}(2, \mathbb{C})$ such that $U(w_j) = T(z_j) = \lambda, 0, 1, \infty$ for $j = 0, 1, 2, 3$ respectively. It follows that $U^{-1}T(z_j) = w_j$.

We have therefore found the following facts:

- Möbius transformations preserve generalized circles.
- $\mathbb{PSL}(2, \mathbb{C})$ acts transitively on triples of distinct points.
- Möbius transformations preserve the cross-ratio.

The distance d_h induced by the hyperbolic metric is called the *hyperbolic distance*. The distance $d_h(z, w)$ between two points z and w is defined as the infimum of the lengths of paths connecting them. Paths that realize the distance between two points are called *geodesics*.

We proceed now to find all the geodesics of the hyperbolic plane. Let first $z = ip$ and $w = iq$ with $p < q$ be two points lying on the imaginary line in $\mathbb{H}$, and let $\gamma(t) = (x(t), y(t)), t \in [0, 1]$, be a differentiable path from z to w (Figure 2.2). Then we have

$$l(\gamma) = \int_0^1 \frac{\sqrt{x'(t)^2 + y'(t)^2}}{y(t)} dt \ \ge \ \int_0^1 \frac{y'(t)}{y(t)} dt$$

$$= \ \log y(1) - \log y(0)$$

$$= \ \log q/p$$

This is in fact an equality if and only if $x(t) = 0$ and $y'(t) \ge 0$ for all $t \in [0, 1]$. Therefore, $d_h(ip, iq) = \log q/p$ and the imaginary axis is a geodesic of $\mathbb{H}$.

Accordingly, the real interval $(-1, 1) \subset \mathbb{D}$ is a geodesic in the unit disc model, since it is the image of the imaginary axis under the isometry $z \mapsto \dfrac{z - i}{z + i}$ used to transport the hyperbolic metric on $\mathbb{H}$ to $\mathbb{D}$.

Let now $z, w \in \mathbb{H}$ be two arbitrary points and choose an element

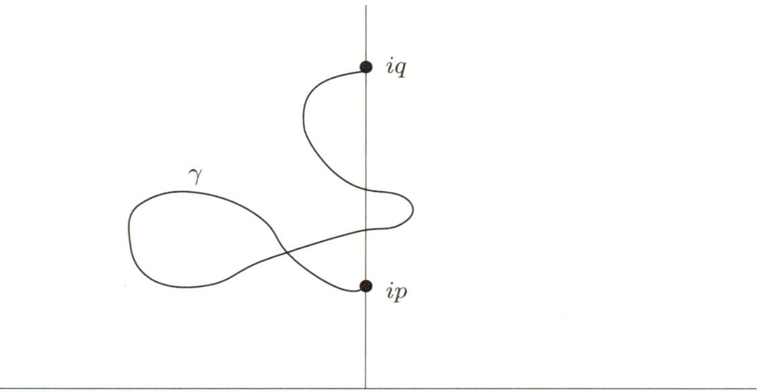

Fig. 2.2. The shortest path from $z = ip$ to $w = iq$ in the hyperbolic metric lies on the imaginary axis.

$T \in \mathbb{PSL}(2, \mathbb{R})$ such that $T(w) = i$. After composing with a suitable hyperbolic rotation around i we can assume that $T(w)$ lies on the imaginary axis. By the above computation the geodesic through z and w is then $T^{-1}(\{z : \mathrm{Re}(z) = 0\})$.

It follows that every geodesic in $\mathbb{H}$ is the image of the imaginary axis by certain isometry. Accordingly, every geodesic in $\mathbb{D}$ is the image of the interval $(-1, 1)$ by some isometry of $\mathbb{D}$. Since Möbius transformations are conformal, it follows the geodesics in $\mathbb{H}$ are generalized circles perpendicular to $\partial \mathbb{H} = \mathbb{R}$. Conversely, if C is a generalized circle perpendicular to $\mathbb{R}$ and $z_1, z_2 \in C$, then as the geodesic L through these points must be a generalized circle perpendicular to $\mathbb{R}$ we necessarily have $L = C$, that is C is a geodesic.

We now summarize some relevant facts about hyperbolic space:

- The generalized circles perpendicular to $\partial \mathbb{H} = \mathbb{R} \cup \{\infty\}$ account for all geodesics of $\mathbb{H}$.

- $\mathbb{PSL}(2, \mathbb{R})$ *acts transitively on the set of all geodesics*, meaning that given an arbitrary pair of geodesics there is always a transformation in $\mathbb{PSL}(2, \mathbb{R})$ sending one to the other.

- Any two arbitrary points $z, w \in \mathbb{H}$ determine a unique geodesic. Two pairs of points z, w and z', w' can be isometrically mapped into each other if and only if $d_h(z, w) = d_h(z', w')$.

Remark 2.4 The above properties of the hyperbolic metric in $\mathbb{H}$ can be easily translated to the unit disc model, since the isometry from $\mathbb{H}$ to $\mathbb{D}$ is realized by a Möbius transformation. The geodesics are now generalized circles perpendicular to $\partial\mathbb{D} = \mathbb{S}^1$.

It is not difficult to find explicit formulas for $d_h(z, w)$ when z and w are arbitrary points in $\mathbb{H}$ (see [Bea83] for the details). One can do it by taking an element $T \in \mathbb{PSL}(2, \mathbb{R})$ such that $T(z)$ and $T(w)$ lie both in the imaginary axis and then computing the hyperbolic distance between $T(z)$ and $T(w)$, which, as we have seen, is the logarithm of its quotient. One obtains

$$\tanh \frac{1}{2} d_h(z, w) = \left| \frac{z - w}{z - \overline{w}} \right| \tag{2.2}$$

The analogous formula in the unit disc $\mathbb{D}$ is

$$\tanh \frac{1}{2} d_h(z, w) = \left| \frac{z - w}{1 - z\overline{w}} \right| \tag{2.3}$$

2.1.2 Groups uniformizing Riemann surfaces of genus $\mathbf{g \geq 2}$

We shall use hyperbolic geometry to construct examples of the kind of subgroups of $\mathbb{PSL}(2, \mathbb{R})$ that occur in Theorem 2.1, that is groups acting freely and properly discontinuously on the hyperbolic plane. The concept of fundamental domain introduced in Examples 1.12 and 1.13 makes perfect sense for these groups. More precisely:

Definition 2.5 Let D be a simply connected closed subset of $\mathbb{H}$ whose boundary ∂D consists of a finite union of differentiable arcs. Then D is said to be a *fundamental domain* for a subgroup $\Gamma < \mathbb{PSL}(2, \mathbb{R})$ if the family $\{\gamma(D); \; \gamma \in \Gamma\}$ tessellates $\mathbb{H}$, i.e.

(i) $\bigcup_{\gamma \in \Gamma} \gamma(D) = \mathbb{H}$; and

(ii) for any $\gamma \in \Gamma \smallsetminus \{\mathrm{Id}\}$ the intersection $D \cap \gamma(D)$ is contained in the boundary of D.

The fundamental domains of the groups we construct in this section will be *hyperbolic polygons*, that is hyperbolically convex subsets of hyperbolic space whose boundary consists of finitely many arcs of hyperbolic geodesics.

Let P be a regular hyperbolic polygon with eight sides and angle $\pi/4$. Note that such a polygon does not exist in Euclidean geometry, since the sum of the angles of an n-gon is forced to be equal to $(n-2)\pi$ in the Euclidean case.

The construction of P can be carried out in the following way: take eight vertices symmetrically distributed around the centre of $\mathbb{D}$, say $v_j = Re^{2(j-1)\pi i/8}$ for $1 \leq j \leq 8$, and let e_j be a geodesic arc joining v_j and v_{j+1} (where subscripts are taken modulo 8).

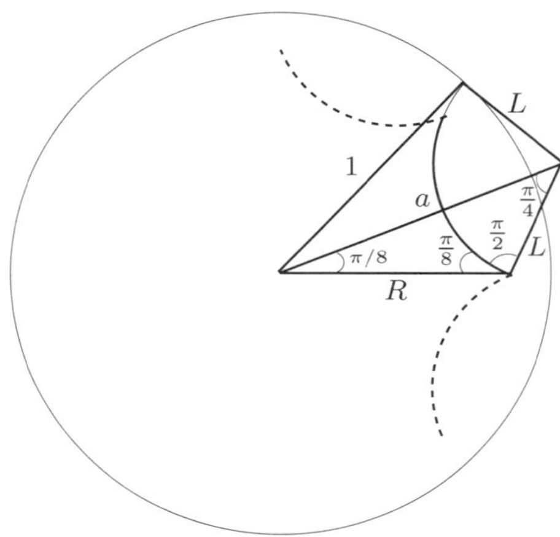

Fig. 2.3. Computation of R.

The angle of the resulting regular polygon at any vertex strictly decreases as R tends to 1. This construction produces a polygon of angle $\pi/4$ exactly for $R = \dfrac{1}{\sqrt[4]{2}}$. To show this, first note that the geodesic passing through v_1 and v_2 is contained in an (Euclidean) circle of centre $ae^{\pi i/8}$ and radius L, where $a > 1$ and $L > 0$ are certain real numbers.

The lower triangle in Figure 2.3 has edges of lengths R, L, a and opposite angles $\pi/4$, $\pi/8$ and $5\pi/8$. By the sine rule we have

$$\frac{L}{\sin \pi/8} = \frac{a}{\sin 5\pi/8}$$

and, using that $a^2 = 1 + L^2$ (the upper triangle of the figure is rectangle) and the addition formulas for sin and cos, we obtain

$$a^2 = \frac{\cos^2 \pi/8}{\cos \pi/4}$$

Again the sine rule gives

$$\frac{R}{\sin \pi/4} = \frac{a}{\sin 5\pi/8}$$

and inserting in this equation the value of a we finally find the announced value

$$R = \sqrt{\cos \frac{2\pi}{8}} = \frac{1}{\sqrt[4]{2}}$$

Let now A_1 be the isometry of $\mathbb{D}$ that sends v_3 to v_2 and v_4 to v_1 (see Figure 2.4). Note that then A_1 is forced to send e_3 to e_1, hence $A_1(P)$ is a regular octogon touching P along the edge e_1. In the same way, let B_1 be determined by $B_1(v_2) = v_5$ and $B_1(v_3) = v_4$ (hence $B_1(e_2) = e_4$), A_2 by $A_2(v_7) = v_6$ and $A_2(v_8) = v_5$ (hence $A_2(e_7) = e_5$), and B_2 by $B_2(v_6) = v_1$ and $B_2(v_7) = v_8$ (hence $B_2(e_6) = e_8$).

Let K be the group generated by this collection of *side-pairing transformations*, as they are called, i.e. let $K = \langle A_1, B_1, A_2, B_2 \rangle$. The images of P by the eight elements $\{A_i^{\pm 1}, B_i^{\pm 1}, \ i = 1, 2\}$ are the eight regular octogons, isometric to P, that meet P along each of its eight edges.

The collection of octogons $\{\gamma(P)\}_{\gamma \in K}$ tessellates $\mathbb{D}$, in the sense that no gap remains uncovered (Figure 2.4). All are hyperbolically isometric to each other, but their Euclidean areas become overwhelmingly smaller as they approach the boundary. Note that, as the angle of P equals $\pi/4$, eight polygons of the tessellation fit perfectly around a given vertex. Moreover, if $\gamma_1 \neq \gamma_2$ the intersection $\gamma_1(P) \cap \gamma_2(P)$ can only be empty, a common vertex or a common edge.

This discussion shows that K acts freely and properly discontinuously on $\mathbb{D}$ with P as fundamental domain (Definition 2.5). In particular, $\mathbb{D}/K$ agrees with $P/\sim$, where $\sim$ is the equivalence relation induced by K on the boundary of P. In other words, $\sim$ is nothing but the equivalence relation we have defined in Section

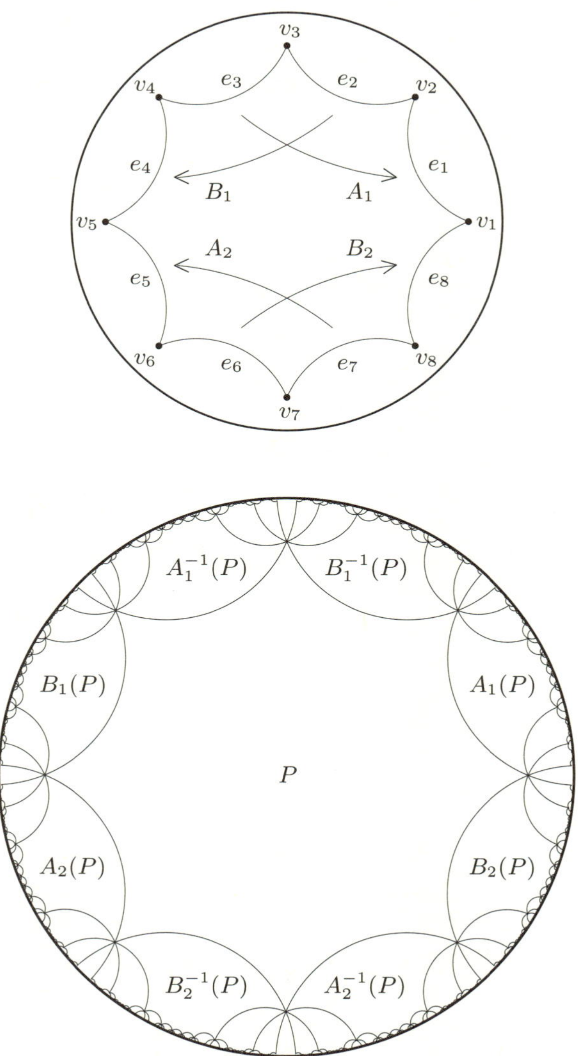

Fig. 2.4. A hyperbolic polygon equipped with a side-pairing identification, and the tessellation produced by the group generated by the side-pairing transformations.

1.2.1 to construct the orientable surface of genus $g = 2$ at a topological level (see Figure 1.7). The eight vertices of P represent the same point $v = [v_i]$ in $\mathbb{D}/K$, as all them are related by some side-pairing transformation. The natural projection $\pi : \mathbb{D} \to \mathbb{D}/K$ is the universal covering of the quotient space $S = \mathbb{D}/K$, which is a genus 2 compact Riemann surface.

One can compute the transformations A_i, B_i explicitly. If we denote

$$\sigma(z) = e^{i\pi/4}z, \ T(z) = -i\frac{z-a}{1-\overline{a}z} \quad \text{with} \quad a = \frac{1+i(\sqrt{2}-1)}{\sqrt[4]{2}}$$

then

$$A_1 = T^{-1}$$
$$B_1 = \sigma \circ T \circ \sigma^{-1}$$
$$A_2 = \sigma^4 \circ T^{-1} \circ \sigma^{-4}$$
$$B_2 = \sigma^5 \circ T \circ \sigma^{-5}$$

In fact this construction can be easily generalized to produce compact Riemann surfaces of arbitrary genus $g \geq 2$ (see [Nar92]). Now K is the group generated by

$$A_1 = T^{-1} \qquad\qquad B_1 = \sigma \circ T \circ \sigma^{-1}$$
$$A_2 = \sigma^4 \circ T^{-1} \circ \sigma^{-4} \qquad B_2 = \sigma^5 \circ T \circ \sigma^{-5}$$
$$A_3 = \sigma^8 \circ T^{-1} \circ \sigma^{-8} \qquad B_3 = \sigma^9 \circ T \circ \sigma^{-9}$$
$$\vdots \qquad\qquad\qquad\qquad \vdots$$
$$A_g = \sigma^{4(g-1)} \circ T^{-1} \circ \sigma^{-4(g-1)} \quad B_g = \sigma^{4g-3} \circ T \circ \sigma^{-(4g-3)}$$

where $\sigma(z) = e^{i2\pi/4g}z$ and T is determined by the conditions $T(R) = Re^{i3\pi/2g}$, $T(Re^{i\pi/2g}) = Re^{i2\pi/2g}$, where $R = \sqrt{\cos\frac{2\pi}{4g}}$.

The quotient space $\mathbb{D}/K$ is a genus g Riemann surface, and a regular hyperbolic polygon P with $4g$ sides is a fundamental polygon. After labelling the vertices of P as above, the transformations A_i are side pairings determined by $A_i(v_{4i}) = v_{4i-3}$ and $A_i(v_{4i-1}) = v_{4i-2}$, whereas B_i are determined by $B_i(v_{4i-1}) = v_{4i}$ and $B_i(v_{4i-2}) = v_{4i+1}$.

As we announced earlier (see Section 1.2.2), we have the following result concerning fundamental groups:

Theorem 2.6 *The fundamental group of a compact orientable* *surface of genus $g > 1$ has generators*

$$\alpha_1, \beta_1, \alpha_2, \beta_2, \ldots, \alpha_g, \beta_g$$

subject to the essentially unique relation

$$\alpha_1\beta_1\alpha_1^{-1}\beta_1^{-1}\alpha_2\beta_2\alpha_2^{-1}\beta_2^{-1} \cdots \alpha_g\beta_g\alpha_g^{-1}\beta_g^{-1} = 1 \qquad (2.4)$$

Proof We prove the result for the particular Riemann surface of genus g constructed in this section. Let α_i, β_i be the loops induced by the paths in P parametrizing the directed edges $e_{1+4(i-1)}$ and $e_{2+4(i-1)}$ counterclockwise oriented (here $i = 1, \ldots, g$). With the notation of Section 1.2.2 the directed edges $e_{3+4(i-1)}$ and $e_{4+4(i-1)}$ counterclockwise oriented correspond to α_i^{-1} and β_i^{-1} respectively. We can assume that $g = 2$, for it will become apparent that the proof is similar in the general case.

Recall (see Theorem 1.69) that K is isomorphic to the fundamental group of $S = \mathbb{D}/K$ via the isomorphism

$$K \xrightarrow{\Phi} \pi_1(S, v)$$
$$A \longmapsto \gamma_A$$

where γ_A is a loop that lifts to a path in $\mathbb{D}$ joining v_1 to its image $A(v_1)$. In particular, $\pi_1(S, v)$ is generated by the elements $\Phi(A_i), \Phi(B_i)$. In turn one can check that these loops can be expressed in terms of α_i, β_i (the precise expressions being the identities (2.7) in Example 2.7). It remains to be shown that

$$\alpha_1\beta_1\alpha_1^{-1}\beta_1^{-1}\alpha_2\beta_2\alpha_2^{-1}\beta_2^{-1} = 1$$

is essentially the unique relation. That is, we have to see that if $w = w(a_i, b_i)$ is a formal word in the kernel of the epimorphism

$$\rho : \mathcal{F}_{2g} \longrightarrow \pi_1(S, v)$$
$$w(a_i, b_i) \longmapsto w(\alpha_i, \beta_i)$$

introduced in Section 1.2.2 then w lies in the smallest normal subgroup containing the word $w_g = a_1b_1a_1^{-1}b_1^{-1}a_2b_2a_2^{-1}b_2^{-1}$.

There is an obvious way to represent a word w as a path γ with initial point v_1 supported on the edges of the fundamental polygons of the tessellation such that $\gamma : [0, 1] \to \mathbb{D}$ is a lift of $\rho(w)$. For instance, the path depicted in the first picture of Figure 2.7 represents the word $w_1 = a_1b_1a_1^{-1}$ whose image under ρ is $\Phi(A_1)$.

When $\rho(w) = 1$, the path γ is closed. Let us denote now by $0 = t_0 < t_1 < \cdots < t_n = 1$ the values at which $\gamma(t_i)$ is a vertex of the tessellation, so that $v_1 = \gamma(0) = \gamma(t_n)$. Note that the condition $\gamma(t_k) = \gamma(t_l) = v$ for $l \neq k$ means that v is a self-intersection point of γ. We argue by induction on the number m of self-intersection points.

If $m = 0$ then $\gamma(0) = \gamma(1) = v_1$ is the only self-intersection point, so that γ is a simple loop enclosing $k \geq 0$ copies of the fundamental polygon P. If $k = 0$, w can only be a trivial word of the form $a_i^{-1} a_i$. If $k = 1$ then obviously w can only be the word w_g or its inverse if the enclosed polygon is P, or the conjugate of one of these two words by $\Phi(A)$ if the enclosed polygon is the transform $A(P)$ of P (we will encounter an explicit example of this situation in Example 2.7). For $k > 1$ we can always write (see Figure 2.5) $\gamma = c'\gamma_1 c$ where both γ_1 and $c'c = \gamma_2$ enclose less than k polygons. Therefore $\gamma = \gamma_2 c^{-1}\gamma_1 c$ and the result follows arguing by induction on k.

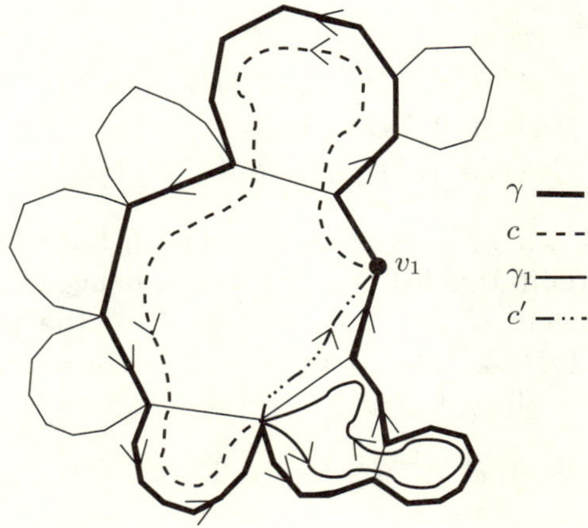

Fig. 2.5. An example of the loops γ and γ_1 and the paths c and c' in the case $k > 1$.

In the case $m > 0$ there must exist a self-intersection vertex $v = \gamma(t_i) = \gamma(t_j)$ with $0 \leq t_i < t_j < 1$. But then we can write $\gamma = \gamma_j \gamma_{ij} \gamma_i$ where $\gamma_i = \gamma([0, t_i])$, $\gamma_{ij} = \gamma([t_i, t_j])$ and $\gamma_j = \gamma[t_j, 1]$. We now observe that both paths $\gamma_j \gamma_i$ and γ_{ij} are loops having each less than m self-intersection points. Now the proof is done.

$\square$

For $g = 1$ Theorem 2.6 also holds. In this case relation (2.4) means that $\pi_1(S)$ is isomorphic to $\mathbb{Z} \oplus \mathbb{Z}$, as was seen in Example 1.73.

Example 2.7 As a further illustration of the correspondence between the fundamental and the uniformizing groups we now check relation (2.4) in terms of the group K. In order to do that we first show that the relation

$$A_1 B_1 A_1^{-1} B_1^{-1} A_2 B_2 A_2^{-1} B_2^{-1} = \mathrm{Id} \tag{2.5}$$

holds in K.

The eight polygons of the tessellation meeting at v_1 are, in cyclic (counterclockwise) order:

$$A_1(P)$$
$$A_1 B_1(P)$$
$$A_1 B_1 A_1^{-1}(P)$$
$$A_1 B_1 A_1^{-1} B_1^{-1}(P)$$
$$A_1 B_1 A_1^{-1} B_1^{-1} A_2(P)$$
$$A_1 B_1 A_1^{-1} B_1^{-1} A_2 B_2(P)$$
$$A_1 B_1 A_1^{-1} B_1^{-1} A_2 B_2 A_2^{-1}(P)$$
$$A_1 B_1 A_1^{-1} B_1^{-1} A_2 B_2 A_2^{-1} B_2^{-1}(P)$$

(see Figure 2.6, where a neighbourhood of v_1 has been magnified). The last polygon in this list is just P itself, hence the claim follows from the fact that K acts freely in $\mathbb{D}$ since the transformation $A_1 B_1 A_1^{-1} B_1^{-1} A_2 B_2 A_2^{-1} B_2^{-1}$ obviously fixes the centre of P.

It follows that the relation

$$\Phi(A_1 B_1 A_1^{-1} B_1^{-1} A_2 B_2 A_2^{-1} B_2^{-1}) = 1 \tag{2.6}$$

must hold in $\pi_1(S, v)$.

If we had $\Phi(A_i) = \alpha_i$, $\Phi(B_i) = \beta_i$, identity (2.6) would agree with identity (2.4). However, this is not the case. In fact with the notation used in the proof of Theorem 2.6 we have (see Figure 2.7)

$$
\begin{aligned}
\Phi(A_1) &= \alpha_1 \beta_1^{-1} \alpha_1^{-1} \\
\Phi(B_1) &= \alpha_1 \beta_1 \alpha_1^{-1} \beta_1^{-1} \alpha_1^{-1} \\
\Phi(A_2) &= \beta_2 \alpha_2 \beta_2^{-1} \alpha_2^{-1} \beta_2^{-1} \\
\Phi(B_2) &= \beta_2 \alpha_2^{-1} \beta_2^{-1}
\end{aligned}
\tag{2.7}
$$

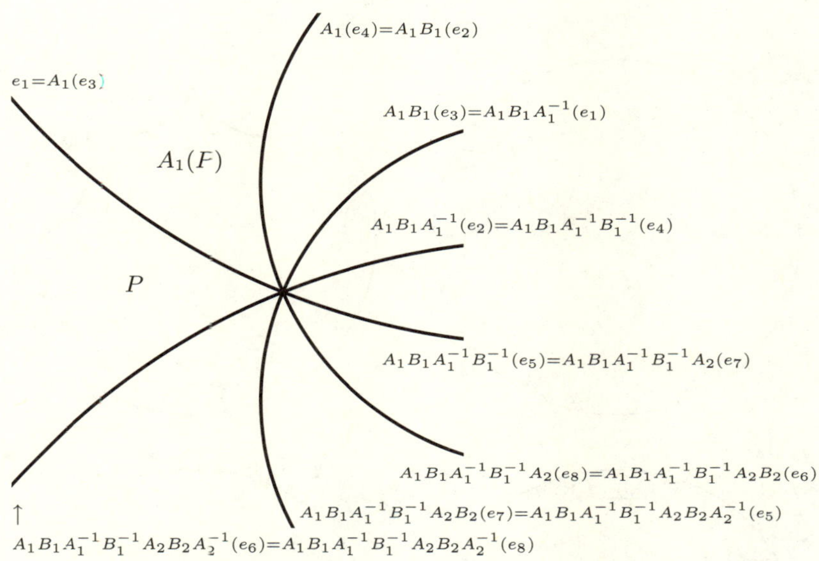

Fig. 2.6. The configuration of the polygons surrounding v_1.

and so (2.6) becomes instead

$$\beta_1\alpha_1\beta_1^{-1}\alpha_1^{-1}\beta_2\alpha_2\beta_2^{-1}\alpha_2^{-1} = 1 \tag{2.8}$$

Nevertheless, it remains true that (2.8) can be derived from (2.4) by first inverting it and then conjugating by $\alpha_2\beta_2\alpha_2^{-1}\beta_2^{-1}$.

It may be illustrative to show how this conjugating element can be determined geometrically. First, equation (2.8) can be rewritten in the form

$$\Phi^{-1}(\beta_1\alpha_1\beta_1^{-1}\alpha_1^{-1}\beta_2\alpha_2\beta_2^{-1}\alpha_2^{-1}) = \mathrm{Id} \in K$$

which in turn is equivalent to saying that the lift of the loop $\beta_1\alpha_1\beta_1^{-1}\alpha_1^{-1}\beta_2\alpha_2\beta_2^{-1}\alpha_2^{-1}$ with starting point at v_1 ends also at v_1. In order to find this lift, note that according to Figure 2.6 the

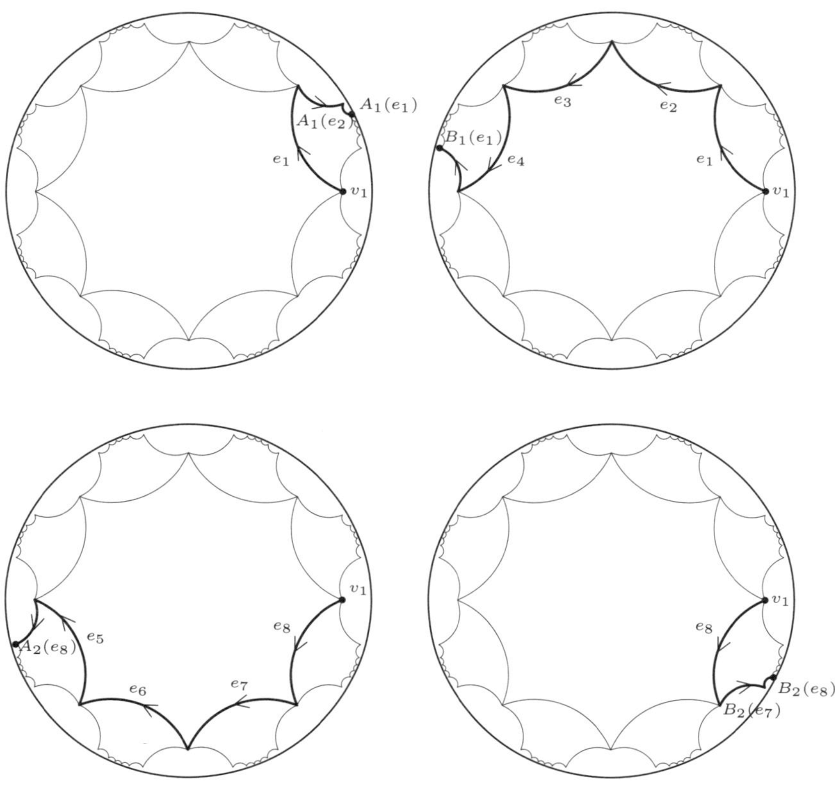

Fig. 2.7. Lifts to $\mathbb{D}$ of the loops corresponding to $\Phi(A_1), \Phi(B_1), \Phi(A_2)$ and $\Phi(B_2)$ in (2.7).

edges of the tessellation meeting v_1 are in clockwise order

$$e_8$$
$$A_1(e_3)$$
$$A_1 B_1(e_2)$$
$$A_1 B_1 A_1^{-1}(e_1)$$
$$A_1 B_1 A_1^{-1} B_1^{-1}(e_4)$$
$$A_1 B_1 A_1^{-1} B_1^{-1} A_2(e_7)$$
$$A_1 B_1 A_1^{-1} B_1^{-1} A_2 B_2(e_6)$$
$$A_1 B_1 A_1^{-1} B_1^{-1} A_2 B_2 A_2^{-1}(e_5)$$

Since our loop begins with β_1 its lift must move first along $A_1 B_1 A_1^{-1} B_1^{-1}(e_4)$ – the other possible candidate, namely $A_1 B_1(e_2)$, is a lift of β_1^{-1}. Now one has again eight possibilities to move forward, and the correct one (the only one which is a lift of α_1) is choosing the next edge of the polygon $A_1 B_1 A_1^{-1} B_1^{-1}(P)$ in clockwise order, namely $A_1 B_1 A_1^{-1} B_1^{-1}(e_3)$. In this way one easily checks (Figure 2.8) that the lift we are looking for is the loop that runs along the boundary of $A_1 B_1 A_1^{-1} B_1^{-1}(P)$ in the clockwise direction, ending back at v_1 as expected.

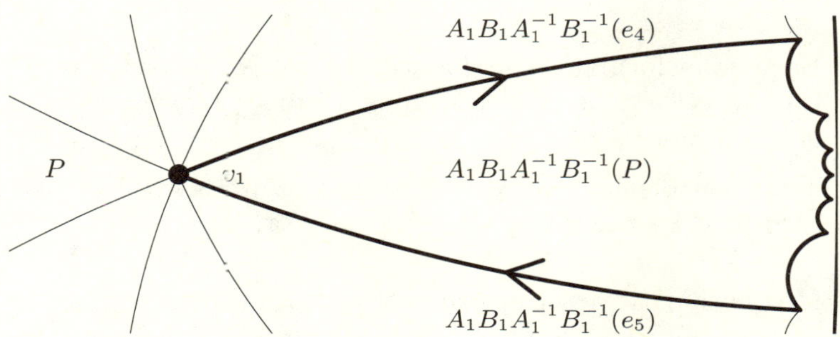

Fig. 2.8. The lift to the unit disc (with base point at v_1) of the path $\beta_1 \alpha_1 \beta_1^{-1} \alpha_1^{-1} \beta_2 \alpha_2 \beta_2^{-1} \alpha_2^{-1} = \Phi(A_1 B_1 A_1^{-1} B_1^{-1} A_2 B_2 A_2^{-1} B_2^{-1})$.

As claimed in the proof of Theorem 2.4 (case $m = 0$, $k = 1$), the element of $\pi_1(S, v)$ conjugating our word $\beta_1 \alpha_1 \beta_1^{-1} \alpha_1^{-1} \beta_2 \alpha_2 \beta_2^{-1} \alpha_2^{-1}$ to w_g^{-1} is precisely $\Phi(A_1 B_1 A_1^{-1} B_1^{-1})$, which by (2.6) is the same as $\Phi(B_2 A_2 B_2^{-1} A_2^{-1})$. By (2.7) this agrees with

$$(\beta_2 \alpha_2^{-1} \beta_2^{-1})(\beta_2 \alpha_2 \beta_2^{-1} \alpha_2^{-1} \beta_2^{-1})(\beta_2 \alpha_2 \beta_2^{-1})(\beta_2 \alpha_2 \beta_2 \alpha_2^{-1} \beta_2^{-1})$$

which equals $\alpha_2 \beta_2 \alpha_2^{-1} \beta_2^{-1}$. Of course this only means that the lift of $\alpha_2 \beta_2 \alpha_2^{-1} \beta_2^{-1}$ starting at v_1 ends at $A_1 B_1 A_1^{-1} B_1^{-1}(v_1)$ (this time this lift goes along one half of $A_1 B_1 A_1^{-1} B_1^{-1}(P)$ counterclockwise).

The following generalization of the construction carried out in this section holds for non-necesarily regular hyperbolic polygons.

Theorem 2.8 (Poincaré's polygon Theorem – first version) *Let P be a convex hyperbolic polygon in $\mathbb{D}$ with sides $l_1, \ldots, l_n$,*

$l'_1, \ldots, l'_n$. *Suppose there exist isometries* $g_1, \ldots, g_n \in \mathrm{Aut}(\mathbb{D})$ *such that* $g_i(l_i) = l'_i$. *Let* K *be the group generated by the elements* g_i. *If the sum of the angles of* P *at every equivalence class of vertices equals* 2π, *then* K *acts freely and properly discontinuously on* $\mathbb{D}$ *and* $\mathbb{D}/K$ *is a compact Riemann surface.*

The proof becomes clear from Figure 2.4, although the details are not easy to write down. In fact it was not until long after Poincaré's time that a complete proof was achieved (see [dR71], [Mas71], [Bea83]).

Here we only emphasize the two essential points of this proof:

(i) Each transformation g_i identifies two of the sides of P, therefore the isometric polygons P and $g_i(P)$ meet along one common edge.

(ii) The sum of the angles of the polygon P at all vertices identified by the side pairings equals 2π.

The idea is that (i) describes the configuration in $\mathbb{D}$ of the successive images of P by the elements of the group K, whereas (ii) implies that these images fit perfectly, leaving no gaps among them but also not overlapping; that is, they *tessellate* $\mathbb{D}$.

2.2 The existence of meromorphic functions

Now the time has come to face the problem of constructing functions on Riemann surfaces of genus 1, i.e. isomorphic to $\mathbb{C}/\Lambda$, and of genus $g \geq 2$, i.e. of the form $\mathbb{H}/K$. Our aim is to show the separation property of the field of meromorphic functions (Theorem 1.90).

The first trial would be choosing a function on $\mathbb{C}$ (or $\mathbb{D}$) and forming the sum $F(z) = \sum_\gamma f(\gamma(z))$ where γ runs along the uniformizing group Λ (or K). By construction the function F will be $\Lambda-$invariant (or $K-$invariant), hence it will induce a well-defined function on the quotient surface. Indeed one of the functions we construct in genus $g = 1$ arises exactly in this way for $f(z) = -2/z^3$. However, the defining series will not converge in general, so we must proceed more carefully.

2.2.1 Existence of functions in genus g = 1

We consider first the existence of functions for surfaces of genus $g = 1$. More information about this question can be found in [Car61], [Ahl78] or [JS87].

Describe S in the form $S = \mathbb{C}/\Lambda$ where $\Lambda = \mathbb{Z}\omega_1 \oplus \mathbb{Z}\omega_2$ for two $\mathbb{R}$-linearly independent complex numbers ω_1 and ω_2.

Lemma 2.9 *The series*

$$\sum_{0 \neq \omega \in \Lambda} \frac{1}{|\omega|^3} \tag{2.9}$$

converges for every lattice Λ as above.

Proof The number of points of Λ belonging to the parallelogram P_n that passes through $n\omega_1$ equals $8n$ (see Figure 2.9). Moreover, if $\omega \in P_n$ and D is the distance from 0 to P_1, we have $|\omega| > nD$. Therefore,

$$\sum_{0 \neq \omega \in \Lambda} \frac{1}{|\omega|^3} \leq \sum_{n=1}^{\infty} \frac{8n}{|Dn|^3} = \frac{8}{D^3} \sum_{n=1}^{\infty} \frac{1}{n^2} = \frac{8}{D^3} \frac{\pi^2}{6}$$

$\square$

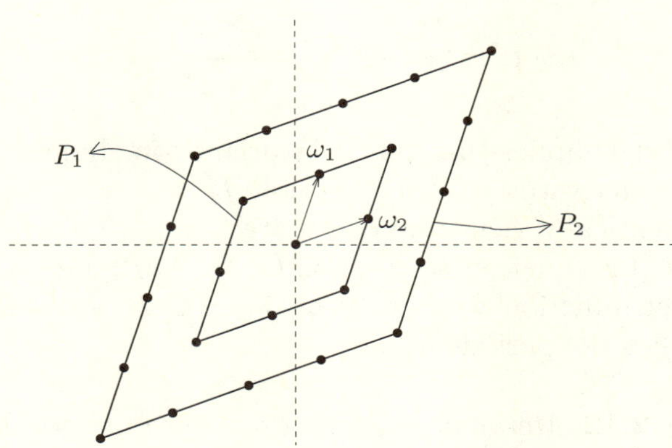

Fig. 2.9. The lattice $\Lambda = \mathbb{Z}\omega_1 \oplus \mathbb{Z}\omega_2$.

Corollary 2.10 (Weierstrass $\wp$ function) *The series*

$$\wp(z) = \frac{1}{z^2} + \sum_{0 \neq \omega \in \Lambda} \left(\frac{1}{(z-\omega)^2} - \frac{1}{\omega^2} \right) \qquad (2.10)$$

defines a meromorphic function in $\mathbb{C}$.

Proof Let $r > 0$ be a fixed positive real number. Take $\omega \in \Lambda$ such that $|\omega| > 2r$ (a restriction that fails to hold for only finitely many lattice points ω). Then, for all $|z| \leq r$, we get

$$\left| \frac{1}{(z-\omega)^2} - \frac{1}{\omega^2} \right| = \left| \frac{z(2\omega - z)}{\omega^2 (z-\omega)^2} \right| = \frac{|z| \, |\omega| \, |2 - (z/\omega)|}{|\omega|^4 \, |1 - (z/\omega)|^2}$$

$$\leq \frac{|z| \, (2 + |z/\omega|)}{|\omega|^3 \, |1 - |z/\omega||^2} \leq \frac{r \, |2 + (1/2)|}{|\omega|^3 \, (1 - 1/2)^2}$$

Now apply Lemma 2.9 and the Weierstrass Theorem on the uniform convergence of analytic functions ([Ahl78])†. $\qquad \square$

Corollary 2.11 *Both $\wp$ and $\wp'$ define meromorphic functions in $S = \mathbb{C}/\Lambda$.*

Proof We have to show that these functions are well-defined modulo Λ, i.e. that they are Λ-invariant. This is obvious for the function

$$\wp'(z) = -2 \sum_{\omega \in \Lambda} \frac{1}{(z-\omega)^3} \qquad (2.11)$$

which in turns implies that $\wp$ is Λ-invariant too. Indeed, for every $\omega \in \Lambda$ the derivative of the function $f(z) = \wp(z + \omega) - \wp(z)$ vanishes identically, hence we can write $\wp(z + \omega) - \wp(z) = c(\omega)$. Evaluating this function at the point $z = -\omega/2$ and noting that $\wp$ is an even function we find $c(\omega) = \wp(\omega/2) - \wp(-\omega/2) = 0$, which implies the periodicity of $\wp$. $\qquad \square$

Corollary 2.12 *Meromorphic functions separate points in Riemann surfaces of genus $g = 1$.*

---

† The Weierstrass Theorem states that if a sequence of analytic functions f_n converges uniformly on compact sets to a function f, then f is also analytic. Moreover, f'_n converges uniformly to f' on compact sets.

Proof Since $\wp\left([z]\right)$ has a unique pole at $[0]$, given two points $[z_1],[z_2] \in \mathbb{C}/\Lambda$ the function

$$f\left([z]\right) = \wp\left([z - z_2]\right) - \wp\left([z_1 - z_2]\right)$$

has a zero at $[z_1]$ and a pole at $[z_2]$. $\qquad\square$

Let us study these functions $\wp$ and $\wp'$ in some more detail. We have

$$\wp(z) = \frac{1}{z^2} + a_2 z^2 + a_4 z^4 + \cdots \tag{2.12}$$

near $z = 0$.

Note that the last series has only terms of even degree ($\wp$ is an even function), and has no constant term since

$$\wp(z) - \frac{1}{z^2} = \sum_{0 \neq \omega \in \Lambda} \left(\frac{1}{(z - \omega)^2} - \frac{1}{\omega^2}\right)$$

vanishes at $z = 0$. Differentiating we find

$$\wp'^2(z) = \frac{4}{z^6} - \frac{8a_2}{z^2} - 16a_4 + \cdots \tag{2.13}$$

while taking third powers in (2.12) yields

$$\wp^3(z) = \frac{1}{z^6} + \frac{3a_2}{z^2} + 3a_4 + \cdots \tag{2.14}$$

It follows that

$$\wp'^2(z) - 4\wp^3(z) = -\frac{20a_2}{z^2} + h(z)$$

where h is a holomorphic function.

If we define now $g_2 = g_2(\Lambda) = 20a_2$, we see that the function $\wp'^2(z) - 4\wp^3(z) + g_2\wp(z)$ is holomorphic everywhere in $\mathbb{C}/\Lambda$, therefore equal to a constant $-g_3 = -g_3(\Lambda)$. Thus we have the following algebraic relation between the functions $\wp$ and $\wp'$

$$\wp'^2 = 4\wp^3 - g_2\wp - g_3.$$

Now, since $\wp$ is a function of degree 2 (it has only one pole and its multiplicity equals 2), Corollary 1.93 implies firstly that $\mathbb{C}(\wp, \wp')$ is the full field of all meromorphic functions in $\mathbb{C}/\Lambda$ and secondly that there is an isomorphism between the transcendental and the algebraic models of the Riemann surface $\mathbb{C}/\Lambda$ given by

$$\mathbb{C}/\Lambda \longrightarrow S_F = \left\{ y^2 = 4x^3 - g_2 x - g_3 \right\} \qquad (2.15)$$

$$[z] \longmapsto (\wp [z], \wp' [z])$$

Remark 2.13 We recall for further use that $\wp'$ is a meromorphic function on $\mathbb{C}/\Lambda$ whose degree equals 3 (it has a single pole of order 3 at $[0]$). Now, since $\wp'(z) = -\wp'(-z)$, the three zeros of $\wp'$ are located at $[\omega_1/2]$, $[\omega_2/2]$ and $[(\omega_1 + \omega_2)/2]$ and they are all simple. The algebraic relation between $\wp'$ and $\wp$ shows that none of these three points is a zero of $\wp$ unless we have $g_3 = 0$. In other words, for the case $g_3 \neq 0$ the (two) zeros of the Weierstrass $\wp$ function are also simple and, since $\wp$ an even function, they are located at $[\pm z_0]$ for certain $z_0 \neq 0$.

2.2.2 Existence of functions in genus g $\geq$ 2

We shall start by constructing a meromorphic function on the Riemann surfaces $S = \mathbb{D}/K$ exhibited in Section 2.1.2. We have:

Lemma 2.14 *Let $K = \{\gamma_n\}_{n=1}^{\infty}$ be the group constructed in Section 2.1.2. The series*

$$\sum \left| \gamma_n' (z) \right|^2 \qquad (2.16)$$

converges uniformly in compact subsets of $\mathbb{D}$.

Proof We split the proof into several steps. Let P be the fundamental polygon of K as described in Section 2.1.2.

(i) For every n we have

$$\int_P \left| \gamma_n' (z) \right|^2 = \int_P (\mathrm{Re}\ \gamma_n)_x^2 + (\mathrm{Im}\ \gamma_n)_x^2 = \int_P \left| \frac{\partial \gamma_n (x, y)}{\partial (x, y)} \right|$$

equals the Euclidean area of $\gamma_n(P)$, and therefore

$$\sum \int_P \left| \gamma_n' (z) \right|^2 = \pi$$

which gives the euclidean area of the unit disc.

(ii) Write $\gamma_n (z) = \dfrac{\overline{a}_n z + \overline{b}_n}{b_n z + a_n}$, where $|a_n|^2 - |b_n|^2 = 1$. Then we have $\gamma_n'(z) = \dfrac{1}{(b_n z + a_n)^2}$. Since $|b_n| < |a_n|$, it follows that for $|z|, |w| \le r < 1$ the following inequalities hold

$$\frac{|b_n z + a_n|}{|b_n w + a_n|} \le \frac{|b_n| \, |z| + |a_n|}{||b_n| \, |w| - |a_n||} \le \frac{|a_n| \, (1 + |z|)}{|a_n| - |b_n| \, |w|}$$

$$\le \frac{|a_n| \, (1 + |z|)}{|a_n| \, (1 - |w|)} \le \frac{1 + r}{1 - r}$$

$$= \frac{1 - r^2}{(1 - r)^2} \le \frac{1}{(1 - r)^2}$$

(iii) Taking $z = 0$ in (ii) yields

$$\frac{|a_n|}{|b_n w + a_n|} \le \frac{1}{(1 - r)^2}$$

hence we conclude that for $|z| \le r$ the following estimate holds

$$\sum |\gamma_n'(z)|^2 = \sum \frac{1}{|b_n z + a_n|^4} \le \frac{1}{(1 - r)^8} \sum \frac{1}{|a_n|^4} \qquad (2.17)$$

(iv) Taking $w = 0$ in (ii) yields

$$\frac{|b_n z + a_n|}{|a_n|} \le \frac{1}{(1 - r)^2}$$

hence

$$\frac{1}{|b_n z + a_n|^4} \ge \frac{(1 - r)^8}{|a_n|^4}$$

and therefore

$$\sum \frac{(1 - r)^8}{|a_n|^4} \le \sum \frac{1}{|b_n z + a_n|^4}$$

(v) If we now choose r such that $P \subset \overline{D(0, r)}$ so that the above estimates hold in P, and we apply the operator $\int_P$ in the last inequality above, we get

$$\text{(Euclidean area of } P) \cdot \sum \frac{(1 - r)^8}{|a_n|^4} \le \sum \int_P \frac{1}{|b_n z + a_n|^4} = \pi$$

and therefore $\sum \dfrac{1}{|a_n|^4} < \infty$. The proof now follows from Equation (2.17).
$\qquad\square$

Corollary 2.15 *Let $a \in \mathbb{D}$, and define $f_a(z) = \left(\dfrac{1}{z-a}\right)$. Then:*

(i) *The series*

$$Q_a(z) = \sum f_a(\gamma_n(z))\,\gamma_n'(z)^2 \qquad (2.18)$$

defines a meromorphic function in $\mathbb{D}$ and, moreover

(ii) *$Q_a(z)$ satisfies the following quadratic relation with respect to the group K*

$$Q_a(\gamma(z)) = \dfrac{1}{\gamma'(z)^2} Q_a(z), \qquad for\ all \quad \gamma = \gamma_n \in K$$

$$(2.19)$$

Proof (i) Since K acts properly discontinuously and fixed point freely, there exists a disc $D(a,\varepsilon)$ such that for all n we have $\gamma_n(D(a,\varepsilon)) \subset \mathbb{D} \smallsetminus D(a,\varepsilon)$. Convergence is now a consequence of the previous lemma together with Weierstrass Theorem on the uniform convergence of analytic functions.

(ii) Direct computation gives

$$\begin{aligned}
Q_a(\gamma(z)) &= \sum f_a(\gamma_n \circ \gamma(z))\,\gamma_n'(\gamma(z))^2 \\
&= \sum f_a(\gamma_m(z)) \dfrac{\gamma_m'(z)^2}{\gamma'(z)^2} \\
&= \dfrac{1}{\gamma'(z)^2} Q_a(z)
\end{aligned}$$

where we have set $\gamma_m = \gamma_n \circ \gamma$.
$\qquad\square$

Proposition 2.16 *The field of meromorphic functions separates points in Riemann surfaces of genus $g \geq 2$.*

Proof Suppose first that S is the Riemann surface of genus g constructed in Section 2.1.2. Let $a, b \in \mathbb{D}$ be points whose images by the projection $\pi : \mathbb{D} \longrightarrow \mathbb{D}/K \simeq S$ are the points $P_1 = [a]$,

$P_2 = [b]$ that we want to become respectively a zero and a pole of some meromorphic function f. Then

$$f(z) = \frac{Q_b(z)}{Q_a(z)} \qquad (2.20)$$

solves our problem, since although Q_a and Q_b do not define functions in $S \simeq \mathbb{D}/K$, their quotient does. This is because relation (2.19) obviously implies that $f(\gamma(z)) = f(z)$.

The same proof would work for any Riemann surface $S' \simeq \mathbb{D}/K'$ with the same genus as S, if we knew that any such group K' has a fundamental domain (maybe less nice than the regular polygon given in Section 2.1.2). This is actually the case as can be shown as follows. By the classification of compact topological surfaces (see Section 1.2.1), S' and S must be homeomorphic and even diffeomorphic to each other. By covering space theory any diffeomorphism $f : S \longrightarrow S'$ (or more precisely the map $f \circ \pi$) lifts to a diffeomorphism $\widetilde{f} : \mathbb{D} \longrightarrow \mathbb{D}$, i.e. we have a commutative diagram as follows:

$$
\begin{array}{ccc}
\mathbb{D} & \xrightarrow{\ \widetilde{f}\ } & \mathbb{D} \\
{\scriptstyle \pi}\downarrow & & \downarrow{\scriptstyle \pi'} \\
S & \xrightarrow{\ f\ } & S'
\end{array}
$$

The fact that $\widetilde{f}$ transforms the group $K = \mathrm{Aut}(\mathbb{D}, \pi)$ into the group $K' = \mathrm{Aut}(\mathbb{D}, \pi')$, that is $K' = \widetilde{f} K \widetilde{f}^{-1}$ (see Section 1.2.5) implies that $\widetilde{f}(P)$ is a fundamental domain for K'.

Although $\widetilde{f}(P)$ will no longer be a hyperbolic polygon, Lemma 2.14 and Corollary 2.15 (and their proofs) remain valid. $\qquad\square$

2.3 Fuchsian groups

Subgroups of $\mathbb{PSL}(2, \mathbb{R})$ acting freely and properly discontinuously on $\mathbb{H}$ arise in a natural way in Riemann surface theory due to the Uniformization Theorem. We are more generally interested in subgroups acting only properly discontinuously (see Definition 1.68) because they still do define a Riemann surface.

Recall that if $\Gamma < \mathbb{PSL}(2, \mathbb{R})$ is such a group, given a point $z \in \mathbb{H}$ there exists a neighbourhood U_z such that $\gamma(U_z) \cap U_z = \emptyset$ for all $\gamma \in \Gamma$ except the finite set of elements $I(z) = \{\gamma_1, \ldots, \gamma_r\}$ fixing the point z. Clearly $I(z)$ is a subgroup of Γ usually called the *stabilizer* or *isotropy group* of z.

Let M be an isomorphism from $\mathbb{H}$ to $\mathbb{D}$ sending z to $0 \in \mathbb{D}$. Then $MI(z)M^{-1}$ is a finite group of rotations around the origin, hence it is a cyclic group generated by the rotation $R_{2\pi/n} : w \to e^{2\pi i/n}w$ where $n = |I(z)|$. It follows that $I(z) = \langle M^{-1} \circ R_{2\pi/n} \circ M \rangle$ itself is also a cyclic group and that the neighbourhood U_z above can be chosen to be $I(z)$-invariant. For this we can simply take $U_z = M^{-1}(D(0, \varepsilon))$. It also follows that the set of fixed points is discrete for otherwise there would be an infinite sequence of elements $\gamma_n \in \Gamma$ and of points $z_n \to z$ such that $\gamma_n(z_n) = z_n$. This would mean that for every $\varepsilon > 0$ there would be infinitely many elements $\gamma \in \Gamma$ such that $\gamma(U_z) \cap U_z \neq \emptyset$.

Now recall that $\mathbb{PSL}(2, \mathbb{R})$, which is naturally equipped with the topology induced by the usual topology of $M_{2\times 2}(\mathbb{R}) \simeq \mathbb{R}^4$, is a topological group. This simply means that the maps $\alpha \mapsto \alpha^{-1}$ and $(\alpha, \beta) \mapsto \alpha\beta$ are continuous.

Definition 2.17 A *Fuchsian group* is a subgroup $\Gamma < \mathbb{PSL}(2, \mathbb{R})$ such that for all $\gamma \in \Gamma$ there is a neighbourhood V of γ in $\mathbb{PSL}(2, \mathbb{R})$ with $V \cap \Gamma = \{\gamma\}$. In other words, the topology induced in Γ is the discrete topology.

An important property regarding the nature of Fuchsian groups is the following:

Lemma 2.18 ([Ive92]) *Let Γ be a Fuchsian group. For every $\alpha \in \mathbb{PSL}(2, \mathbb{R})$ (not necessarily in Γ) there is a neighbourhood $V \subset \mathbb{PSL}(2, \mathbb{R})$ such that $V \cap \Gamma$ is a finite set. In particular, Γ is a closed subset of $\mathbb{PSL}(2, \mathbb{R})$.*

Proof By definition of Fuchsian group the result holds trivially if $\alpha \in \Gamma$. In particular, we can take an open neighbourhood W of the identity such that $W \cap \Gamma = \{\mathrm{Id}\}$.

For arbitrary α the set $\alpha \cdot W^{-1} = \{\alpha\beta^{-1} \mid \beta \in W\}$ is an open neighbourhood of α. Now, if $\alpha \cdot W^{-1} \cap \Gamma = \emptyset$ we are done. If not, let $\gamma \in \alpha \cdot W^{-1} \cap \Gamma$, so that $\alpha = \gamma\beta$ for some $\beta \in W$. Take $V = \gamma W$, which is obviously an open neighbourhood of α as well as of γ. Then, since $W \cap \Gamma = \{\mathrm{Id}\}$ we have $\gamma^{-1}(V \cap \Gamma) = \{\mathrm{Id}\}$ and therefore $V \cap \Gamma = \{\gamma\}$.

In particular, if $\gamma_n \in \Gamma$ is a sequence converging to α, the existence of a neighbourhood V as above implies that all but finitely many elements of the sequence agree with α. Thus Γ is a closed subset of $\mathbb{PSL}(2, \mathbb{R})$. $\qquad\square$

Fuchsian groups and groups acting properly discontinuously are equivalent concepts.

Proposition 2.19 $\Gamma < \mathbb{PSL}(2, \mathbb{R})$ *is a Fuchsian group if and only if it acts properly discontinuously on* $\mathbb{H}$.

Proof Suppose that Γ is not a Fuchsian group. Then there exists an infinite sequence of distinct elements $\gamma_n \in \Gamma$ such that $\lim_{n \to \infty} \gamma_n = \gamma \in \Gamma$, hence if we put $\beta_n = \gamma^{-1}\gamma_n$ we find that $\beta_n \to \mathrm{Id}$ and therefore $\beta_n(z) \to z$. Thus for any neighbourhood U of z there are infinitely many elements $\beta_n \in \Gamma$ such that $\beta_n(U) \cap U \neq \emptyset$ and so Γ does not act properly discontinuously.

On the other hand, assume Γ is Fuchsian. If Γ does not act properly discontinuously, then there exists $z_0 \in \mathbb{H}$ and an infinite sequence $\{\gamma_k\} \subset \Gamma \smallsetminus I(z_0)$ of distinct elements such that $\gamma_k(D_h(z_0, 1/k)) \cap D_h(z_0, 1/k) \neq \emptyset$ where D_h stands for hyperbolic disc. In particular, $\{\gamma_k\}$ is contained in the set $C = \{\alpha \in \mathbb{PSL}(2, \mathbb{R}) \mid d_h(z_0, \alpha(z_0)) \leq 1\}$, which is compact (see Lemma 2.20 below). Therefore, $\{\gamma_k\} = \{\gamma_k\} \cap C$ is contained in $\Gamma \cap C$, which by Lemma 2.18 is a closed (hence compact) subset of C and being discrete it must be finite. Contradiction. $\qquad\square$

Lemma 2.20 *Given* $z_0 \in \mathbb{H}$ *and* $\varepsilon > 0$, *the set*

$$C_\varepsilon = \{\alpha \in \mathbb{PSL}(2, \mathbb{R}) \mid d_h(z_0, \alpha(z_0)) \leq \varepsilon\}$$

is a compact subset of $\mathbb{PSL}(2, \mathbb{R})$.

Proof We can work with $\mathrm{SL}(2, \mathbb{R}) \subset M_{2 \times 2}(\mathbb{R}) \simeq \mathbb{R}^4$. For an element $\alpha = \begin{pmatrix} a & b \\ c & d \end{pmatrix}$ we denote $|\alpha|^2 = a^2 + b^2 + c^2 + d^2$.

Since the set C_ε is clearly closed we only need to show that it is bounded. By conjugation by an element of $\mathbb{PSL}(2, \mathbb{R})$ sending z_0 to $i \in \mathbb{H}$, we can assume that $z_0 = i$.

A straightforward computation using formula (2.2) shows that

$$\tanh \frac{1}{2} d_h(i, \alpha(i)) = \sqrt{\frac{|\alpha|^2 - 2}{|\alpha|^2 + 2}}$$

therefore if $|\alpha| \to \infty$ then $\tanh \frac{1}{2} d_h(i, \alpha(i)) \to 1$, which implies $d_h(i, \alpha(i)) \to \infty$. $\qquad\square$

The most obvious example of Fuchsian group is $\mathbb{PSL}(2, \mathbb{Z})$. We will show in Section 2.4.4 that the quotient space $\mathbb{H}/\mathbb{PSL}(2, \mathbb{Z})$ is a Riemann surface isomorphic to $\mathbb{C}$. In fact the quotient of $\mathbb{H}$ by the action of an arbitrary Fuchsian group always gives a Riemann surface.

Proposition 2.21 *If Γ is a Fuchsian group then the quotient $\mathbb{H}/\Gamma$ has a Riemann surface structure for which the canonical projection $\pi : \mathbb{H} \longrightarrow \mathbb{H}/\Gamma$ is holomorphic.*

Proof We first observe that $\mathbb{H}/\Gamma$ is Hausdorff. Indeed if $\pi(x)$, $\pi(y)$ are two different points in $\mathbb{H}/\Gamma$ then there must exist hyperbolic discs $D_h(x, \varepsilon)$ and $D_h(y, \varepsilon)$ such that $\pi(D_h(x, \varepsilon)) \cap \pi(D_h(y, \varepsilon)) = \emptyset$. Otherwise there would be an infinite sequence of distinct elements $\gamma_n \in \Gamma$ such that $\gamma_n(D_h(x, 1/n)) \cap D_h(y, 1/n) \neq \emptyset$. This implies that if $\alpha \in \mathbb{PSL}(2, \mathbb{R}) \smallsetminus \Gamma$ is such that $\alpha(x) = y$ we have $\alpha^{-1}\gamma_n(D_h(x, 1/n)) \cap D_h(x, 1/n) \neq \emptyset$ for every n. By Lemma 2.20 the sequence $\alpha^{-1}\gamma_n$ is contained in a compact set and therefore, passing to a subsequence if necessary, we can assume that $\alpha^{-1}\gamma_n \to \beta$, hence $\gamma_n \to \alpha\beta$, which by Lemma 2.18 is an element of Γ. This cannot happen for a Fuchsian group.

The quotient map $\pi : \mathbb{H} \longrightarrow \mathbb{H}/\Gamma$ restricted to a small neighbourhood provides a coordinate function around the points where Γ acts freely. If U and V are two such neighbourhoods and $\pi(U) \cap \pi(V) \neq \emptyset$ then the transition function looks like

$$\left(\pi|_V^{-1} \circ \pi|_U\right)(z) = \gamma(z)$$

where $\gamma \in \Gamma$ is the element which sends $z \in U$ to $\gamma(z) \in V$ (as in Example 1.13).

Now, suppose $z_0 \in \mathbb{H}$ is a fixed point of some element. Let U be $I(z_0)$-stable and small enough so that $U/\Gamma = U/I(z_0)$. The map $\pi : U \smallsetminus \{z_0\} \longrightarrow U \smallsetminus \{z_0\}/\Gamma$ is a covering of Riemann surfaces with degree n and therefore π is of the form $z \mapsto z^n$ (Example 1.72). More precisely we have a commutative diagram

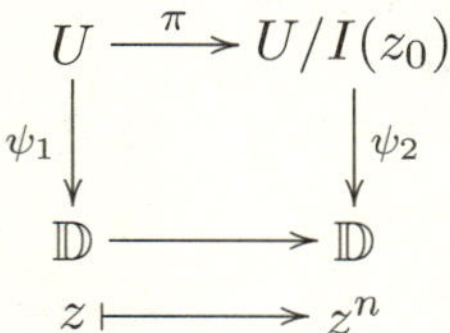

where the maps ψ_i are holomorphic isomorphisms. We can therefore take ψ_2 as a chart around $[z_0]$, and the resulting transition function $\psi_2 \circ \pi = \psi_1^n$ is certainly holomorphic. $\qquad\square$

Remark 2.22 We observe for further use that along the previous proof we have obtained the identity $\mathrm{ord}_z(\pi) = |I(z)|$.

Proposition 2.23 *A Fuchsian group Γ acts freely on $\mathbb{H}$ if and only if Γ is torsion free (i.e. if it has no non-trivial elements of finite order).*

Proof An element $\gamma \in \mathbb{PSL}(2, \mathbb{R})$ fixes a point if and only if it is conjugate to a rotation $R_\theta(z) = e^{2\pi i \theta} z$ in the unit disc.

Now if $\theta = m/n \in \mathbb{Q}$ then the order of R_θ (hence of γ) is n, whereas if $\theta \notin \mathbb{Q}$ then $\left\{ R_\theta^k = R_{k\theta} \right\}_k$ is an infinite sequence of different rotations which must have limit points since its elements are parametrized by their rotation angles and these lie in the compact set $[0, 2\pi]$. $\qquad\square$

Corollary 2.24 *A subgroup of $\mathbb{PSL}(2, \mathbb{R})$ acts freely and properly discontinuously on $\mathbb{H}$ if and only if it is discrete and torsion free.*

Of course different Fuchsian groups may produce the same Riemann surface. In the case of torsion free groups it is easy to characterize when this occurs.

Proposition 2.25 *Let $S_1 = \mathbb{H}/\Gamma_1$ and $S_2 = \mathbb{H}/\Gamma_2$ be two Riemann surfaces (compact or not) uniformized by freely acting Fuchsian groups Γ_1 and Γ_2. Then S_1 and S_2 are isomorphic if and only if there exists $T \in \mathbb{PSL}(2, \mathbb{R})$ such that $T \circ \Gamma_1 \circ T^{-1} = \Gamma_2$.*

Proof The *if* part is trivial. The isomorphism $T : \mathbb{H} \to \mathbb{H}$ induces a well-defined isomorphism between S_1 and S_2. Indeed, with the obvious notation, if $[z]_1 = [w]_1$ then $w = \gamma_1(z)$ for some $\gamma_1 \in \Gamma_1$. Therefore, $T(w) = T \circ \gamma_1(z) = \gamma_2 \circ T(z)$ for some $\gamma_2 \in \Gamma_2$, hence $[T(w)]_2 = [T(z)]_2$.

Conversely, an isomorphism $\phi : S_1 \to S_2$ determines a commutative diagram

$$
\begin{array}{ccc}
\mathbb{H} & \xrightarrow{\;\tilde{\phi}\;} & \mathbb{H} \\
\downarrow{\scriptstyle p_1} & & \downarrow{\scriptstyle p_2} \\
\mathbb{H}/\Gamma_1 & \xrightarrow{\;\phi\;} & \mathbb{H}/\Gamma_2
\end{array}
$$

where $\tilde{\phi} = T \in \mathrm{Aut}(\mathbb{H}) = \mathbb{PSL}(2, \mathbb{R})$ is a lift of ϕ (more precisely, of $\phi \circ p_1$). In turn this implies that $\Gamma_2 = T \circ \Gamma_1 \circ T^{-1}$. $\qquad\square$

2.4 Fuchsian triangle groups

In this section we study the Fuchsian groups whose fundamental domain has the minimum possible number of sides. They are called *triangle groups* and admit fundamental domains with only four hyperbolic sides, which are obtained by glueing two copies of some hyperbolic triangle.

2.4.1 Triangles in hyperbolic space

A hyperbolic triangle in $\mathbb{H}$ is a topological triangle whose edges are hyperbolic geodesic segments. We admit in this definition the possibility of triangles with edges of infinite (hyperbolic) length, a case in which some vertex lies in $\mathbb{R} \cup \{\infty\}$ and the corresponding angle is zero.

What characterizes hyperbolic geometry is the fact that the sum of angles of a hyperbolic triangle is less than π. In fact we have the following:

Proposition 2.26 *If T is a hyperbolic triangle with angles α, β and γ, then the hyperbolic area of T equals $a(T) = \pi - \alpha - \beta - \gamma$.*

Proof (Following [JS87]). Suppose we have proved the statement for triangles with (at least) one null angle. Then, if T is a triangle with angles α, β and γ we can construct T' as in Figure 2.10 in such a way that both T' and $T \cup T'$ are triangles with one null angle.

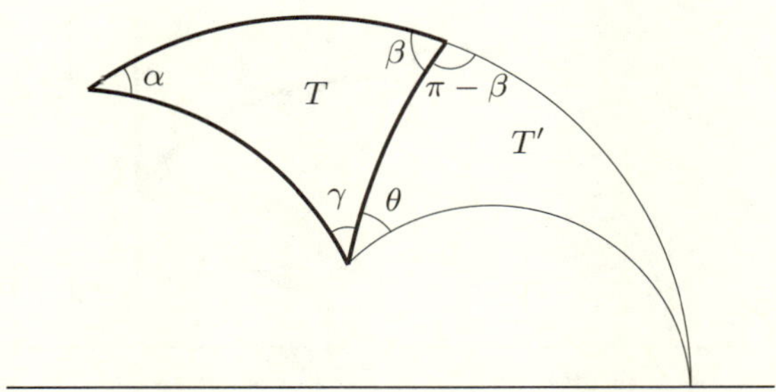

Fig. 2.10. The hyperbolic triangles T, T' and $T \cup T'$ in $\mathbb{H}$.

Then we would have

$$
\begin{aligned}
a(T) &= a(T \cup T') - a(T') \\
&= (\pi - \alpha - (\gamma + \theta) - 0) - (\pi - (\pi - \beta) - \theta - 0) \\
&= \pi - \alpha - \beta - \gamma
\end{aligned}
$$

as we wanted to show.

In order to prove the result for a triangle with exactly one null angle, we can restrict ourselves to the case in which T has a vertex at ∞, one edge (of finite length) contained in $\{|z| = r\}$ and two straight edges (of infinite edges) L_a and L_b contained in $\{\mathrm{Re}(z) = a\}$ and $\{\mathrm{Re}(z) = b\}$ respectively (see Figure 2.11). The reason is that we can always achieve this configuration applying transformations of the form $z \longmapsto z + t$ and $z \longmapsto -1/z$.

Let now α and β be the angles of T shown in Figure 2.11. By elementary Euclidean geometry we see that $b = r \cos \beta$ and $a = -r \cos \alpha = r \cos(\pi - \alpha)$ respectively.

We can therefore compute

$$
a(T) = \int_T \frac{dxdy}{y^2} = \int_a^b \int_{\sqrt{r^2 - x^2}}^\infty \frac{dy}{y^2} dx = \int_a^b \frac{dx}{\sqrt{r^2 - x^2}}
$$

$$
= \int_{\pi - \alpha}^\beta \frac{-r \sin \theta}{r \sin \theta} d\theta = \pi - \alpha - \beta
$$

where in the second line we have employed the change of variables $x = r \cos \theta$ with $\theta \in (0, \pi)$. The proof is done. $\qquad \square$

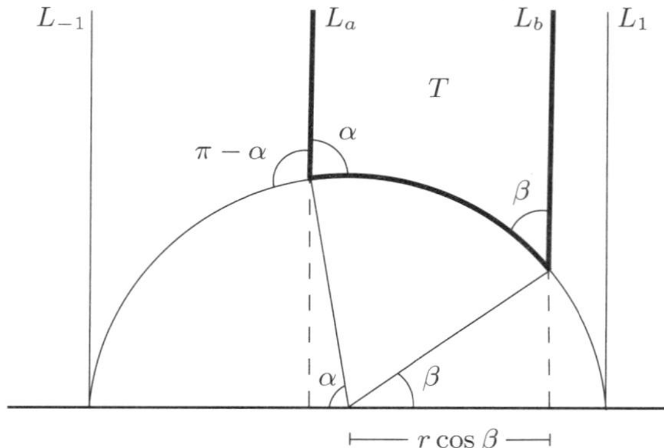

Fig. 2.11. A hyperbolic triangle with a vertex at ∞.

Given three non-negative real numbers α, β and γ such that $\alpha + \beta + \gamma < \pi$ there exists a hyperbolic triangle with angles α, β and γ. Figure 2.11 illustrates the construction of such a triangle in the case $\gamma = 0$.

We begin with the (degenerate) triangle with sides are L_{-1}, L_1 and $C \subset \{|z| = R\}$, whose three angles equal 0. If we move continuously the geodesic L_a from L_{-1} towards L_1 the angle between C and L_a measured in positive sense goes from zero to π. There must be intermediate positions where the line L_a meets C with angles α and $\pi - \beta$ (see Figure 2.11).

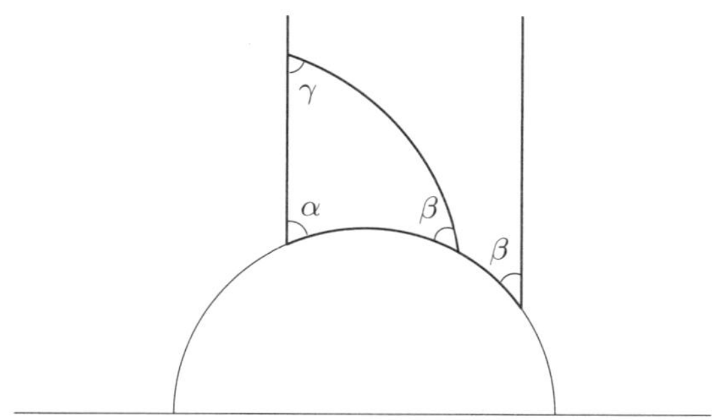

Fig. 2.12. A hyperbolic triangle with angles α, β, γ.

Now, Figure 2.12 is meant to llustrate the existence of a triangle with given angles $\alpha, \beta, \gamma > 0$. The proof relies again on a continuity argument. Starting from Figure 2.11, we see that if we now move the geodesic L_b from L_1 to the left, keeping the angle β constant in the process, the angle γ formed with L_a goes from 0 to $\pi - (\alpha + \beta)$, since in the limit we would encounter a triangle with zero area.

2.4.2 Reflections

The *reflection R_L in the geodesic L* is the unique non-trivial isometry that fixes every point of L. The reflection in the imaginary axis L_0 is $R_0(z) = -\bar{z}$, and the reflection in an arbitrary geodesic L can be described as a conjugate of this one. Namely, $R_L = M^{-1} \circ R_0 \circ M$, with $M \in \mathbb{PSL}(2, \mathbb{R})$ such that $M(L) = L_0$. Note that reflections are *anticonformal isometries* (they reverse orientation).

If R is an arbitrary reflection, it is clear that $R_0 \circ R$ is an orientation-preserving isometry, hence $R_0 \circ R \in \mathbb{PSL}(2, \mathbb{R})$. Therefore, every reflection can be expressed in the form

$$R(z) = \frac{a\bar{z} + b}{c\bar{z} + d}, \quad \text{where } a, b, c, d \in \mathbb{R} \text{ and } ad - bc = -1$$

2.4.3 Construction of triangle groups

Let R_i be the reflection in the side L_i of the hyperbolic triangle T of Figure 2.13, with vertices v_1, v_2, v_3 and angles $\dfrac{\pi}{n}, \dfrac{\pi}{m}$ and $\dfrac{\pi}{l}$ respectively (here we denote by L_i the edge from v_i to v_{i+1}, with subscripts taken modulo 3). The numbering of the vertices v_1, v_2, v_3 is understood to follow counterclockwise order, as in Figure 2.13, and n, m, l are either positive integers or ∞ (the latter meaning that the angle at the corresponding vertex equals zero, see for instance Figure 2.17). In what follows we assume that $\dfrac{1}{n} + \dfrac{1}{m} + \dfrac{1}{l} < 1$. This condition guarantees the existence in hyperbolic space of such a triangle T, as was seen in Section 2.4.1.

Let us now consider the images of T under the following se-

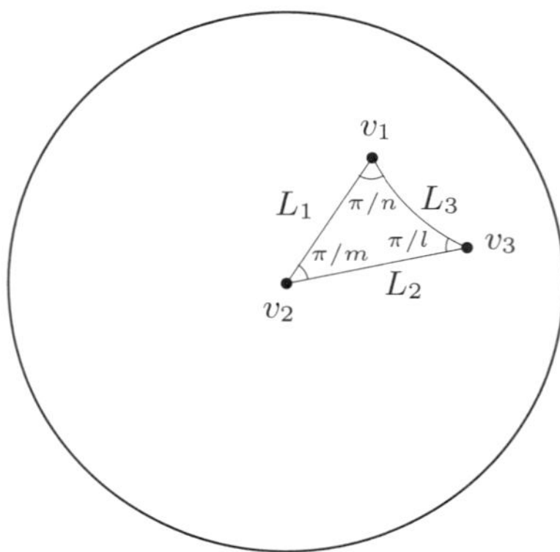

Fig. 2.13. The starting triangle T in the case $(n, m, l) = (3, 4, 5)$.

quence of compositions of the reflections R_1 and R_2

$$
\begin{aligned}
&T \\
&R_1\,(T) \\
&R_1 \circ R_2\,(T) \\
&R_1 \circ R_2 \circ R_1\,(T) \\
&(R_1 \circ R_2)^2\,(T) \\
&\qquad\vdots \\
&(R_1 \circ R_2)^{k-1}\,(T) \\
&(R_1 \circ R_2)^{k-1} \circ R_1\,(T) \\
&(R_1 \circ R_2)^k\,(T) \\
&\qquad\vdots
\end{aligned}
\tag{2.21}
$$

Since $v_2 \in L_1 \cap L_2$, this vertex remains fixed along the whole process, hence all the triangles have a common vertex at v_2. Moreover, (2.21) is a sequence of *adjacent triangles.*

Indeed:

- $R_1\,(T) =: T^-$ is the reflection of T in the side L_1.
- $R_1 \circ R_2\,(T) = R_1 \circ R_2 \circ R_1^{-1}\,(R_1\,(T))$ is the reflection of $R_1\,(T)$ in its side $R_1\,(L_2)$.
- $R_1 \circ R_2 \circ R_1\,(T) = (R_1 \circ R_2) \circ R_1 \circ (R_1 \circ R_2)^{-1}\,R_1 \circ R_2\,(T)$ is the reflection of $R_1 \circ R_2\,(T)$ in its side $R_1 \circ R_2\,(L_1)$, etc.

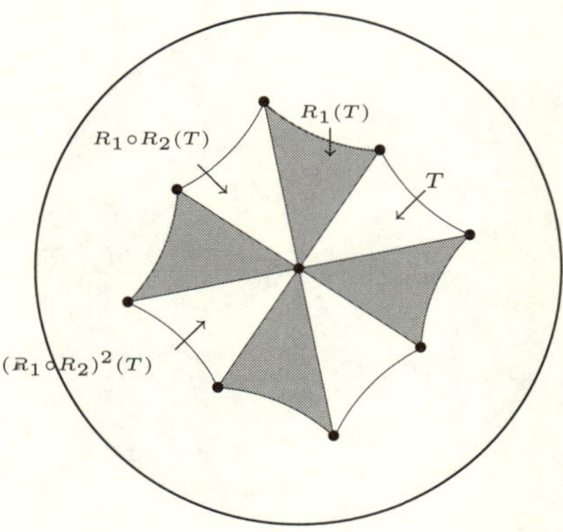

Fig. 2.14. D_2 is a neighbourhood of v_2 constructed as a reunion of triangles.

Since the angle at v_2 of each of these $2m$ triangles equals π/m, we see that the reunion of the first $2m$ triangles of the sequence (2.21) produces a complete neighbourhood D_2 of v_2 (see Figure 2.14). It also follows that $(R_1 \circ R_2)^m = \mathrm{Id}$.

Alternatively, this same neighbourhood D_2 can be also obtained as the reunion of images of the quadrilateral

$$Q = T \cup T^- = T \cup R_1(T)$$

since

$$D_2 = Q \cup (R_1 \circ R_2)(Q) \cup \cdots \cup (R_1 \circ R_2)^{m-1}(Q)$$

as can be seen by reordering in the appropiate way the terms of sequence (2.21).

It becomes apparent that by reflecting T in every possible direction, that is translating T by all the elements of the group $\langle R_1, R_2, R_3 \rangle$, we obtain a tessellation of $\mathbb{D}$ (see Figure 2.15).

Alternatively, a tessellation of $\mathbb{D}$ is obtained also translating the quadrilateral Q by all the transformations in the group

$$\Gamma = \Gamma_{n,m,l} = \langle x_1, x_2, x_3 \rangle$$

where $x_1 = R_3 \circ R_1$, $x_2 = R_1 \circ R_2$ and $x_3 = R_2 \circ R_3$. Clearly the relations

$$x_1^n = x_2^m = x_3^l = x_1 x_2 x_3 = 1$$

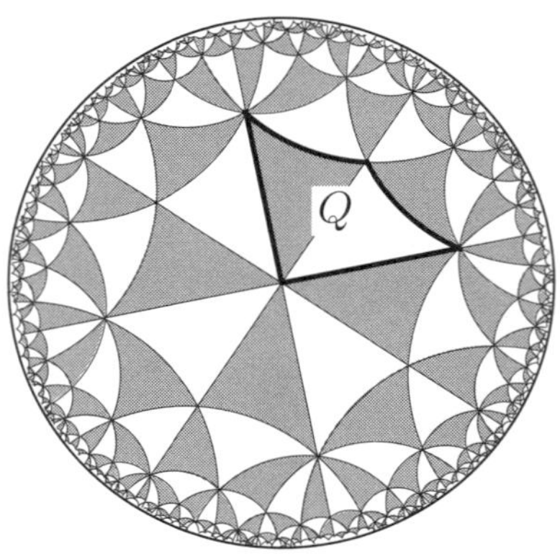

Fig. 2.15. The tessellation of $\mathbb{D}$ corresponding to the triangle group $\Gamma_{n,m,l}$ with $(n, m, l) = (3, 4, 5)$.

hold in Γ.

Note that now the transformations x_1, x_2 and x_3 are certainly conformal isometries, therefore $\Gamma < \mathrm{Aut}(\mathbb{D}) \simeq \mathbb{PSL}\,(2, \mathbb{R})$. In fact x_1, x_2 and x_3 are hyperbolic rotations around v_1, v_2 and v_3 through angles $2\pi/n, 2\pi/m$ and $2\pi/l$ respectively.

The fact that the quadrilaterals $\{\gamma(Q) \mid \gamma \in \Gamma\}$ provide a tessellation of the unit disc is a manifestation of the fact that Γ acts properly discontinuously and that Q is a fundamental domain. Of course this is not a fixed point free action. The set of points of $\mathbb{D}$ fixed by elements of Γ agrees with the set of vertices of the tessellation. Accordingly, the finite order (torsion) elements of Γ are the conjugates of powers of x_1, x_2 and x_3.

According to Proposition 2.21, $\mathbb{D}/\Gamma$ carries a Riemann surface structure. From the side identifications in Q it is easy to deduce that the genus of this surface is 0, therefore $\mathbb{D}/\Gamma \simeq \widehat{\mathbb{C}}$ (or $\widehat{\mathbb{C}}$ minus one, two or three points in case one, two or three of the angles equal zero).

These last statements are nothing but a particular application of a more general version of Poincaré's theorem. The difference with the situation we studied in Section 2.1.2 is the fact that the sum of the angles of all the vertices of Q equivalent to, say, v_2 is no longer 2π but rather $2\pi/m$. Therefore, the reunion of the small circle sectors around these vertices does not provide a complete neighbourhood of the vertex. We need now to apply a mapping of the form $z \to z^m$ to get a complete neighbourhood. According to the proof of Proposition 2.21 this only reflects the fact that the stabilizer of v_2 is a cyclic group of order m, namely $I(v_2) = \langle R_1 \circ R_2 \rangle = \langle x_2 \rangle$.

The version of Poincaré's polygon Theorem that covers the case of our group $\Gamma_{n,m,l}$ is:

Theorem 2.27 (Poincaré's polygon Theorem) *Let P be a convex hyperbolic polygon in $\mathbb{D}$ with sides $l_1, \ldots, l_n, l'_1, \ldots, l'_n$. Suppose there exist $g_1, \ldots, g_n \in \mathrm{Aut}(\mathbb{D})$ such that $g_i(l_i) = l'_i$. Let G be the group generated by the g_i. If the sum of the angles of P at the j-th equivalence class of vertices equals $2\pi/n_j$ for an integer n_j, then G is a Fuchsian group and $\mathbb{D}/G$ is a compact Riemann surface.*

Summarizing, the projection

$$\pi : \mathbb{H} \longrightarrow \mathbb{H}/\Gamma$$

$$z \longmapsto [z]_\Gamma$$

ramifies at all points z fixed by some element of $\Gamma = \Gamma_{n,m,l}$. These are the vertices v_i and their translates by Γ. According to Remark 2.22 each of these points has branching order equal to n, m or l.

Since $x_1 \circ x_2 \circ x_3 = \mathrm{Id}$, we can also write

$$\Gamma_{n,m,l} = \langle x_1, x_2 \rangle = \langle x_1, x_3 \rangle = \langle x_2, x_3 \rangle$$

Definition 2.28 The group $\Gamma_{n,m,l}$ described above is called the *triangle group* of signature (n, m, l).

Remark 2.29 The triangle group of signature (n, m, l) is unique up to conjugation in $\mathbb{PSL}(2, \mathbb{R})$.

Given two hyperbolic triangles T_1, T_2 with the same angles there exists an isometry γ of hyperbolic space that maps T_1 to T_2. The two triangle groups Γ_1, Γ_2 corresponding to T_1, T_2 are conjugated by γ, that is $\gamma^{-1} \circ \Gamma_2 \circ \gamma = \Gamma_1$.

As an abstract group, $\Gamma_{n,m,l}$ can be defined in terms of generators and relations as follows (see [JS87], Appendix 2)

$$\Gamma_{n,m,l} = \langle x_1, x_2, x_3; \ x_1^n = x_2^m = x_3^l = x_1 x_2 x_3 = 1 \rangle$$

where x_j^∞ means no relation, e.g. the group

$$\Gamma_{\infty,\infty,\infty} = \langle x_1, x_2, x_3; \ x_1 x_2 x_3 = 1 \rangle = \langle x_1, x_2 \rangle$$

is the free group with two generators.

Remark 2.30 When $\dfrac{1}{n} + \dfrac{1}{m} + \dfrac{1}{l} = 1$ or $\dfrac{1}{n} + \dfrac{1}{m} + \dfrac{1}{l} > 1$ one can construct the triangle group $\Gamma_{n,m,l}$ as a subgroup of isometries of Euclidean 2-space or the sphere respectively. The starting triangle T with angles $\dfrac{\pi}{n}, \ \dfrac{\pi}{m}$ and $\dfrac{\pi}{l}$ exists in Euclidean or spherical geometry rather than in hyperbolic space, but the construction of such an Euclidean or spherical triangle group goes exactly along the same lines as that of a hyperbolic triangle group. The resulting group of transformations is still infinite in the Euclidean case, but is finite in the spherical case. For instance, in Figure 2.16 the tessellations corresponding to $\Gamma_{2,2,2}$ and $\Gamma_{2,3,6}$ are depicted.

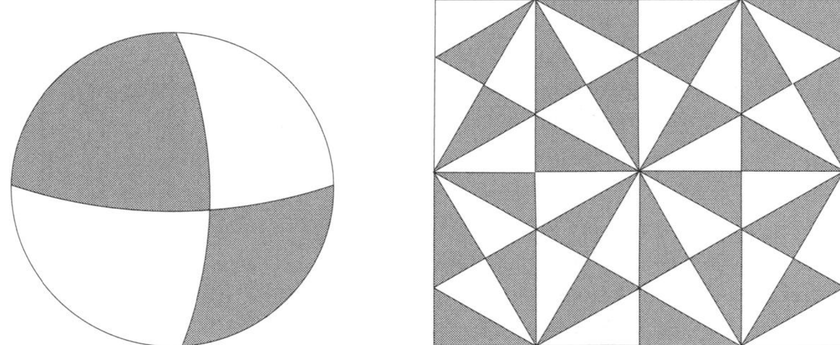

Fig. 2.16. Tessellation of the sphere and the complex plane associated to a triangle group $\Gamma_{2,2,2}$ and $\Gamma_{2,3,6}$ respectively.

From the presentation

$$\Gamma_{2,2,2} = \langle x_1, x_2, x_3 \ : \ x_1^2 = x_2^2 = x_3^2 = x_1 x_2 x_3 = 1 \rangle$$

we see that $\Gamma_{2,2,2}$ is generated by two elements of order 2 which commute (their product has again order 2), hence $\Gamma_{2,2,2}$ is Klein's group of order 4.

2.4.4 The modular group $\mathrm{PSL}(2,\mathbb{Z})$

In this section we introduce the most important triangle group of all. We now work with the upper half plane model of the hyperbolic plane.

Let T be the hyperbolic triangle with vertices located at $v_1 = i$, $v_2 = e^{\pi i/3}$ and $v_3 = \infty$. The edges L_i $(i = 1,2,3)$ of T lie on $|z| = 1$, $\mathrm{Re}(z) = 1/2$ and $\mathrm{Re}(z) = 0$ respectively (see the shadded region in Figure 2.17). The angles of T at its vertices are $\pi/2$, $\pi/3$ and 0 respectively, hence the construction of the previous section produces a triangle group $\Gamma_{2,3,\infty}$. We recall that the quotient $\mathbb{H}/\Gamma_{2,3,\infty}$ is the sphere minus one point.

With the same notation as in Section 2.4.3, one can easily check that

$$R_1(z) = \frac{1}{\bar{z}}, \quad R_2(z) = -\bar{z} + 1, \quad R_3(z) = -\bar{z}$$

and therefore any chosen pair of the following three elements of $\mathrm{PSL}(2,\mathbb{R})$

$$\begin{array}{rcl}
x_1(z) & = & R_3 \circ R_1(z) = -1/z \\
x_2(z) & = & R_1 \circ R_2(z) = 1/(-z+1) \\
x_3(z) & = & R_2 \circ R_3(z) = z + 1
\end{array}$$

generate $\Gamma_{2,3,\infty}$.

If we take $Q = T \cup R_3(T)$ as fundamental domain, the images of Q by the elements of $\Gamma_{2,3,\infty}$ provide the well-known tessellation of hyperbolic space shown in Figure 2.17.

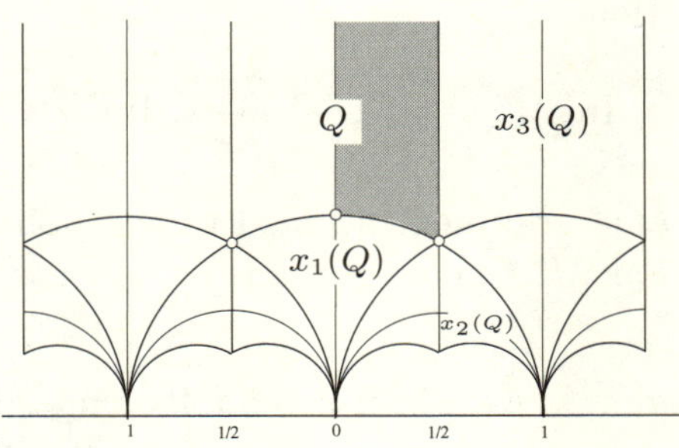

Fig. 2.17. Fundamental domain of $\Gamma_{2,3,\infty} = \mathrm{PSL}(2,\mathbb{Z})$.

This group is specially relevant in number theory, because it consists of the elements of $\mathrm{PSL}(2, \mathbb{R})$ with integer entries.

Theorem 2.31 *The group $\Gamma_{2,3,\infty}$ above agrees with*

$$\mathrm{PSL}(2, \mathbb{Z}) = \frac{\mathrm{SL}(2, \mathbb{Z})}{\langle \pm \mathrm{Id} \rangle}$$

Proof Obviously $\Gamma_{2,3,\infty} \leq \mathrm{PSL}(2, \mathbb{Z})$.

Conversely, suppose that $\mathrm{PSL}(2, \mathbb{Z})$ strictly contains $\Gamma_{2,3,\infty}$. Let $A \in \mathrm{PSL}(2, \mathbb{Z}) \smallsetminus \Gamma_{2,3,\infty}$ and take a point z in the interior $\overset{\circ}{Q}$ of the fundamental domain Q above such that $A(z)$ lies also in the interior of another fundamental domain, say $A(z) \in \gamma\left(\overset{\circ}{Q}\right)$ for some $\gamma \in \Gamma_{2,3,\infty}$. Then the transformation $M = \gamma^{-1} \circ A$ will send z back to $\overset{\circ}{Q}$. This way we will have in $\overset{\circ}{Q}$ the two points z and $M(z) = \dfrac{az + b}{cz + d}$ with $a, b, c, d \in \mathbb{Z}$, $ad - bc = 1$ and $M \neq \pm \mathrm{Id}$.

We first observe that c has to be non-zero. Indeed if $c = 0$ then $a = d = \pm 1$, hence the transformation M would be a translation by a non-zero integer and so z and $M(z)$ could not both lie in Q.

On the other hand, since $|z| > 1$ and $|\mathrm{Re}\, z| < 1/2$, we have

$$\begin{aligned}
|cz + d|^2 &= c^2 |z|^2 + 2cd\,\mathrm{Re}\, z + d^2 \\
&> c^2 + d^2 - |cd| \\
&= (|c| - |d|)^2 + |cd|
\end{aligned}$$

and this is a positive integer since otherwise $c = d = 0$, which is a contradiction. Therefore, $|cz + d|^2 > 1$ (strict inequality), and it follows by (2.1) that

$$\mathrm{Im}\, M(z) = \frac{\mathrm{Im}\, z}{|cz + d|^2} < \mathrm{Im}\, z$$

Exactly the same argument applied to the point $M(z)$ and to the transformation M^{-1} shows that $\mathrm{Im}\, z < \mathrm{Im}\, M(z)$, which is a contradiction. $\qquad\square$

In the next example we will use the following observation:

Lemma 2.32 *Let $\Gamma_1 < \Gamma_2$ be an inclusion of Fuchsian groups of index n, and $\gamma_1, \ldots, \gamma_n \in \Gamma_2$ a set of representatives of $\Gamma_1 \backslash \Gamma_2$, the* *right cosets of Γ_2 modulo Γ_1. Let Q and D be hyperbolic polygons such that Q is a fundamental domain for Γ_2 and $D = \bigcup_{i=1}^{n} \gamma_i(Q)$. Then D is a fundamental domain for Γ_1.*

Proof (1) Note first that

$$\bigcup_{\gamma \in \Gamma_1} \gamma(D) = \bigcup_{\gamma \in \Gamma_1} \gamma \left(\bigcup_{i=1}^{n} \gamma_i(Q) \right)$$
$$= \bigcup_{\gamma \in \Gamma_1} \bigcup_{i=1}^{n} \gamma\gamma_i(Q)$$
$$= \bigcup_{\beta \in \Gamma_2} \beta(Q)$$
$$= \mathbb{H}$$

(2) For any $\gamma \in \Gamma_1$ one clearly has

$$D \cap \gamma(D) = \bigcup_{i,j=1}^{n} \gamma_i(Q) \cap \gamma\gamma_j(Q)$$

Now, D being a convex hyperbolic polygon, if $\gamma_i(Q) \cap \gamma\gamma_j(Q)$ contains an interior point of D then $\gamma\gamma_j(Q)$ must equal one of the polygons $\gamma_k(Q)$ that intersect $\gamma_i(Q)$, or equivalently $\gamma\gamma_j = \gamma_k$. Since $\gamma_1,\ldots,\gamma_n$ are distinct representatives of Γ_2 modulo Γ_1 the last equality occurs only if $\gamma = \mathrm{Id}$. Our conclusion is that if $\gamma \in \Gamma_1 \setminus \{\mathrm{Id}\}$ then $D \cap \gamma(D)$ is contained in the boundary of D. $\square$

Example 2.33 The modular group $\mathbb{PSL}(2,\mathbb{Z})$, also denoted $\Gamma(1)$, contains with index 6 the so-called *principal congruence subgroup of level* 2, defined as

$$\Gamma(2) = \left\{ \begin{pmatrix} a & b \\ c & d \end{pmatrix} \in \mathbb{PSL}(2,\mathbb{Z}) \ : \ \begin{pmatrix} a & b \\ c & d \end{pmatrix} \equiv \mathrm{Id} \bmod 2 \right\}$$

From the exact sequence

$$1 \to \Gamma(2) \to \Gamma(1) \to \mathbb{PSL}(2,\mathbb{Z}/2\mathbb{Z}) \to 1$$

we deduce that $\Gamma(2) \lhd \Gamma(1)$ with index 6.

The group $\Gamma(2)$ is itself a triangle group. In fact we claim that $\Gamma(2)$ is the triangle group $\Gamma_{\infty,\infty,\infty}$ corresponding to the triangle T with vertices at $\infty, 0$ and 1 and angle 0 at every vertex. This can be seen quite clearly after a good choice of the fundamental domain for $\Gamma(1)$ and representatives for the six equivalency classes of $\Gamma(2)\backslash\Gamma(1)$. Instead of Q it will be more convenient to choose the quadrilateral $Q' = T \cup R_2(T)$ with vertices at i, $e^{i\pi/3}$, $1+i$ and ∞ as fundamental domain for $\Gamma(1)$ (see Figure 2.18).

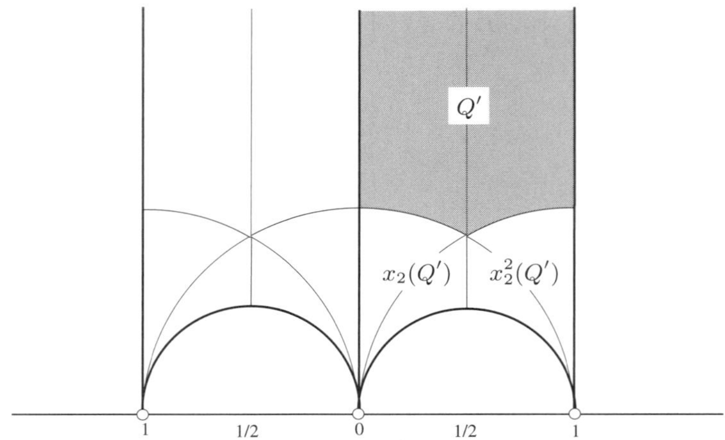

Fig. 2.18. The group $\Gamma(2)$ agrees with the triangle group $\Gamma_{\infty,\infty,\infty}$.

For a set of representatives we take

$$\mathrm{Id} = \begin{pmatrix} 1 & 0 \\ 0 & 1 \end{pmatrix}, \quad x_2 = \begin{pmatrix} 0 & -1 \\ 1 & -1 \end{pmatrix}, \quad x_2^2 = \begin{pmatrix} 1 & -1 \\ 1 & 0 \end{pmatrix},$$

$$x_3^{-1} = \begin{pmatrix} 1 & -1 \\ 0 & 1 \end{pmatrix}, \quad x_3^{-1}x_2 \begin{pmatrix} -1 & 0 \\ 1 & -1 \end{pmatrix}, \quad x_3^{-1}x_2^2 = \begin{pmatrix} 0 & -1 \\ 1 & 0 \end{pmatrix}$$

By Lemma 2.32 the reunion

$$D = Q' \cup x_2(Q') \cup x_2^2(Q') \cup x_3^{-1}(Q') \cup x_3^{-1}x_2(Q') \cup x_3^{-1}x_2^2(Q')$$

is a fundamental domain for $\Gamma(2)$ (see Figure 2.18). Now, D can be also described as the reunion of the two adjacent triangles $\widetilde{T} = Q' \cup x_2(Q') \cup x_2^2(Q')$ (with vertices at $0, 1, \infty$) and $\widetilde{T}^- = R_1(\widetilde{T})$ (with vertices at $0, -1$ and ∞).

Since $\Gamma(2)$ identifies the sides of D in the same way as the triangle group corresponding to $\widetilde{T}$, it follows that $\Gamma(2)$ equals $\Gamma_{\infty,\infty,\infty}$.

This identification allows us to find explicit generators for $\Gamma(2)$.

With the notation as above, we have

$$\widetilde{R}_1(z) = -\overline{z}, \quad \widetilde{R}_2(z) = \frac{\overline{z}}{2\overline{z}-1}, \quad \widetilde{R}_3(z) = -\overline{z}+2$$

hence

$$\widetilde{x}_1(z) = \widetilde{R}_3\widetilde{R}_1(z) \;=\; z+2 \quad \text{(which fixes } \infty\text{)}$$

$$\widetilde{x}_2(z) = \widetilde{R}_1\widetilde{R}_2(z) \;=\; \frac{-z}{2z-1} \quad \text{(which fixes } 0\text{)}$$

$$\widetilde{x_3}(z) = \widetilde{R}_2\widetilde{R}_3(z) \;=\; \frac{-z+2}{-2z+3} \quad \text{(which fixes } 1\text{)}$$

One obviously has $\widetilde{x}_1\widetilde{x}_2\widetilde{x}_3 = \mathrm{Id}$, as mentioned at the end of the previous section.

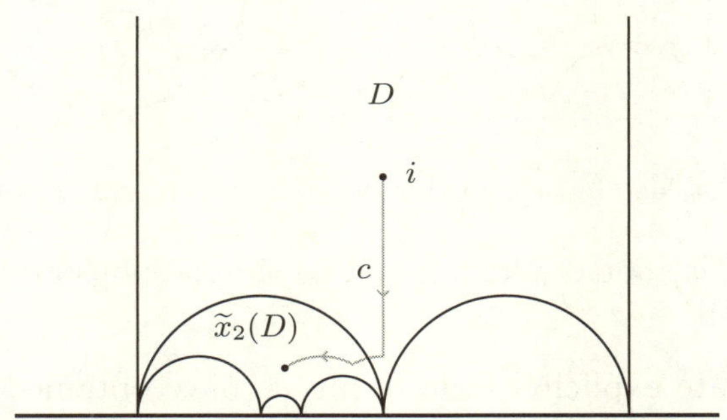

Fig. 2.19. A path in $\mathbb{H}$ joining i to $\widetilde{x}_2(i)$.

Now, being torsion free, the group $\Gamma(2) = \Gamma_{\infty,\infty,\infty}$ acts freely on $\mathbb{H}$ and so $\Gamma(2)$ must be isomorphic to the fundamental group of $\mathbb{H}/\Gamma(2) = \mathbb{H}/\Gamma_{\infty,\infty,\infty}$, which clearly is isomorphic to the sphere with three points deleted corresponding to the vertices $0,1,\infty$ of the domain D. Since $\mathbb{PSL}(2,\mathbb{C})$ acts transitively on triples of points we can choose this isomorphism in such a way that these vertices correspond respectively to the points $0,1,\infty \in \widehat{\mathbb{C}}$.

The identification of $\Gamma(2)$ with $\pi_1(\widehat{\mathbb{C}} \smallsetminus \{0,1,\infty\}, P)$ can be made very explicit. Let us choose as base point P the projection of the point $z = i$ onto $\mathbb{H}/\Gamma(2)$. According to Theorem 1.69, the loop corresponding to $\widetilde{x}_2$ is the projection to $\mathbb{H}/\Gamma(2)$ of any path c in $\mathbb{H}$ connecting $z = i$ to $\widetilde{x}_2(i) = -\frac{2}{5} + \frac{i}{5}$. In Figure 2.19 we have chosen one such c as a piecewise geodesic path, connecting first $z = i$ with a point z' in the imaginary axis near the real line, followed by a geodesic arc orthogonal to ∂D that leaves D traveling within $\widetilde{x}_2(D)$ towards the point $z'' = \widetilde{x}_2(z')$, and from here heading towards $\widetilde{x}_2(i)$ along the arc $\widetilde{x}_2([z', i])$.

Taking into account the side-identifications in D, we see that our choice of c projects to a loop in $\widehat{\mathbb{C}} \smallsetminus \{0, 1, \infty\}$ that goes near the missing point 0 through a path c_0, turns once (counterclockwise) around it, and goes back to P following c_0 in the reverse direction (see Figure 2.20).

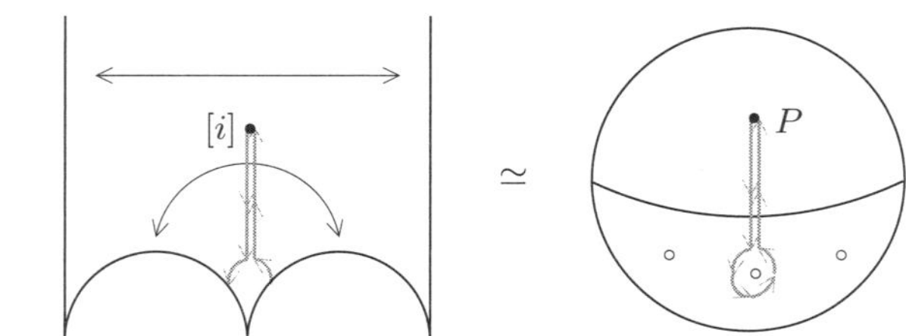

Fig. 2.20. The loop in the sphere minus three points corresponding to $\widetilde{x}_2$.

Similarly, $\widetilde{x}_1$ and $\widetilde{x}_3$ correspond to simple loops around ∞ and 1 respectively.

Let us state explicitly the result we have obtained in this example.

Theorem 2.34 *The triangle group $\Gamma_{\infty,\infty,\infty}$ agrees with the principal congruence subgroup of level two $\Gamma(2)$ and it is isomorphic to the fundamental group of the thrice punctured sphere.*

In the same way one can introduce the *principal congruence subgroups* $\Gamma(n)$ for arbitrary level $n > 1$ as the kernel of the epimorphism

$$\Gamma(1) \longrightarrow \mathbb{PSL}(2, \mathbb{Z}/n\mathbb{Z})$$

defined by reducing the entries modulo n. These groups will be studied in the next section (Example 2.40).

2.5 Automorphisms of Riemann surfaces

As a consequence of Proposition 2.25 we get:

Proposition 2.35 *If Γ is a Fuchsian group acting freely on $\mathbb{H}$, then*

$$\mathrm{Aut}\,(\mathbb{H}/\Gamma) = N\,(\Gamma)\,/\Gamma,$$

where $N\,(\Gamma) = \{T \in \mathbb{PSL}\,(2,\mathbb{R}) : T \circ \Gamma \circ T^{-1} = \Gamma\}$ *is the* normal-izer *of* Γ *in* $\mathbb{PSL}\,(2,\mathbb{R})$.

Proof Taking $\Gamma_1 = \Gamma_2$ in Proposition 2.25, we get a group epi-morphism

$$N\,(\Gamma) \longrightarrow \mathrm{Aut}\,(\mathbb{H}/\Gamma)$$

whose kernel is precisely Γ. $\qquad\square$

Proposition 2.36 *If Γ is a non-cyclic Fuchsian group, then the normalizer $N\,(\Gamma)$ is also a Fuchsian group.*

Proof If $N\,(\Gamma)$ is not Fuchsian there is a sequence $\gamma_n \in N\,(\Gamma)$ such that $\gamma_n \to \mathrm{Id}$. Therefore, for any pair of elements $\gamma, \beta \in \Gamma$ we have $\gamma_n \circ \gamma \circ \gamma_n^{-1} \to \gamma$ and $\gamma_n \circ \beta \circ \gamma_n^{-1} \to \beta$. Since Γ is discrete, it follows that for n sufficiently large $\gamma, \beta \in \mathrm{Comm}(\gamma_n)$. By Remark 2.3 γ and β must commute. We conclude that Γ is abelian, which contradicts Lemma 2.2. $\qquad\square$

Corollary 2.37 *The automorphism group of a compact Riemann surface (or, equivalently, of an algebraic curve) of genus $g \geq 2$ is finite.*

Proof Let $S = \mathbb{H}/\Gamma$. By Proposition 2.35, the quotient map

$$f : S \longrightarrow S/\mathrm{Aut}(S)$$

can be described in terms of Fuchsian groups as the following morphism of compact Riemann surfaces

$$\mathbb{H}/\Gamma \xrightarrow{\ f\ } \mathbb{H}/N\,(\Gamma)$$
$$[z]_\Gamma \longmapsto [z]_{N(\Gamma)}$$

The commutativity of the obvious diagram

$$\begin{array}{ccc} \mathbb{H} & & \\ \pi \downarrow & \searrow^{\phi} & \\ \mathbb{H}/\Gamma & \xrightarrow{f} & \mathbb{H}/N(\Gamma) \end{array}$$

implies that the map f is holomorphic. Therefore, $d = \deg(f)$ is a finite number. But we can show that this number is precisely the order of the group $N(\Gamma)/\Gamma$. If $\{\gamma_n\} \subset N(\Gamma)$ is a complete set of representatives of the quotient group, then

$$f^{-1}\left([z]_{N(\Gamma)}\right) = \{([\gamma_n(z)]_\Gamma)\}$$

where the points $[\gamma_n(z)]_\Gamma$ are all distinct except for the case in which z is a fixed point of some element of $N(\Gamma)$. $\qquad\square$

From the proof of the above theorem we extract the following result.

Corollary 2.38 *Let $H \le K$ be an inclusion of Fuchsian groups. Then the degree of the corresponding morphism of Riemann surfaces*

$$\mathbb{H}/H \longrightarrow \mathbb{H}/K$$

equals the index $[K : H]$.

We now give a formula that relates the genus of a Riemann surface and the genus of a finite quotient.

Lemma 2.39 *Let S be a compact Riemann surface of genus $g \ge 2$ and $G \le \mathrm{Aut}(S)$. Then*

$$2g - 2 = |G|(2\overline{g} - 2) + \sum_{p \in S}(|I(p)| - 1)$$

where $\overline{g}$ is the genus of S/G and $I(p) \le G$ denotes the stabilizer of p in G. Equivalently

$$2g - 2 = |G|\left(2\overline{g} - 2 + \sum_{i=1}^{n}\left(1 - \frac{1}{|I(p_i)|}\right)\right)$$

where $\{p_1, \ldots, p_n\}$ is a maximal set of fixed points of G inequivalent under the action of G.

Proof The result follows from the application of the Riemann–Hurwitz formula to the quotient morphism $f : S \longrightarrow S/G$. Recall first that the degree of f equals $|G|$ and that a point $p \in S$ is a branching point of f if and only if it is a fixed point of some automorphism of S. As for the computation of the ramification indices, we shall consider again the commutative diagram

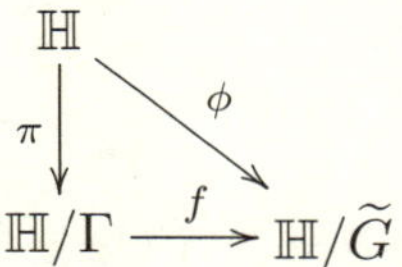

where Γ is the Fuchsian group uniformizing S and $\widetilde{G}$ is the subgroup of $N(\Gamma)$ such that $\widetilde{G}/\Gamma \simeq G$ (that is, $\widetilde{G}$ consists of all lifts of G to $\mathbb{H}$). The ramification index of ϕ at z equals $|\widetilde{I}(z)|$, where $\widetilde{I}$ denotes the stabilizer of z in $\widetilde{G}$. As π is an unramified map, the ramification index of f at $[z]_\Gamma$ equals also $|\widetilde{I}(z)|$. But $|\widetilde{I}(z)|$ can be naturally identified with the stabilizer $I(p)$ of $p = [z]_\Gamma$ in $\widetilde{G}/\Gamma \simeq G$ since Γ does not fix points.

In order to obtain the second formula, note that if p and q are two fixed points of G such that $q = h(p)$ for some $h \in G$ then $I(p)$ and $I(q)$ are conjugate subgroups (by means of h). It follows that, given a branching value b of f, the partial sum

$$\sum_{p\in\{f^{-1}(b)\}} (|I(p)| - 1)$$

equals

$$\frac{|G|}{|I(p_j)|}(|I(p_j)| - 1) = |G|\left(1 - \frac{1}{|I(p_j)|}\right)$$

where p_j is any representative of the fibre $f^{-1}(b)$. $\qquad\square$

As an application of this lemma, we now compute the genus of the Riemann surface uniformized by the principal congruence subgroup of level $n > 1$.

Example 2.40 Recall that $\Gamma(n)$ is the kernel of the epimorphism

$$\Gamma(1) \longrightarrow \mathbb{PSL}(2, \mathbb{Z}/n\mathbb{Z})$$

defined by reducing the entries modulo n. We observe that $\Gamma(n)$ is torsion free (equivalently it does not fix points, see Proposition

2.23) because the generators $z \mapsto -1/z$ and $z \mapsto 1/(1-z)$ of $\Gamma(1)$ do not belong to $\Gamma(n)$ and neither do any of their conjugates, since the inclusion $\Gamma(n) < \Gamma(1)$ is normal.

We would like to compute the genus of the quotient Riemann surfaces $\mathbb{H}/\Gamma(n)$ (or rather their compactifications).

There is an obvious commutative diagram

$$\begin{array}{c} \mathbb{H} \\ \downarrow \searrow \\ \mathbb{H}/\Gamma(n) \xrightarrow{\ f\ } \mathbb{H}/\Gamma(1) \simeq \widehat{\mathbb{C}} \smallsetminus \{\infty\} \end{array} \qquad (2.22)$$

where the horizontal morphism is a quotient map given by the obvious action of the group $\mathbb{PSL}(2, \mathbb{Z}/n\mathbb{Z}) = \Gamma(1)/\Gamma(n)$ on $\mathbb{H}/\Gamma(n)$. Since clearly this action extends to the compactification $\widehat{\mathbb{H}/\Gamma(n)}$ we can use Lemma 2.39 to compute its genus g_n. We have

$$2g_n - 2 = [\Gamma(1) : \Gamma(n)] \left(-2 + \left(1 - \frac{1}{m_1} - \frac{1}{m_2} - \frac{1}{m_3} \right) \right) \qquad (2.23)$$

where m_1, m_2, m_3 stand for the multiplicities above the three branching values in $\mathbb{H}/\Gamma(1)$, namely $[i]$, $[\xi_6]$ and ∞.

As $\Gamma(n)$ is torsion free and $\Gamma(1) = \Gamma_{2,3,\infty}$ we know that $m_1 = 2$, $m_2 = 3$ and therefore only m_3 remains to be worked out. For this we need to understand the compactification of $\mathbb{H}/\Gamma(n)$ and the behaviour of (the extension of) f near the compactifying points over ∞. In order to do that we consider the open set $U = \{z \in \mathbb{H} : \operatorname{Im}(z) > 1\}$. As in Example 1.71 we have an obvious commutative diagram of coverings

$$\begin{array}{c} U \\ \downarrow \searrow \\ U/\langle z \mapsto z + n \rangle \longrightarrow U/\langle z \mapsto z + 1 \rangle \end{array} \qquad (2.24)$$

which is equivalent to

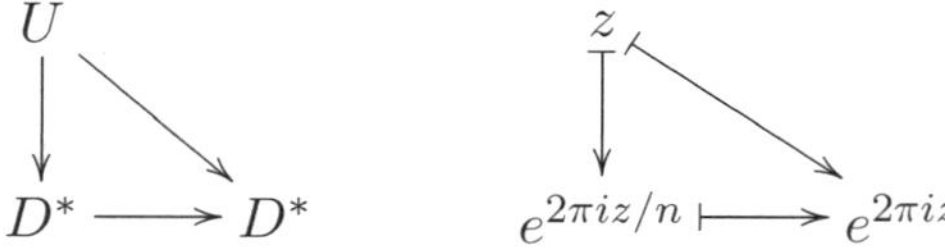

where D^* is a punctured disc of radius $e^{-2\pi}$ around the origin.

Moreover (2.24) is precisely the restriction of diagram (2.22) to U. This is because $\langle z \mapsto z + n \rangle$ is the subgroup of $\Gamma(n)$ that stabilizes U, and any other element $\gamma = \begin{pmatrix} a & b \\ c & d \end{pmatrix} \in \Gamma(1)$ maps U into a disc $\gamma(U)$, which by formula (2.1) does not intersect U. Actually, such discs are called *horocycles*. Their boundary $\gamma(\{\mathrm{Im}(z) = 1\})$ is a circle passing through the rational point a/c.

It follows that $m_3 = n$ and so formula (2.23) gives the genus g_n in terms of the order of the group $\mathbb{PSL}(2, \mathbb{Z}/n\mathbb{Z})$ (cf. [Gun62], [JS87]). For instance, when n is a prime number p we have $\mathbb{PSL}(2, \mathbb{Z}/p\mathbb{Z}) = p^2(p-1)/2$ and therefore

$$g_p = \begin{cases} 0 & \text{if } p = 2 \\ 1 + \frac{(p^2-1)(p-6)}{24} & \text{if } p \neq 2 \end{cases}$$

Another consequence of the above discussion is that the action of $\Gamma(1)$ on $\mathbb{H}$ extends to an action on $\widehat{\mathbb{H}} := \mathbb{H} \cup \mathbb{Q} \cup \{\infty\}$ in such a way that $\widehat{\mathbb{H}}/\Gamma(n)$ agrees with the compactification of $\mathbb{H}/\Gamma(n)$. This is why in what follows we will make no distinction between $\widehat{\mathbb{H}}/\Gamma(n)$ and $\widehat{\mathbb{H}/\Gamma(n)}$.

A deeper application of Lemma 2.39 is the following theorem:

Theorem 2.41 (Hurwitz) *The number of automorphisms of a compact Riemann surface of genus $g \geq 2$ never exceeds $84(g-1)$.*

Proof We include here the proof given in [FK92].

Let S be a compact Riemannn surface of genus $g \geq 2$. By the previous lemma we have

$$2g - 2 = N \left(2\overline{g} - 2 + \sum_{i=1}^{n} \left(1 - \frac{1}{m_i} \right) \right) \tag{2.25}$$

where we have denoted $N = |\mathrm{Aut}(S)|$, $m_i = |I(p_i)|$, and $\overline{g}$ is the genus of $S/\mathrm{Aut}(S)$. Note that $m_i \geq 2$, and therefore we have $\frac{1}{2} \leq 1 - \frac{1}{m_i} < 1$. The proof consists of a case-by-case analysis:

(a) If $\overline{g} \geq 2$ then from (2.25) we obtain $2g - 2 \geq 2N$, that is $N \leq g - 1$.

(b) If $\overline{g} = 1$ then $n > 0$ since otherwise $g = 1$ (see part (iii) of Corollary 1.77). Therefore, from (2.25) we get the inequality $2g - 2 \geq \frac{N}{2}$, or equivalently $N \leq 4(g-1)$.

(c) Let $\bar{g} = 0$. Then $n \geq 3$ since otherwise we would have $2g - 2 < -2N + 2N = 0$.

Now, if $n \geq 5$ then we get $2g - 2 \geq N\left(-2 + \frac{5}{2}\right) = \frac{N}{2}$, therefore $N \leq 4(g-1)$. On the other hand, if $n = 4$ one of the multiplicities m_i must be grater than 2 for if not again $2g - 2 < -2N + 2N$. Thus $2g - 2 \geq -2N + N\left(\frac{3}{2} + \frac{2}{3}\right)$, that is $N \leq 12(g-1)$.

Suppose finally that $n = 3$ and that $2 \leq m_1 \leq m_2 \leq m_3$. It can be easily seen that $m_2 > 2$ and $m_3 > 3$. Now we argue as follows:

- If $m_3 \geq 7$ we get $2g - 2 \geq N\left(-2 + \frac{1}{2} + \frac{2}{3} + \frac{6}{7}\right)$, therefore $N \leq 84(g-1)$.
- If $m_3 = 6$ and $m_1 = 2$ then $m_2 \geq 4$ and $N \leq 24(g-1)$.
- If $m_3 = 6$ and $m_1 \geq 3$ then $N \leq 12(g-1)$.
- If $m_3 = 5$ and $m_1 = 2$ then $m_2 \geq 4$ and $N \leq 40(g-1)$.
- If $m_3 = 5$ and $m_1 \geq 3$ then $N \leq 15(g-1)$.
- Finally, if $m_3 = 4$ then $m_1 \geq 3$ and $N \leq 24(g-1)$.

$\square$

From the proof we see that the Hurwitz bound is attained if and only if $\bar{g} = 0$, $n = 3$ and $(m_1, m_2, m_3) = (2, 3, 7)$. Therefore we have

Corollary 2.42 *A compact Riemann surface S of genus $g > 1$ reaches the Hurwitz bound if and only if the quotient $S/\mathrm{Aut}(S)$ has genus 0, and the projection $S \longrightarrow S/\mathrm{Aut}(S)$ ramifies over three values with branching orders $2, 3$ and 7.*

Remark 2.43 The group $\mathrm{Aut}(S)$ is not even finite when the genus g of S is lower than 2. As we know, $\widehat{\mathbb{C}}$ is the only compact Riemann surface of genus 0, and its automorphism group agrees with $\mathbb{PSL}(2, \mathbb{C})$. On the other hand, if $g = 1$ then S can be described as $\mathbb{C}/\Lambda$, where Λ is a lattice. For any translation τ we have $\tau \Lambda \tau^{-1} = \Lambda$ (translations always commute), and therefore τ induces an automorphism of S.

2.5.1 The action of the automorphism group on the function field

We remind the reader that any subgroup $G \leq \mathrm{Aut}(S)$ acts on the function field $\mathcal{M}(S)$ by $\tau^*(\varphi) = \varphi \circ \tau$, where $\tau \in G$ and $\varphi \in \mathcal{M}(S)$ (see Proposition 1.95). We shall denote by G^* the subgroup of $\mathrm{Aut}(\mathcal{M}(S))$ defined by $G^* = \{\tau^* \mid \tau \in G\}$. The next proposition identifies the function field of S/G with the subfield $\mathcal{M}(S)^{G^*}$ of $\mathcal{M}(S)$ consisting of the elements fixed by G^*.

Proposition 2.44 *If $G \leq \mathrm{Aut}(S)$ then $\mathcal{M}(S)^{G^*} \simeq \mathcal{M}(S/G)$.*

Proof Let $\pi : S \longrightarrow S/G$ denote the canonical projection. We claim that the mapping

$$\mathcal{M}\left(\frac{S}{G}\right) \xrightarrow{\ \pi^*\ } \mathcal{M}(S)$$

$$f \longmapsto \pi^*(f) = f \circ \pi$$

defines a field isomorphism over $\mathbb{C}$ whose image equals precisely $\mathcal{M}(S)^{G^*}$.

Note first that for $\tau \in G$ we have

$$\tau^*(\pi^*(f)) = f \circ \pi \circ \tau = f \circ \pi = \pi^*(f)$$

and therefore $\mathrm{Im}(\pi^*) \subset \mathcal{M}(S)^{G^*}$. On the other hand, any given $h \in \mathcal{M}(S)^{G^*}$ induces a well-defined function on $\dfrac{S}{G}$ by means of the rule $[P] \longmapsto h(P)$. $\qquad\square$

As an application of this result, we next describe the algebraic equation of an arbitrary prime Galois cover of $\widehat{\mathbb{C}}$. In order to do that, it will be convenient to introduce the concept of rotation number. Let $\tau \in \mathrm{Aut}(S)$ be an automorphism of order p fixing a point $Q \in S$ and (U, φ) a τ-invariant chart around Q centred at the origin. Then, of course, $\varphi \circ \tau \circ \varphi^{-1}$ is an order p automorphism of the disc $D = \varphi(U)$ which fixes the origin. Hence, it is of the form

$$\varphi \circ \tau \varphi^{-1}(z) = \xi_p^k z$$

The integer k will be referred to as the *rotation number* of τ at Q. This number is well defined modulo p because if (V, ψ) is another such chart and $\psi \circ \tau \circ \psi^{-1}(z) = \xi_p^l z$ we can write

$$
\begin{aligned}
\xi_p^l &= (\psi \circ \tau \circ \psi^{-1})'(0) \\
&= ((\varphi \circ \psi^{-1})^{-1})'(0) \cdot (\varphi \circ \tau \circ \varphi^{-1})'(0) \cdot (\varphi \circ \psi^{-1})'(0) \\
&= \xi_p^k
\end{aligned}
$$

Furthermore, even if (V, ψ) is not centred at the origin, say $\psi(Q) = z_0$, it is still true that $(\psi \circ \tau \circ \psi^{-1})'(z_0) = \xi_p^k$. This can be seen by choosing a local isomorphism with $\alpha(z_0) = 0$ and noting that

$$
\begin{aligned}
\xi_p^k &= (\alpha \circ \psi \circ \tau \circ \psi^{-1} \circ \alpha^{-1})'(0) \\
&= \alpha'(z_0) \cdot (\psi \circ \tau \circ \psi^{-1})'(z_0) \cdot (\alpha')^{-1}(0) \\
&= (\psi \circ \tau \circ \psi^{-1})'(z_0)
\end{aligned}
$$

Proposition 2.45 *Let S be a compact Riemann surface admiting an automorphism τ of prime order p so that $S/\langle\tau\rangle$ has genus zero. Assume that τ fixes $r+1$ points $P_1, \ldots, P_{r+1}$ with rotation numbers $d_1, \ldots, d_{r+1}$. Then S is isomorphic to the Riemann surface of an algebraic curve of the form*

$$
y^p = (x - a_1)^{m_1} \cdots (x - a_r)^{m_r}
$$

where $1 \le m_i < p$ and $m_1 + \cdots + m_r$ is prime to p.

Moreover, there is an isomorphism

$$
\Phi : S \longrightarrow \{y^p = (x - a_1)^{m_1} \cdots (x - a_r)^{m_r}\}
$$

under which $\tau \in \mathrm{Aut}(S)$ corresponds to $(x, y) \longmapsto (x, \xi_p y)$, the points $P_1, \ldots P_r, P_{r+1}$ to $\tilde{P}_1 = (a_1, 0), \ldots, \tilde{P}_r = (a_r, 0), \tilde{P}_{r+1} = \infty$ and the integers $m_1, \ldots, m_r, m_{r+1} := -\sum_{i=1}^r m_i$ are the inverses of $d_1, \ldots, d_{r+1}$ modulo p.

Proof By hypothesis we have the following Galois extension with group $G^* = \langle \tau^* \rangle$

$$
\mathbb{C}(\mathbf{x}) \simeq \mathcal{M}(S/G) = \mathcal{M}(S)^{G^*} \hookrightarrow \mathcal{M}(S)
$$

Viewed as an endomorphism of the vector space $\mathcal{M}(S)$ over the field $\mathcal{M}(S)^{G^*}$, the automorphism τ^* satisfies the polynomial $P(X) = X^p - 1$, hence its minimal polynomial is a divisor of $P(X)$ and, in particular, the eigenvalues of τ^* are p-th roots of unity.

Let $\mathbf{y} \in \mathcal{M}(S) \smallsetminus \mathcal{M}(S)^{G^*}$ be an eigenvector with $\tau^*(\mathbf{y}) = \eta_p \mathbf{y}$, where η_p is some non-trivial p-th root of unity. Then clearly

$\tau^*(\mathbf{y}^k) = \eta_p^k \mathbf{y}^k$. This has two consequences. The first one is that by changing $\mathbf{y}$ by a suitable power $\mathbf{y}^k$ we can assume that $\tau^*(\mathbf{y}) = \xi_p \mathbf{y}$ and the second one is that $\tau^*(\mathbf{y}^p) = \mathbf{y}^p$ and so $\mathbf{y}^p \in \mathcal{M}(S/G) = \mathbb{C}(\mathbf{x})$, that is

$$\mathbf{y}^p = \frac{a(\mathbf{x})}{b(\mathbf{x})} \in \mathbb{C}(\mathbf{x})$$

Clearing out denominators we get $(b(\mathbf{x}) \cdot \mathbf{y})^p = a(\mathbf{x})\,(b(\mathbf{x}))^{p-1}$. From here the proof of the first part follows replacing $\mathbf{y}$ with $b(\mathbf{x})\cdot\mathbf{y}$ and arguing as in the first step of Section 1.3.2.

Moreover, the fact that $\tau^*(\mathbf{x}) = \mathbf{x}$ and $\tau^*(\mathbf{y}) = \xi_p \mathbf{y}$ means that under the isomorphism

$$\Phi : S \longrightarrow \{y^p = (x - a_1)^{m_1} \cdots (x - a_r)^{m_r}\}$$

defined by $\Phi = (\mathbf{x}, \mathbf{y})$, the automorphism τ is given by the rule $(x, y) \longmapsto (x, \xi_p y)$.

More precisely, for a generic point $(x, y) = (\mathbf{x}(P), \mathbf{y}(P))$ one has

$$
\begin{aligned}
\Phi \circ \tau \circ \Phi^{-1}(x, y) &= \Phi \circ \tau(P) = (\mathbf{x} \circ \tau(P), \mathbf{y} \circ \tau(P)) \\[2mm]
&= (\tau^* \mathbf{x}(P), \tau^* \mathbf{y}(P)) = (\mathbf{x}(P), \xi_p \mathbf{y}(P)) \\[2mm]
&= (x, \xi_p y)
\end{aligned}
$$

We finally check that the rotation number of τ at the point P_i, that is the rotation number of $\tilde{\tau} = \Phi \circ \tau \circ \Phi^{-1}$ at $\tilde{P}_i = (a_i, 0)$, is the inverse of m_i. We recall from Example 1.11 that a parametrization around the point $\tilde{P}_i$ can be defined by

$$\varphi^{-1}(t) = \left(t^p + a_i,\, t^{m_i} \sqrt[p]{\prod_{j \neq i} (t^p + a_i - a_j)^{m_j}} \right)$$

Its inverse is given by

$$\varphi(x, y) = y^{1/m_i} \left(\sqrt[p]{\prod_{i \neq j} (x - a_j)^{m_j}} \right)^{-1/m_i}$$

Now

$$
\begin{aligned}
\varphi \circ \tilde{\tau} \circ \varphi^{-1}(t) &= \varphi\left(t^p + a_i,\, \xi_p t^{m_i} \sqrt[p]{\prod_{i \neq j} (t^p + a_i - a_j)^{m_j}} \right) \\
&= \xi_p^{1/m_i} t
\end{aligned}
$$

as required.

As for the point $\tilde{P}_{r+1} = \infty$ the corresponding calculation is

$$\varphi \circ \tilde{\tau} \circ \varphi^{-1}(t) = \varphi\left(\frac{1}{t^p}, \xi_p t^{-(m_1+\cdots+m_r)} \sqrt[p]{\prod(1 - a_i t^p)^{m_i}}\right)$$
$$= \xi_p^{-1/(m_1+\cdots+m_r)} t$$

$\square$

When $p = 2$ we say that S is a *hyperelliptic Riemann surface*. Proposition 2.45 yields the following:

Corollary 2.46 *If S is a Riemann surface of genus $g \geq 2$ admitting an involution J such that $S/\langle J \rangle$ has genus zero, then S is a hyperelliptic Riemann surface with equation of the form $y^2 = (x - a_1) \cdots (x - a_{2g+1})$ and $J(x, y) = (x, -y)$.*

An important result relative to the hyperelliptic involution J is:

Proposition 2.47 *The hyperelliptic involution J of a hyperelliptic Riemann surface S is the only automorphism of order 2 such that $S/\langle J \rangle \simeq \mathbb{P}^1$.*

Proof Two such involutions J_1 and J_2 would give rise to two meromorphic functions

$$f_i : S \longrightarrow S/\langle J_i \rangle \simeq \mathbb{P}^1 \quad (i = 1, 2)$$

of degree 2. Since $[\mathcal{M}(S) : \mathbb{C}(f_1)] = \deg(f_1) = 2$, we see that either $\mathcal{M}(S) = \mathbb{C}(f_1, f_2)$ or $f_2 \in \mathbb{C}(f_1)$.

The first case is impossible by Proposition 1.97. In the second case we have $f_2 = M \circ f_1$, where M is certain rational function. Since $\deg(f_2) = \deg(f_1) = 2$, this rational function must have degree 1, hence it is a Möbius transformation.

Therefore, we can write

$$J_1(P) = Q$$
$$\Updownarrow$$
$$f_1(P) = f_1(Q)$$
$$\Updownarrow$$
$$M \circ f_1(P) = M \circ f_1(Q)$$
$$\Updownarrow$$
$$f_2(P) = f_2(Q)$$
$$\Updownarrow$$
$$J_2(P) = Q$$

We conclude that $J_1 = J_2$. $\qquad\square$

This result has a number of consequences.

Corollary 2.48 *The hyperelliptic involution J of a hyperelliptic Riemann surface lies in the centre of* $\mathrm{Aut}(S)$.

Proof For any $\tau \in \mathrm{Aut}(S)$ the automorphism $\tau \circ J \circ \tau^{-1}$ must agree with J by Proposition 2.47. $\qquad\square$

Corollary 2.49 *Two hyperelliptic curves*

$$\{y^2 = \prod_{i=1}^{2g+2} (x - a_i)\} \quad and \quad \{y^2 = \prod_{i=1}^{2g+2} (x - b_i)\}$$

are isomorphic if and only if there is a Möbius transformation M such that $M(\{a_1, \ldots, a_{2g+2}\}) = \{b_1, \ldots, b_{2g+2}\}$.

Proof The *if* part was proved in Example 1.83.

For the *only if* part we can argue as follows. If there is an isomorphism $f : S_1 \longrightarrow S_2$ and J_1 and J_2 denote the corresponding hyperelliptic involutions, then by Proposition 2.47 one has

$$J_2 = f \circ J \circ f^{-1} \tag{2.26}$$

This yields an obvious commutative diagram as follows

$$
\begin{array}{ccc}
S_1 & \xrightarrow{\;\;f\;\;} & S_2 \\
{\scriptstyle x}\downarrow & & {\scriptstyle x}\downarrow \\
S_1/\langle J_1 \rangle \simeq \widehat{\mathbb{C}} & \xrightarrow{\;M\;} & S_2/\langle J_2 \rangle \simeq \widehat{\mathbb{C}}
\end{array}
$$

By the commutativity of the diagram the degree of M has to be 1, hence it is a Möbius transformation. Moreover, denoting $\mathrm{Fix}(J_i)$ the set of fixed points of J_i, identity (2.26) implies

$$\{b_i\} = \mathbf{x}(\mathrm{Fix}(J_2)) = \mathbf{x} \circ f(\mathrm{Fix}(J_1)) = M(\{a_i\})$$

as claimed. $\qquad\square$

2.5.2 Uniformization of Klein's curve of genus three

We devote this section to study again Klein's quartic (cf. Proposition 1.44), now from the point of view of Fuchsian groups.

Example 2.50 The principal congruence group $\Gamma(7)$ is a normal subgroup of index 168 of the modular group $\Gamma(1)$ (Example 2.40). Hence there is a natural action of $\mathbb{PSL}(2, \mathbb{Z}/7\mathbb{Z}) \simeq \Gamma(1)/\Gamma(7)$ on $\mathbb{H}/\Gamma(7)$. This action extends to its compactification $\widehat{\mathbb{H}}/\Gamma(7)$.

As the genus of $\widehat{\mathbb{H}}/\Gamma(7)$ is three (see Example 2.40), and the Hurwitz bound for the number of automorphisms in the genus three case equals precisely 168, we deduce that the normalizer $N(\Gamma(7))$ of $\Gamma(7)$ in $\mathbb{PSL}(2, \mathbb{R})$ equals $\Gamma(1)$, and moreover that the group of automorphisms of both $\mathbb{H}/\Gamma(7)$ and $\widehat{\mathbb{H}}/\Gamma(7)$ is isomorphic to $\mathbb{PSL}(2, \mathbb{Z}/7\mathbb{Z})$.

Now observe that the element $z \longmapsto z + 1$ induces an automorphism τ of order 7 on $\widehat{\mathbb{H}}/\Gamma(7)$. The Riemann–Hurwitz formula applied to the group $\langle \tau \rangle$ reads

$$2 \cdot 3 - 2 = 7(2g' - 2) + 6r$$

where g' is the genus of the quotient Riemann surface $S/\langle \tau \rangle$ and r is the number of branch values. Simple inspection shows that necessarily $g' = 0$ and $r = 3$. Now, a combination of Proposition 2.45 and Proposition 1.96 implies that $\widehat{\mathbb{H}}/\Gamma(7)$ is isomorphic to the Riemann surface of either $y^7 = x(x - 1)$ or $y^7 = x(x - 1)^2$, and in both cases τ is given by the formula $\tau(x, y) = (x, \xi_7 y)$.

We claim that in fact $\widehat{\mathbb{H}}/\Gamma(7) \simeq \{y^7 = x(x - 1)^2\}$. To rule out the other possibility we recall from the last lines in Section 1.3.2 that the Riemann surface $\{y^7 = x(x - 1)\}$ is hyperelliptic, and therefore the centre of its automorphism group is non-trivial (Corollary 2.48), while $\mathbb{PSL}(2, \mathbb{Z}/7\mathbb{Z})$ is known to be a simple group.

Example 2.51 Figure 2.21 describes a very well-known Fuchsian group that was studied first by Felix Klein (cf. [Kle78]). The regular 14-gon P of angle $2\pi/7$ is the fundamental domain of a Fuchsian group K generated by the identification of the sides e_i and e_j of P for the following pairs

$$(i,j) \in \{(1,6), (2,11), (3,8), (4,13), (5,10), (7,12), (9,14)\}$$

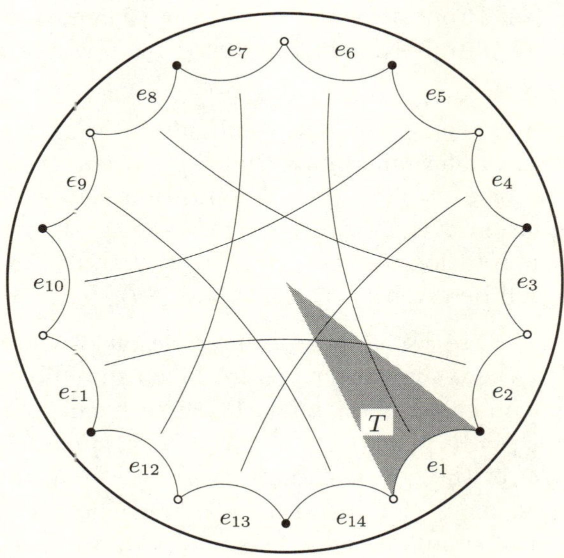

Fig. 2.21. Klein's Riemann surface is uniformized by a subgroup $K < \Gamma_{7,7,7}$. The lines indicate the side-pairing transformations that generate K.

From the point of view of plain topology these identifications are the same as those in Example 1.51, therefore we already know that the genus of the resulting surface is $g = 3$. The group K is discrete by Poincaré's polygon Theorem, as the set of vertices splits into two classes of seven identified vertices each, yielding two points $Q_1, Q_2 \in \mathbb{D}/K$ (depicted in Figure 2.21 as a white and a black point respectively).

Let $\Gamma_{7,7,7}$ be the triangle group determined by the triangle T with vertices at the origin and two consecutive vertices of P (in particular T is an equilateral triangle of angle $\pi/7$, see Figure 2.21). The group K is clearly contained in $\Gamma_{7,7,7}$, since its generators preserve the tessellation corresponding to this triangle group. Note that P consists of seven copies of the fundamental domain of $\Gamma_{7,7,7}$ related by the order 7 rotation $\widetilde{\tau} : z \longmapsto \xi_7 z$; therefore $[\Gamma_{7,7,7} : K] = 7$ by Lemma 2.32. The inclusion is normal, since $\widetilde{\tau}$ clearly normalizes K.

It follows (see Proposition 2.35) that the Riemann surface $\mathbb{D}/K$ has an order 7 automorphism τ induced by $\widetilde{\tau}$ whose quotient is $S/\langle \tau \rangle \simeq \widehat{\mathbb{C}}$.

As in the previous example, a combination of Proposition 2.45 and Proposition 1.96 would imply that $\mathbb{D}/K$ is isomorphic to either $y^7 = x(x-1)$ or $y^7 = x(x-1)^2$. Proposition 2.45 will show that the right equation is the second one provided that we are able to check that the rotation numbers of τ are 1, 2 and 4. In any of the two cases τ will be given by $\tau(x,y) = (x, \xi_7 y)$.

For $j = 1, \ldots, 14$ let v_j be the intersection of the edges e_{j-1} and e_j of P (where the subscripts are taken modulo 14), and let $v_0 = 0$. One can easily check that the three fixed points of τ are $Q_0 = [v_0]_K$, $Q_1 = [v_1]_K$ and $Q_2 = [v_2]_K$. Now, since K acts freely, the restriction of the quotient map $\pi : \mathbb{D} \longrightarrow \mathbb{D}/K$ to a suitable neighbourhood of v_j (for instance, a hyperbolic disc centred at v_j with radius ε small enough) defines a parametrization around Q_j whose inverse ψ_j is a τ-invariant chart around Q. Moreover, if γ_j is any automorphism of the hyperbolic disc sending v_j to 0, the composition $\varphi_j = \gamma_j \circ \psi_j$ is even a τ-invariant chart around Q centred now at the origin, and we can use it to compute the rotation number of τ at Q_j.

In the case of Q_0 we see that $\varphi_0 \circ \tau \circ \varphi_0^{-1}(z) = \xi_7 z$ by construction, and therefore that the rotation number of τ at Q_0 equals 1. We claim that $\varphi_1 \circ \tau \circ \varphi_1^{-1}(z) = \xi_7^2 z$ and $\varphi_2 \circ \tau \circ \varphi_2^{-1}(z) = \xi_7^4 z$, and therefore that the rotation number of τ at Q_1 (resp. at Q_2) equals 2 (resp. 4).

The collection of 14-gons $\{\gamma(P),\ \gamma \in K\}$ is a tessellation of the hyperbolic disc $\mathbb{D}$ such that seven polygons meet at every vertex. The seven polygons meeting at v_2 are, in cyclic (counterclockwise)

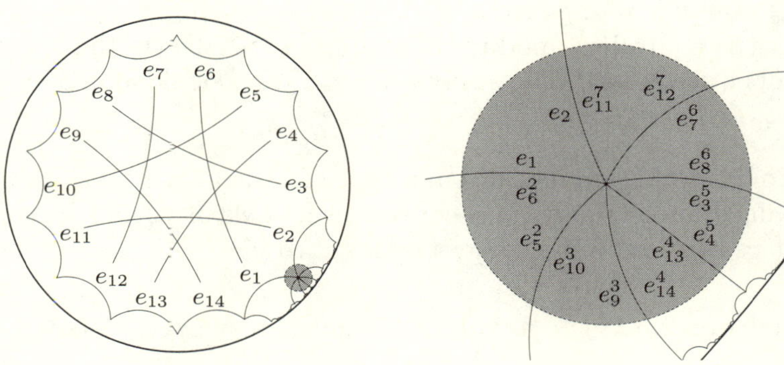

Fig. 2.22. Computation of the rotation number of τ at $Q_2 = [v_2]_K$. The lines on the left-hand side indicate the side pairings. The picture on the right-hand side shows a magnified detail.

order

$$P_1 = P$$
$$P_2 = \gamma_{6,1}(P)$$
$$P_3 = \gamma_{6,1} \circ \gamma_{10,5}(P)$$
$$P_4 = \gamma_{6,1} \circ \gamma_{10,5} \circ \gamma_{14,9}(P)$$
$$P_5 = \gamma_{6,1} \circ \gamma_{10,5} \circ \gamma_{14,9} \circ \gamma_{4,13}(P)$$
$$P_6 = \gamma_{6,1} \circ \gamma_{10,5} \circ \gamma_{14,9} \circ \gamma_{4,13} \circ \gamma_{8,3}(P)$$
$$P_7 = \gamma_{6,1} \circ \gamma_{10,5} \circ \gamma_{14,9} \circ \gamma_{4,13} \circ \gamma_{8,3} \circ \gamma_{12,7}(P)$$

where $\gamma_{i,j}$ stands for the side-pairing transformation that sends e_i to e_j (compare with Example 2.7). If γ is the element of K sending P to P_k we set $e_i^k = \gamma(e_k)$. In other words, e_i^k is the edge of P_k which is K-equivalent to the edge e_i of P. With this notation we find that the eight edges meeting at v_2 are in cyclic (counterclockwise) order

$$e_1 = \gamma_{6,1}(e_6) = e_6^2$$
$$e_5^2 = \gamma_{6,1}(e_5) = \gamma_{6,1} \circ \gamma_{10,5}(e_{10}) = e_{10}^3$$
$$e_9^3 = \gamma_{6,1} \circ \gamma_{10,5}(e_9) = \gamma_{6,1} \circ \gamma_{10,5} \circ \gamma_{14,9}(e_{14}) = e_{14}^4$$
$$e_{13}^4 = \gamma_{6,1} \circ \gamma_{10,5} \circ \gamma_{14,9}(e_{13}) = \gamma_{6,1} \circ \gamma_{10,5} \circ \gamma_{14,9} \circ \gamma_{4,13}(e_4) = e_4^5$$
$$e_3^5 = \gamma_{6,1} \circ \gamma_{10,5} \circ \gamma_{14,9} \circ \gamma_{4,13}(e_3)$$

$$\vdots$$

(see Figure 2.22) and we already see that we must turn an angle $4 \cdot \frac{2\pi}{7}$ around v_2 to go from e_1 to $\gamma_{4,13} \circ \gamma_{14,9} \circ \gamma_{10,5} \circ \gamma_{6,1}(e_3) = e_3^5$, which is equivalent modulo K to $e_3 = \widetilde{\tau}(e_1)$, hence the rotation number at Q_2 is 4. The study around v_1 can be done in a similar way.

To finish this section, we summarize all the information we have obtained about the various equivalent models of Klein's Riemann surface of genus 3 in the next classical result.

Theorem 2.52 *The following Riemann surfaces are isomorphic:*

(i) $\{y^7 = x(x-1)^2\}$.

(ii) *The Riemann surface of genus $g = 3$ uniformized by the Fuchsian group generated by the side-pairing identifications in Figure 2.21.*

(iii) $\dfrac{\widehat{\mathbb{H}}}{\Gamma(7)}$.

(iv) *Klein's plane quartic $X_0^3 X_1 + X_1^3 X_2 + X_2^3 X_0 = 0$.*

Moreover the automorphism group is isomorphic to the group $\mathbb{PSL}(2, \mathbb{Z}/7\mathbb{Z})$ *of order 168. In particular, it reaches Hurwitz's bound for a curve of genus 3.*

Proof The equivalence of the first three models has been shown in the last two examples, whereas the equivalence with Klein's quartic was established in Section 1.1.3.

The order of the group $\mathbb{PSL}(2, \mathbb{Z}/7\mathbb{Z}) \simeq \Gamma(1)/\Gamma(7)$, equals precisely $168 = 84 \cdot (3-1)$, the maximal number of automorphisms of a Riemann surface of genus three. It follows that the automorphism group of Klein's Riemann surface is indeed isomorphic to $\mathbb{PSL}(2, \mathbb{Z}/7\mathbb{Z})$ and in particular all the automorphisms can be described in terms of the model in part (iii) as

$$\frac{\widehat{\mathbb{H}}}{\Gamma(7)} \longrightarrow \frac{\widehat{\mathbb{H}}}{\Gamma(7)}$$
$$[z] \longmapsto [\gamma(z)]$$

for some $\gamma \in \Gamma(1)$. $\qquad\square$

Remark 2.53 Any triple $\{\sigma_2, \sigma_3, \sigma_7\}$ of elements of order 2, 3 and 7 respectively generates the automorphism group of Klein's Riemann surface, since the order of $\langle \sigma_2, \sigma_3, \sigma_7 \rangle$ must be divisible by $2 \cdot 3 \cdot 7$ and $\mathbb{PSL}(2, \mathbb{Z}/7\mathbb{Z})$ does not contain any proper subgroup of such order (see [CCN$^+$85]). In the algebraic model (i) we can take for instance

$$\sigma_3(x, y) = \left(\frac{1}{1-x}, \frac{y^2}{1-x} \right)$$

and

$$\sigma_7(x, y) = (x, \xi_7 y)$$

but the order 2 automorphisms have a more complicated expression. As it is stated in [KK79], denoting $\omega_1 = \xi_7 + \xi_7^2 + \xi_7^5 + \xi_7^6$ and $y_1 = \xi_7^2 + \xi_7^5$ the map σ induced in projective space by the matrix

$$\begin{pmatrix} \omega_1 & 1 & y_1 \\ 1 & y_1 & \omega_1 \\ y_1 & \omega_1 & 1 \end{pmatrix}$$

is an automorphism of order 2 of Klein's plain quartic. Since we already know (Proposition 1.44) that the two algebraic models are related by the isomorphism

$$\{y^7 = x(x-1)^2\} \xrightarrow{\Phi} \{x_0^3 x_1 + x_1^3 x_2 + x_2^3 x_0 = 0\}$$

$$(x, y) \longmapsto \left(\frac{1}{y^3} : \frac{x-1}{y^5} : \frac{1-x}{y^6} \right)$$

whose inverse is

$$\{y^7 = x(x-1)^2\} \xleftarrow{\Phi^{-1}} \{x_0^3 x_1 + x_1^3 x_2 + x_2^3 x_0 = 0\}$$

$$\left(1 + \frac{x_1^3}{x_0 x_2^2}, -\frac{x_1}{x_2} \right) \longleftarrow (x_0 : x_1 : x_2)$$

we can take

$$\sigma_2(x, y) = \Phi^{-1} \sigma \Phi(x, y) = \left(1 + \frac{B^3}{A \cdot C^2}, -\frac{B}{C} \right)$$

where

$$\begin{aligned}
A(x, y) &= \omega_1 y^3 - y_1(x-1) + (x-1)y \\
B(x, y) &= y^3 - \omega_1(x-1) + y_1(x-1)y \\
C(x, y) &= y_1 y^3 - (x-1) + \omega_1(x-1)y
\end{aligned}$$

2.6 The moduli space of compact Riemann surfaces

Uniformization theory allows us to gain insight on the size of the set of isomorphism classes of Riemann surfaces of given genus g (the *moduli space* $\mathcal{M}_g$). We have already said that there is only one isomorphism class of Riemann surfaces of genus 0 (Theorem 2.1), therefore the moduli space of genus 0 reduces to a single point. We shall consider now the case $g > 0$.

2.6.1 The moduli space $\mathcal{M}_1$

Theorem 2.1 along with the observation that for $\tau = \pm\omega_2/\omega_1$ the automorphism of $\mathbb{C}$ defined by $T(z) = z/\omega_1$ induces an isomorphism

$$\mathbb{C}/(\omega_1\mathbb{Z} \oplus \omega_2\mathbb{Z}) \xrightarrow{\overline{T}} \mathbb{C}/(\mathbb{Z} \oplus \tau\mathbb{Z})$$

shows that every compact Riemann surface of genus $g = 1$ is isomorphic to one of the form $\mathbb{C}/\Lambda_\tau$ for some $\tau \in \mathbb{H}$, where Λ_τ stands for the lattice $\mathbb{Z} \oplus \tau\mathbb{Z}$ acting by translations. Hence, only one complex parameter is needed to describe $\mathcal{M}_1$. In fact:

Proposition 2.54 $\mathcal{M}_1 \simeq \mathbb{H}/\mathbb{PSL}(2,\mathbb{Z}) \simeq \mathbb{C}.$

Proof The content of the statement is that for $\tau_1, \tau_2 \in \mathbb{H}$ the Riemann surfaces $\mathbb{C}/\Lambda_{\tau_1}$ and $\mathbb{C}/\Lambda_{\tau_2}$ are isomorphic if and only if $\tau_1 = \dfrac{a\tau_2 + b}{c\tau_2 + d}$ for some $\begin{pmatrix} a & b \\ c & d \end{pmatrix} \in \mathbb{PSL}(2,\mathbb{Z})$.

Just as in the proof of Proposition 2.25, covering space theory shows that the existence of an isomorphism

$$\mathbb{C}/\Lambda_{\tau_1} \xrightarrow{\overline{T}} \mathbb{C}/\Lambda_{\tau_2}$$

(which we can always assume to send $[0] \in \mathbb{C}/\Lambda_{\tau_1}$ to $[0] \in \mathbb{C}/\Lambda_{\tau_2}$) is equivalent to the existence of $T \in \mathrm{Aut}(\mathbb{C})$ (which we can always assume to be of the form $T(z) = \omega \cdot z$) such that

$$T \circ \Lambda_{\tau_1} = \Lambda_{\tau_2} \circ T$$

that is

$$\omega(\mathbb{Z} \oplus \tau_1\mathbb{Z}) = (\mathbb{Z} \oplus \tau_2\mathbb{Z})$$

This is equivalent to the existence of matrices $A = \begin{pmatrix} d & c \\ b & a \end{pmatrix}$ and $A' = \begin{pmatrix} d' & d' \\ b' & a' \end{pmatrix}$ with integer coefficients and determinant $\det(A) = \det(A') = \pm 1$ such that

$$\begin{pmatrix} \omega \\ \omega\tau_1 \end{pmatrix} = A \begin{pmatrix} 1 \\ \tau_2 \end{pmatrix} = A' \begin{pmatrix} \omega \\ \omega\tau_1 \end{pmatrix}$$

It follows that $\tau_1 = A\tau_2 = \dfrac{a\tau_2 + b}{c\tau_2 + d}$ and that in fact A belongs to $\mathbb{PSL}(2, \mathbb{R})$, hence to $\mathbb{PSL}(2, \mathbb{Z})$, since both τ_1 and τ_2 lie in $\mathbb{H}$.

Conversely, if $\tau_1 = \dfrac{a\tau_2 + b}{c\tau_2 + d}$ the isomorphism $\mathbb{C}/\Lambda_{\tau_1} \simeq \mathbb{C}/\Lambda_{\tau_2}$ is induced by $T(z) = (c\tau_2 + d) \cdot z$. $\qquad\qquad\square$

The identification $\mathcal{M}_1 \simeq \mathbb{C}$ can be also understood from the point of view of algebraic curves. We know that any Riemann surface of genus 1 can be regarded as an algebraic curve of the form $y^2 = 4x^3 - g_2 x - g_3$, see equation (2.15). Now, by Example 1.83 this curve can be written in the canonical form $C_\lambda : y^2 = x(x-1)(x-\lambda)$ for some $\lambda \in \mathbb{C} \setminus \{0, 1\}$.

The next step is to determine when C_λ and $C_{\lambda'}$ produce isomorphic Riemann surfaces. For this question we will need the following auxiliary result:

Lemma 2.55 *Let C_λ be the compact Riemann surface corresponding to the curve $y^2 = x(x - 1)(x - \lambda)$, $\lambda \neq 0$, and let $P \in C_\lambda$ be a given point. Then C_λ has exactly one automorphism of order 2 fixing P. When $P = (0, 0)$ (or $(1, 0)$, or $(\lambda, 0)$, or ∞) this automorphism is given by*

$$J_\lambda(x, y) = (x, -y)$$

Proof Let $C_\lambda \simeq \mathbb{C}/\Lambda$ for certain lattice Λ. An automorphism f of $\mathbb{C}/\Lambda$ is induced by an automorphism of $\mathbb{C}$, say $\widetilde{f}(z) = az + b$, such that $a\Lambda \subset \Lambda$. Assume that P corresponds to the point $[0]_\Lambda \in \mathbb{C}/\Lambda$, something we can always achieve by conjugation by a translation. Then we can choose our lift $\widetilde{f}$ to be of the form $\widetilde{f}(z) = az$.

If f is going to be of order 2 we must have

$$[(\widetilde{f})^2(z)]_\Lambda = [a^2 z]_\Lambda = [z]_\Lambda$$

and, as Λ is a discrete set, we deduce that $a^2 z - z = \left(a^2 - 1\right) z$ is a constant value in Λ. This can only happen if $a = \pm 1$. Therefore, f has to be given by $f\left([z]\right) = [-z]$.

The last statement about J_λ follows from the uniqueness of f.

$\square$

Proposition 2.56 *For $\lambda \neq 0, 1$ let C_λ be the compact Riemann surface $\{y^2 = x(x-1)(x-\lambda)\}$. Then C_λ is isomorphic to C_μ if and only if $\mu \in \Sigma(\lambda)$, where*

$$\Sigma(\lambda) = \left\{ \lambda,\ 1 - \lambda,\ \frac{1}{\lambda},\ \frac{\lambda}{\lambda - 1},\ \frac{1}{1 - \lambda},\ \frac{\lambda - 1}{\lambda} \right\}$$

Proof For $i = 1, \ldots, 6$ let $\mu_i = \mu_i\left(\lambda\right)$ be the six values in $\Sigma(\lambda)$, and let M_i be the obvious Möbius transformation that satisfies $\mu_i = M_i\left(\lambda\right)$ for all λ. The required isomorphism is given by

$$C_\lambda \longrightarrow C_\mu$$
$$(x, y) \longmapsto \left(M_i(x), R(x, y)\right)$$

where R is a rational function uniquely determined up to sign that can be obtained by forcing the point $\left(M_i(x), R(x, y)\right)$ to lie in C_μ (just as we did in Example 1.83). For instance, in the case $M_i(x) = \dfrac{x}{x - 1}$ this means that

$$R(x, y)^2 = \left(\frac{x}{x-1}\right)\left(\frac{x}{x-1} - 1\right)\left(\frac{x}{x-1} - \frac{\lambda}{\lambda-1}\right)$$

$$= \frac{-x\left(x - \lambda\right)}{(x-1)^3\left(\lambda - 1\right)}$$

$$= \frac{-y^2}{\left(\lambda - 1\right)(x-1)^4}$$

hence we can take

$$R\left(x, y\right) = i\frac{y}{(x-1)^2\sqrt{\left(\lambda - 1\right)}}$$

The other cases can be easily handled in a similar way.

Conversely, let

$$\Phi = \left(\Phi_1, \Phi_2\right) : C_\lambda \longrightarrow C_\mu$$

be an isomorphism. We can assume $\Phi(\lambda,0) = (\mu,0)$, again by composition with a translation in $\mathbb{C}/\Lambda \simeq C_\mu$.

By the previous lemma, $J_\lambda = \Phi^{-1} \circ J_\mu \circ \Phi$, that is $\Phi \circ J_\lambda = J_\mu \circ \Phi$. It follows that for all $(x,y) \in C_\lambda$ we have

$$(\Phi_1(x,-y),\Phi_2(x,-y)) = (\Phi_1(x,y),-\Phi_2(x,y))$$

In other words, we have a commutative diagram

$$
\begin{array}{ccc}
C_\lambda & \xrightarrow{\ \Phi\ } & C_\mu \\
{\scriptstyle\pi_\lambda}\downarrow & & \downarrow{\scriptstyle\pi_\mu} \\
\mathbb{P}^1 & \xrightarrow{\ M\ } & \mathbb{P}^1
\end{array}
$$

where π_λ, π_μ are the respective projections onto the first coordinate and $M(x) = \Phi_1(x,y) = \Phi_1(x,-y)$ is an isomorphism, its inverse being $x \longmapsto \left(\Phi^{-1}\right)_1(x)$. It follows that M is a Möbius transformation that, due to the commutativity of the diagram, must send the branching values $\{0,1,\infty,\lambda\}$ of π_λ to the branching values $\{0,1,\infty,\mu\}$ of π_μ. Moreover, since $\Phi(\lambda,0) = (\mu,0)$ we even know that $M(\lambda) = \mu$ and $M(\{0,1,\infty\}) = \{0,1,\infty\}$. But the Möbius transformations that permute $0,1,\infty$ form a group isomorphic to Σ_3 consisting of the six matrices M_i defined above. By definition, the orbit $\Sigma_3(\lambda)$ of the point λ is precisely $\Sigma(\lambda)$. $\square$

Corollary 2.57 *The Riemann surfaces C_λ and C_μ are isomorphic if and only if $j(\lambda) = j(\mu)$, where j stands for the so-called j-invariant, given by*

$$j(\lambda) = \frac{\left(1-\lambda+\lambda^2\right)^3}{\lambda^2\left(\lambda-1\right)^2}$$

Proof A direct computation shows that $j(\lambda) = j(\lambda')$ for every $\lambda' \in \Sigma(\lambda)$. Since, as a rational function, $j(x)$ has degree 6, it follows that $j^{-1}(j(\lambda)) = \Sigma(\lambda)$. Therefore, $j(\lambda) = j(\mu)$ if and only if $\mu \in \Sigma(\lambda)$, as was to be seen. $\square$

2.6.2 The moduli space $\mathcal{M}_\mathbf{g}$ for $\mathbf{g} > 1$

As in the case $g=1$, for $g \geq 2$ there are also two ways to define the moduli space $\mathcal{M}_g$; the algebraic one due to Mumford and others (see [MF82], [HM98]) and the trascendental one due to Fricke,

Teichmüller, Ahlfors, Bers and others (see e.g. [Abi80], [Ber72], [Bus92], [Gar87], [Har74], [Nag88]). In this case $\mathcal{M}_g$ is a variety of complex dimension $3g - 3$. A precise description of any of the two approaches is far beyond the scope of this book. However, we can at least give an argument to show that the given dimension is right. In view of Theorem 2.6 the Uniformization Theorem for genus $g \geq 2$ tells us that describing a Riemann surface of genus $g \geq 2$ amounts to specifying $2g$ real 2×2 matrices $\{\gamma_i\}_{i=1}^{2g}$ satisfying:

(1) $\det(\gamma_i) = 1$

(2) $\begin{pmatrix} s_{11} & s_{12} \\ s_{21} & s_{22} \end{pmatrix} = \prod_{i=1}^{g}[\gamma_i, \gamma_{g+i}] = \begin{pmatrix} 1 & 0 \\ 0 & 1 \end{pmatrix}$

By (1) each γ_i depends only of three real parameters, which makes a total amount of $3 \cdot 2g = 6g$ real parameters. Now (2) imposes three relations among the entries of the matrices γ_i (the fourth one, $s_{22} = 1$, is a consequence of the fact that the determinant equals one). This reduces the number of needed parameters to $6g - 3$.

Finally, since for any $\gamma \in \mathbb{PSL}(2, \mathbb{R})$ the groups $\Gamma = \langle \gamma_i \rangle$ and $\gamma \Gamma \gamma^{-1} = \langle \gamma \gamma_i \gamma^{-1} \rangle$ uniformize isomorphic Riemann surfaces (see Proposition 2.25), we conclude that the true number of real parameters needed is $6g - 6$. In fact it is known that the set of isomorphism classes of Riemann surfaces of genus $g \geq 2$ can be endowed with a complex analytic structure of (complex) dimension $3g - 3$.

2.7 Monodromy

Let $f : S_1 \to S$ be a morphism of degree d ramified over the values $y_1, \ldots, y_n \in S$. If $y \in S$ is a chosen regular value we can associate to f a group homomorphism

$$M_f : \pi_1\left(S \smallsetminus \{y_1, \ldots, y_n\}, y\right) \longrightarrow \mathrm{Bij}\left(f^{-1}(y)\right)$$

$$\gamma \longmapsto M_f(\gamma) = \sigma_\gamma^{-1}$$

where σ_γ is defined as follows.

Since $f : S_1 \smallsetminus f^{-1}\{y_1, \ldots, y_n\} \longrightarrow S \smallsetminus \{y_1, \ldots, y_n\}$ is a covering map and γ is a loop based on y, we can lift γ to a path $\widetilde{\gamma}$ with initial point at any given point $x \in f^{-1}(y)$ and endpoint at certain $x' \in f^{-1}(y)$. We then set $\sigma_\gamma(x) = x'$.

The reason why M_f is a homomorphism is that if $\gamma = \alpha\beta$ is a composition of two loops a lift $\widetilde{\gamma}$ can be viewed as a lift of α followed by a lift of β. Notice that if we had chosen to define $M_f(\gamma) = \sigma_\gamma$ we would have obtained an anti-homomorphism.

Of course if we number the points in $f^{-1}(y)$, i.e. if we choose a bijection $\varphi : \{1, \ldots, d\} \longrightarrow f^{-1}(y)$ such that $\varphi(j) = x_j$ then M_f, or more precisely M_f^φ defined by

$$M_f^\varphi(\gamma) = \varphi^{-1} \circ M_f(\gamma) \circ \varphi$$

becomes a group homomorphism from the fundamental group to the symmetric group Σ_d.

In the sequel we will refer to this representation of the fundamental group $\pi_1 \left(S \smallsetminus \{y_1, \ldots, y_n\}\right)$ in the symmetric group Σ_d as the *monodromy* of f. The homomorphism M_f depends on the ordering of the points in $f^{-1}(y)$ only up to conjugation in the symmetric group. This is obvious; if $\psi : \{1, \ldots, d\} \longrightarrow f^{-1}(y)$ is another numbering then

$$M_f^\psi(\gamma) = \tau^{-1} \circ M_f^\varphi(\gamma) \circ \tau$$

where $\tau \in \Sigma_d$ is the permutation $\varphi^{-1} \circ \psi$. Similarly, the role played by the base point $y \in S$ is also irrelevant, in the sense that different choices give rise to conjugate monodromies. Let z be another point of S with $f^{-1}(z) = \{x_1', \ldots, x_d'\}$ and consider the corresponding monodromy homomorphism

$$M_f' : \pi_1 \left\{S \smallsetminus \{y_1, \ldots, y_n\}, z\right\} \longrightarrow \Sigma_d$$

$$\beta \longmapsto (\sigma_\beta')^{-1}$$

Then if c is a path in $S \smallsetminus \{y_1, \ldots, y_n\}$ joining y to z, we have

$$\pi_1 \left(S \smallsetminus \{y_1, \ldots, y_n\}, z\right) = c^{-1} \circ \pi_1 \left(S \smallsetminus \{y_1, \ldots, y_n\}, y\right) \circ c$$

and also

$$\widetilde{\gamma}_c = \widetilde{c}^{-1} \circ \widetilde{\gamma} \circ \widetilde{c} \qquad \text{if} \qquad \gamma_c = c^{-1} \circ \gamma \circ c$$

hence $\sigma_\gamma' = \sigma_{c^{-1}\gamma c} = \sigma_c \sigma_\gamma \sigma_c^{-1}$ where $\sigma_c \in \Sigma_d$ is the permutation defined by $\sigma_c(i) = j$ when the lift of c with initial point x_i has endpoint x_j'.

When we refer to the *monodromy group* of f, which we will denote by $\mathrm{Mon}(f)$, we will mean the image of M_f inside Σ_d, namely

$$\mathrm{Mon}(f) = \{\sigma_\gamma \mid \gamma \in \pi_1\left(S \smallsetminus \{y_1,\ldots,y_n\}\right)\} \leq \Sigma_d$$

The connectedness of the surface implies that the monodromy group is always a *transitive subgroup* of Σ_d. Any pair of given points $x_i, x_j \in f^{-1}\{y\}$ can be joined by a path $\widetilde{\gamma}$, which is obviously the lift, with initial point x_i, of the path $\gamma = f(\widetilde{\gamma})$ (a loop based on y).

2.7.1 Monodromy and Fuchsian groups

From the point of view of Fuchsian groups, monodromy can be rephrased as follows.

Consider the unramified covering

$$f : S_1 \smallsetminus f^{-1}(\{y_1,\ldots,y_n\}) \longrightarrow S \smallsetminus \{y_1,\ldots,y_n\}$$

and let

$$\pi : \mathbb{H}/\Gamma_1 \longrightarrow \mathbb{H}/\Gamma$$

be the Fuchsian group representation of this map provided by covering space theory. That is, $\Gamma_1 < \Gamma$ and there is a commutative diagram as follows:

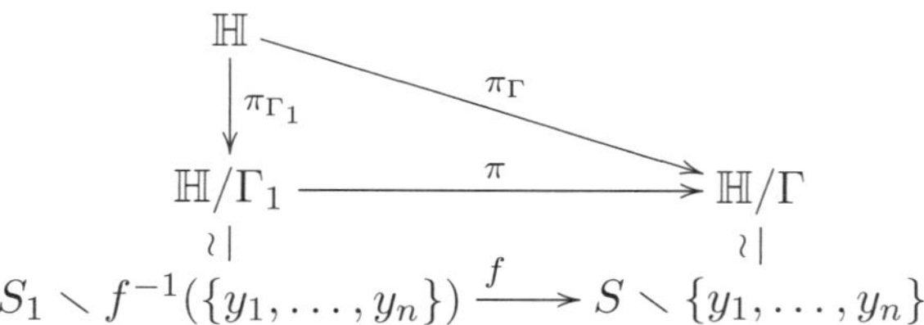

We now have the following identifications:

- $y = [z_0]_\Gamma$ for some $z_0 \in \mathbb{H}$.
- $\pi_1(S \smallsetminus \{y_1,\ldots,y_n\}, y) \simeq \Gamma$.
- $f^{-1}(y) = \{[\beta(z_0)]_{\Gamma_1}\}$, where β runs along a set of representatives of $\Gamma_1 \backslash \Gamma$, the set of right cosets of Γ_1 in Γ. We recall that $\Gamma_1\beta_1 = \Gamma_1\beta_2 \Leftrightarrow \beta_1\beta_2^{-1} \in \Gamma_1$.

Consider the bijection

$$\Phi : \Gamma_1\backslash\Gamma \longrightarrow f^{-1}(y)$$
$$\Gamma_1\beta \longmapsto [\beta(z_0)]_{\Gamma_1}$$

We would like to understand what the monodromy map looks like as a group homomorphism $M_f : \Gamma \longrightarrow \mathrm{Bij}(\Gamma_1\backslash\Gamma)$. To describe $M_f(\gamma)$ for $\gamma \in \Gamma$ as a bijection of $\Gamma_1\backslash\Gamma$ one proceeds as follows. According to the identification $\pi_1(S \smallsetminus \{y_1, \ldots, y_n\}, y) \simeq \Gamma$ the element γ corresponds to the loop $\pi_\Gamma([z_0, \gamma(z_0)])$, where $[z_0, \gamma(z_0)]$ is a path in $\mathbb{H}$ connecting z_0 to $\gamma(z_0)$ (see Section 1.2.5). The lift of this loop to $\mathbb{H}/\Gamma_1$ with initial point $[\beta(z_0)]_{\Gamma_1}$ is clearly the path $\pi_{\Gamma_1}(\beta[z_0, \gamma(z_0)])$, whose endpoint is $[\beta\gamma(z_0)]_{\Gamma_1}$, which corresponds to $\Gamma_1\beta\gamma$. This means that $\sigma_\gamma(\Gamma_1\beta) = \Gamma_1\beta\gamma$. Thus, we arrive at the following neat description of the monodromy map

$$M_f : \Gamma \longrightarrow \mathrm{Bij}(\Gamma_1\backslash\Gamma)$$
$$\gamma \longmapsto M_f(\gamma)$$

where

$$M_f(\gamma) : \Gamma_1\backslash\Gamma \longrightarrow \Gamma_1\backslash\Gamma \qquad (2.27)$$
$$\Gamma_1\beta \longmapsto M_f(\gamma)(\Gamma_1\beta) := \Gamma_1\beta\gamma^{-1}$$

In fact this description of the monodromy map of

$$\mathbb{H}/\Gamma_1 \longrightarrow \mathbb{H}/\Gamma$$

remains valid even when the groups are not torsion free (cf. formula (4.5) in Section 4.4).

We note in passing that the stabilizer of $\Gamma_1\beta \in \Gamma_1\backslash\Gamma$, that is the subgroup of Γ consisting of those $\gamma \in \Gamma$ such that $\Gamma_1\beta = \Gamma_1\beta\gamma^{-1}$, agrees with $\beta^{-1}\Gamma_1\beta$. This discussion yields the following crucial remark:

Remark 2.58 The subgroup $\Gamma_1 < \Gamma$ which defines the covering f is recovered (up to conjugation) from the monodromy map as the stabilizer of a point of the fibre of an unbranched value.

Another interesting consequence of our previous discussion is the following:

Corollary 2.59 *The monodromy map $M_\pi : \Gamma \longrightarrow \mathrm{Bij}(\Gamma_1 \backslash \Gamma)$ induces an isomorphism*

$$\frac{\Gamma}{\bigcap_{\beta \in \Gamma} \beta^{-1}\Gamma_1\beta} \simeq \mathrm{Mon}(\pi)$$

Proof Indeed $\gamma \in \ker(M_\pi)$ if and only if for every $\beta \in \Gamma$ we have

$$\Gamma_1\beta\gamma^{-1} = M_\pi(\gamma)(\Gamma_1\beta) = \Gamma_1\beta$$

that is if and only if $\gamma \in \beta^{-1}\Gamma_1\beta$ for every $\beta \in \Gamma$, as was to be seen. $\qquad\square$

Remark 2.60 If $\Gamma_1 \lhd \Gamma$ then the quotient $G = \Gamma_1 \backslash \Gamma = \Gamma/\Gamma_1$ is a group, thus we are allowed to write

$$M_\pi(\gamma)(\beta\Gamma_1) = (\beta\Gamma_1)(\gamma^{-1}\Gamma_1)$$

We can conclude that in this situation $\mathrm{Mon}(f)$ agrees with the group G viewed as a subgroup of $\mathrm{Bij}(G)$ via the monomorphism

$$g \longmapsto \sigma_g$$

where for any $h \in G$ we have $\sigma_g(h) = hg^{-1}$.

2.7.2 Characterization of a morphism by its monodromy

The relevance of the monodromy can be clearly perceived in the following result:

Theorem 2.61 *Let $f_i : S_i \longrightarrow S$, for $i = 1, 2$ be two morphisms of some degree d, with the same branch values $\{y_k\} \subset S$. Then f_1 and f_2 have conjugate monodromies if and only if they are isomorphic coverings, i.e. if there exists an isomorphism $\phi : S_1 \longrightarrow S_2$ such that the diagram*

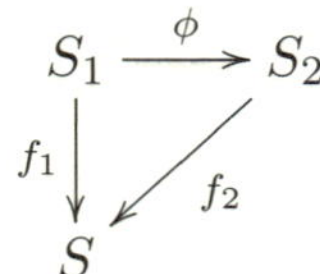

is commutative.

Proof The result is obvious in one of the directions. If the coverings are isomorphic, then the monodromies coincide, provided the elements of $f_2^{-1}(y) = \phi\left(f_1^{-1}(y)\right)$ have been labelled in the right way (meaning that the labels of $f_1^{-1}(y)$ have been transferred to $f_2^{-1}(y)$ via ϕ).

According to (the extension) Lemma 1.80, for the converse we only need to show that if the monodromies are conjugate there is a commutative diagram as follows

$$\mathbb{H}/\Gamma_1 = S_1 \setminus f_1^{-1}\{y_k\} \xrightarrow{\ \ \phi\ \ } S_2 \setminus f_2^{-1}\{y_k\} = \mathbb{H}/\Gamma_2$$
$$\Big\downarrow{\scriptstyle f_1} \qquad\qquad\qquad\qquad\qquad \Big\downarrow{\scriptstyle f_2}$$
$$\mathbb{H}/\Gamma = S \setminus \{y_k\} \xrightarrow{\ \ \text{Id}\ \ } S \setminus \{y_k\} = \mathbb{H}/\Gamma$$

where $\Gamma_1, \Gamma_2 < \Gamma$ are Fuchsian groups such that the covering maps $f_i : S_i \setminus f_i^{-1}\{y_k\} \longrightarrow S \setminus \{y_k\}$ are represented by the natural projections $\mathbb{H}/\Gamma_i \longrightarrow \mathbb{H}/\Gamma$, $i = 1, 2$.

Now if the monodromies of f_1 and f_2 are conjugate then there must be a bijection $\varphi : \Gamma_1 \backslash \Gamma \simeq \Gamma_2 \backslash \Gamma$ such that the stabilizer of M_{f_1} at a point $\Gamma\beta_1$ ($\beta_1 \in \Gamma_1$) agrees with the stabilizer of M_{f_2} at the point $\Gamma\beta_2 = \varphi(\Gamma\beta_1)$ for some $\beta_2 \in \Gamma_2$. But, by the observation made above, these stabilizers are subgroups of Γ conjugate to Γ_1 and Γ_2 respectively, therefore Γ_1 and Γ_2 must be conjugate subgroups. Finally, if $\Gamma_2 = \gamma\Gamma_1\gamma^{-1}$ for some $\gamma \in \Gamma$ then the required isomorphism ϕ is induced by the element γ. $\qquad\square$

Remark 2.62 This is the result upon which the classical representation of a Riemann surface as a union of a number of copies of $\widehat{\mathbb{C}}$ suitably glued together is based.

For instance, a hyperelliptic curve

$$S = \left\{y^2 = (x - \lambda_1) \cdots (x - \lambda_{2g+2})\right\}$$

can be obtained by glueing two copies of $\widehat{\mathbb{C}}$. One begins by drawing in $\widehat{\mathbb{C}}$ a collection of $g + 1$ segments c_i connecting the points λ_{2i-1} and λ_{2i}, for $i = 1, \ldots, g + 1$. Then one cuts $\widehat{\mathbb{C}}$ open along these segments to obtain a space Y isomorphic to a sphere with $g + 1$ holes each of them with an upper and lower boundary c_i^+ and c_i^-, corresponding to the two sides of c_i (see Figure 2.23).

Finally, one glues two copies of Y together by identifying the lower bound c_i^- of the first copy to the upper bound c_i^+ of the

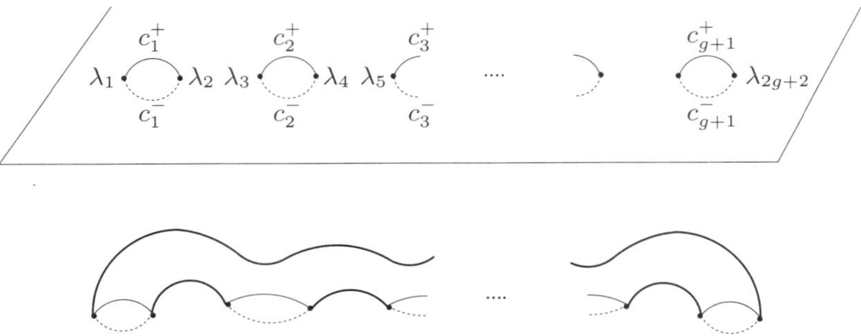

Fig. 2.23. Cutting $\widehat{\mathbb{C}}$ along c_i to obtain Y, a sphere with $g+1$ holes.

second copy (and vice versa) to obtain a new Riemann surface X. The lift of a small loop γ encircling λ_1, say, jumps from the first copy to the second one (see Figure 2.24). Thus X carries in an obvious way a $2:1$ map to $\widehat{\mathbb{C}}$ ramified over $\lambda_1,\ldots,\lambda_{2g+2}$, which has the same monodromy as the **x** function in S. In particular, S and X must be isomorphic.

We finish this section with the following useful result:

Proposition 2.63 *Let S be a compact Riemann surface and let $B = \{a_1,\ldots,a_n\} \subset S$ be a finite subset. Given a natural number $d \geq 1$ there are only finitely many pairs $(\widetilde{S}, f)$ where $\widetilde{S}$ is a compact Riemann surface and $f : \widetilde{S} \longrightarrow S$ is a morphism of degree d and branching value set B.*

Proof Assume first that $S = \mathbb{P}^1$ and $n = 3$, a special case that will be very important in the next chapters. With the notation as above the Fuchsian group Γ in this case is the principal congruence subgroup $\Gamma(2)$ (see Theorem 2.34), a group generated by two elements γ_1 and γ_2. Therefore, any possible monodromy map

$$M_f : \Gamma(2) \longrightarrow \Sigma_d$$

is determined by the images of γ_1 and γ_2 and hence there can only be finitely many of them.

The same argument can be applied in the general case, since the group uniformizing $S \setminus B$ is always finitely generated. This

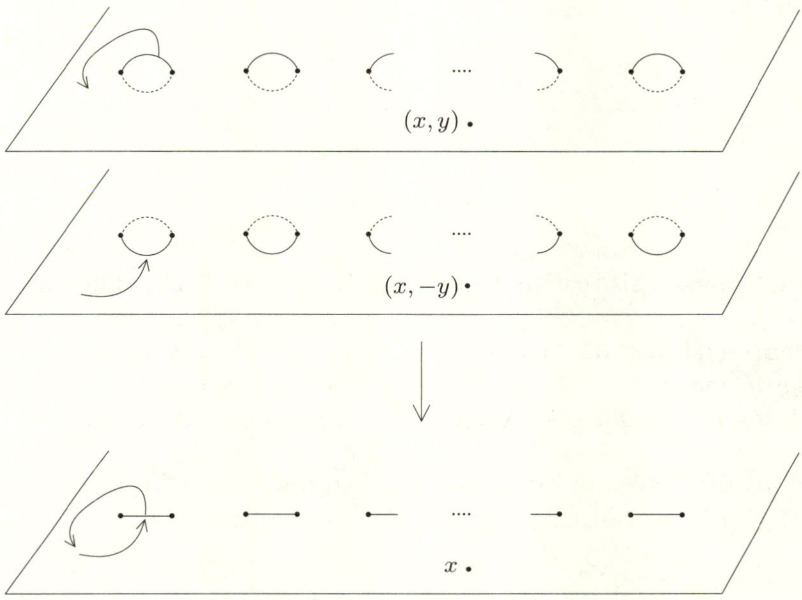

Fig. 2.24. The monodromy of the obvious $2 : 1$ cover from X to $\widehat{\mathbb{C}}$

is a well-known fact we will not prove (or need) in the rest of the book. $\qquad\square$

2.8 Galois coverings

Definition 2.64 A covering $f : S_1 \longrightarrow S_2$ is called a *Galois* (or *normal*, or *regular*) covering if the covering group

$$\mathrm{Aut}(S_1, f) = \{h \in \mathrm{Aut}(S_1) \: : f \circ h = f\}$$

acts transitively on each fibre.

A Galois covering can be thought of as a quotient $S_1 \longrightarrow S_1/G$ where $G = \mathrm{Aut}(S_1, f)$. More precisely, there is a commutative

diagram

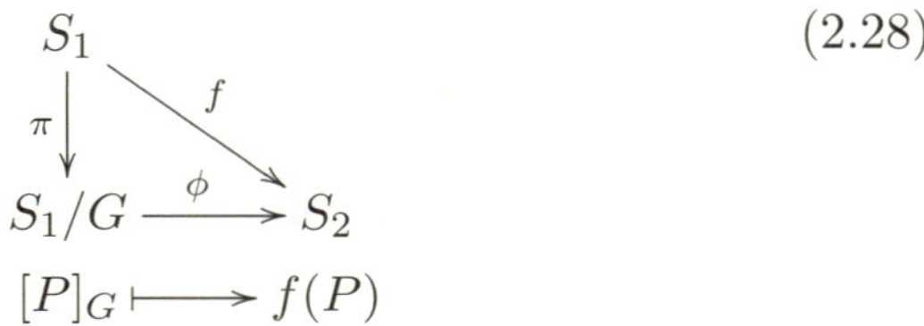

$$(2.28)$$

where ϕ is an isomorphism.

Of course this terminology comes from field theory. One has:

Proposition 2.65 *A morphism $f : S_1 \longrightarrow S_2$ is a Galois covering if and only if $f^* : \mathcal{M}(S_2) \longrightarrow \mathcal{M}(S_1)$ is a Galois extension. Moreover, in that case $\mathrm{Aut}(S_1, f) \simeq \mathrm{Gal}(\mathcal{M}(S_1)/\mathcal{M}(S_2))$.*

Proof If we denote by G^* the subgroup of $\mathrm{Aut}(\mathcal{M}(S_1))$ given by $G^* = \{h^* \mid h \in \mathrm{Aut}(S_1, f)\}$ then by Proposition 2.44

$$f^*(\mathcal{M}(S_2)) = \pi^*(\mathcal{M}(S_1/G)) = \mathcal{M}(S_1)^{G^*}$$

and this is one of the ways to say that $f^* : \mathcal{M}(S_2) \hookrightarrow \mathcal{M}(S_1)$ is a Galois extension with Galois group $G^* \simeq \mathrm{Aut}(S_1, f)$.

Conversely, if $f^*(\mathcal{M}(S_2)) = \mathcal{M}(S_1)^H$ for some group H of field automorphisms of $\mathcal{M}(S_1)$ over $\mathbb{C}$ then, due to the equivalence between the categories of Riemann surfaces and function fields (Proposition 1.95), the group H agrees with G^* for some subgroup $G \leq \mathrm{Aut}(S_1)$ and, again by Proposition 2.44, there is a field isomorphism

$$(\pi^*)^{-1}\big|_{\mathcal{M}(S_1)^{G^*}} \circ f^* : \mathcal{M}(S_2) \longrightarrow \mathcal{M}(S_1/G)$$

which must agree with ϕ^* for some isomorphism $\phi : S_1/G \longrightarrow S_2$, so we get a diagram like (2.28). $\qquad\square$

By now we have encountered several examples of Galois coverings, the simplest ones being the hyperelliptic covers of $\mathbb{P}^1$ given by

$$\mathbf{x} : S = \{y^2 = \prod(x - a_i)\} \longrightarrow \mathbb{P}^1$$

$$(x, y) \longmapsto x$$

whose covering group G is the order 2 group generated by the hyperelliptic involution $J(x, y) = (x, -y)$. Indeed we have the following commutative diagram:

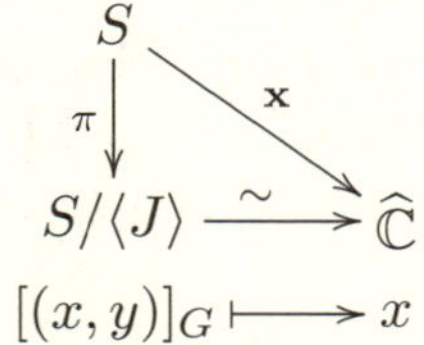

More generally, all covers of degree 2 are normal. This is only because the corresponding statement holds for function fields.

Another example of normal covering is given in Example 1.21, where the covering group is the order d group generated by the automorphism $(x, y) \longmapsto (x, \xi_d y)$ of the Fermat curve. Also normal are the coverings given explicitly as quotient maps, such as the covering of $\mathbb{P}^1$ by Klein's curve of genus three induced by the (normal) inclusion of $\Gamma(7)$ in the modular group (see Example 2.50). The covering group is in this case the quotient group $\mathbb{P}\mathrm{SL}(2, \mathbb{Z})/\Gamma(7) \simeq \mathbb{P}\mathrm{SL}(2, \mathbb{Z}/7\mathbb{Z})$.

Galois coverings are easily characterized in terms of their monodromy groups, as shown in the following proposition:

Proposition 2.66 *A covering* $f : S_1 \longrightarrow S_2$ *is normal if and only if* $\deg(f) = |\mathrm{Mon}(f)|$.

Proof If f is a normal covering with covering group G then certainly $\deg(f) = |G| = |\mathrm{Mon}(f)|$, the last equality by Remark 2.60.

Conversely, since the monodromy group $\mathrm{Mon}(f)$ is a transitive subgroup of Σ_d with $d = \deg(f)$, the fact that $d = |\mathrm{Mon}(f)|$ implies that for any $j \in \{1, \dots, d\}$ there is a unique element in $\mathrm{Mon}(f)$ sending 1 to j. When applied to $j = 1$ this means that $M_f(\gamma)(1) = 1$ is equivalent to $M_f(\gamma) = \mathrm{Id} \in \Sigma_d$. In other words, the stabilizer $M_f^{-1}(I(1))$ agrees with the kernel of M_f and so our covering is equivalent to the compactification of a covering of the form $\mathbb{H}/\Gamma_1 \longrightarrow \mathbb{H}/\Gamma$ with $\Gamma_1 \lhd \Gamma$, which is a Galois covering with covering group $G \simeq \Gamma/\Gamma_1$. $\square$

2.9 Normalization of a covering of $\mathbb{P}^1$

Let $f : S \longrightarrow \mathbb{P}^1$ be a covering of degree $d > 0$ with monodromy group $\mathrm{Mon}(f) \subset \Sigma_d$. In this section we will prove that $\mathrm{Mon}(f)$

is isomorphic to the group $\mathrm{Aut}(\widetilde{S}, \widetilde{f})$ of covering transformations of certain morphism $\widetilde{f} : \widetilde{S} \longrightarrow \mathbb{P}^1$ associated to $f : S \longrightarrow \mathbb{P}^1$, called the *normalization* of f and characterized by the fact that the corresponding field extension $\widetilde{f}^* : \mathcal{M}(\mathbb{P}^1) \hookrightarrow \mathcal{M}(\widetilde{S})$ is the normalization of the extension $f^* : \mathcal{M}(\mathbb{P}^1) \hookrightarrow \mathcal{M}(S)$ in the sense of Galois theory. We recall that the normalization of a field extension $i : K \hookrightarrow L$ is a Galois extension of K of the lowest possible degree among those containing L. In other words, it is a Galois extension $\widetilde{j} : K \hookrightarrow \widetilde{L}$ of minimum degree such that there is a field extension $j : L \hookrightarrow \widetilde{L}$ making commutative the diagram

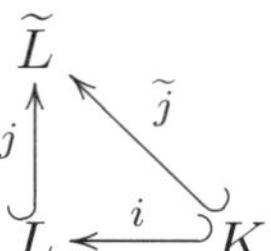

Accordingly, we make the following definition:

Definition 2.67 The *normalization* of a covering $f : S \longrightarrow \mathbb{P}^1$ of degree $d > 0$ is a Galois covering $\widetilde{f} : \widetilde{S} \longrightarrow \mathbb{P}^1$ of the lowest possible degree with the property that there is a third morphism $\pi : \widetilde{S} \to S$ making the diagram

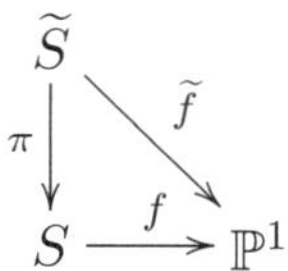

commutative.

Let $\mathcal{M}(S) = \mathbb{C}(f, h)$ and let $F(X, Y) \in \mathbb{C}[X, Y]$ be an irreducible polynomial such that $F(f, h) = 0$. The results in Section 1.3.1 tell us that the ramified coverings $f : S \longrightarrow \mathbb{P}^1$ and $\mathbf{x} : S_F \longrightarrow \widehat{\mathbb{C}}$ (or, equivalently, the corresponding field extensions $f^* : \mathcal{M}(\mathbb{P}^1) \hookrightarrow \mathcal{M}(S)$ and $\mathbb{C}(\mathbf{x}) \subset \mathbb{C}(\mathbf{x}, \mathbf{y}) = \mathcal{M}(S_F)$) are isomorphic. Moreover, the minimal polynomial of $\mathbf{y}$ over $\mathbb{C}(\mathbf{x})$ is precisely $F(\mathbf{x}, Y) \in \mathbb{C}(\mathbf{x})[Y]$.

In these terms we can construct the Riemann surface $\widetilde{S}$ as follows. Consider the normalization of the field extension $\mathbb{C}(\mathbf{x}) \hookrightarrow \mathbb{C}(\mathbf{x}, \mathbf{y})$. It is known that the normalization is the splitting field of the polynomial $F(\mathbf{x}, Y)$ over $\mathbb{C}(\mathbf{x})$. Recall that this is an abstract field $\widetilde{\mathcal{M}}$ enjoying the following two properties:

(i) There is a $\mathbb{C}$-embedding $\varphi : \mathbb{C}(\mathbf{x}, \mathbf{y}) \hookrightarrow \widetilde{\mathcal{M}}$ such that the polynomial $F(\varphi(\mathbf{x}), Y) \in \mathbb{C}(\varphi(\mathbf{x}))[Y]$ has all its d roots in $\widetilde{\mathcal{M}}$.

(ii) If these roots are denoted by $\widetilde{y}_1, \ldots, \widetilde{y}_d$ with, say, $\widetilde{y}_1 = \varphi(\mathbf{y})$ and $\varphi(\mathbf{x})$ is denoted by $\widetilde{x}$, then $\widetilde{\mathcal{M}} = \mathbb{C}(\widetilde{x}, \widetilde{y}_1, \ldots, \widetilde{y}_d)$.

By the Primitive Element Theorem $\widetilde{\mathcal{M}} = \mathbb{C}(\widetilde{x}, \widetilde{y})$ where

$$\widetilde{y} = a_1(\widetilde{x})\widetilde{y}_1 + \cdots + a_d(\widetilde{x})\widetilde{y}_d$$

for certain polynomials $a_1, \ldots, a_d \in \mathbb{C}[X]$.

Now, Proposition 1.95 implies that there is a commutative diagram of field embeddings over $\mathbb{C}$

$$
\begin{array}{ccc}
\mathcal{M}(S_F) & \xrightarrow{\;\widetilde{\varphi}\;} & \mathcal{M}(S_{\widetilde{F}}) \\
{\scriptstyle \alpha_2 = \mathrm{Id}}\big\downarrow & & \big\downarrow{\scriptstyle \alpha_1} \\
\mathbb{C}(\mathbf{x}, \mathbf{y}) & \xrightarrow{\;\varphi\;} & \widetilde{\mathcal{M}}
\end{array}
$$

where:

- $\widetilde{F} \in \mathbb{C}[X, Y]$ is an irreducible polynomial satisfying $\widetilde{F}(\widetilde{x}, \widetilde{y}) = 0$.
- α_1 is a field isomorphism defined by $\alpha_1^{-1}(\widetilde{x}) = \widetilde{\mathbf{x}}$, $\alpha_1^{-1}(\widetilde{y}) = \widetilde{\mathbf{y}}$ where $\widetilde{\mathbf{x}}, \widetilde{\mathbf{y}}$ stand for the coordinate functions on the Riemann surface $S_{\widetilde{F}}$. In particular $\mathcal{M}(S_{\widetilde{F}}) = \mathbb{C}(\widetilde{\mathbf{x}}, \widetilde{\mathbf{y}})$.
- $\widetilde{\varphi} : \mathcal{M}(S_F) \hookrightarrow \mathcal{M}(S_{\widetilde{F}})$ agrees with the pull-back Φ^* of a morphism $\Phi : S_{\widetilde{F}} \longrightarrow S_F$ defined by $\Phi(x, y) = (R_1(x, y), R_2(x, y))$ where $R_1, R_2 \in \mathbb{C}(X, Y)$ are rational functions that express the image of $\mathbf{x}$ and $\mathbf{y}$ under $\alpha_1^{-1} \circ \varphi$ in terms of the coordinate functions $\widetilde{\mathbf{x}}, \widetilde{\mathbf{y}}$ of $S_{\widetilde{F}}$.

Since $\alpha_1^{-1} \circ \varphi(\mathbf{x}) = \widetilde{\mathbf{x}}$ we see that Φ is actually of the form $\Phi(x, y) = (x, R_2(x, y))$. In other words, there is a commutative diagram as follows:

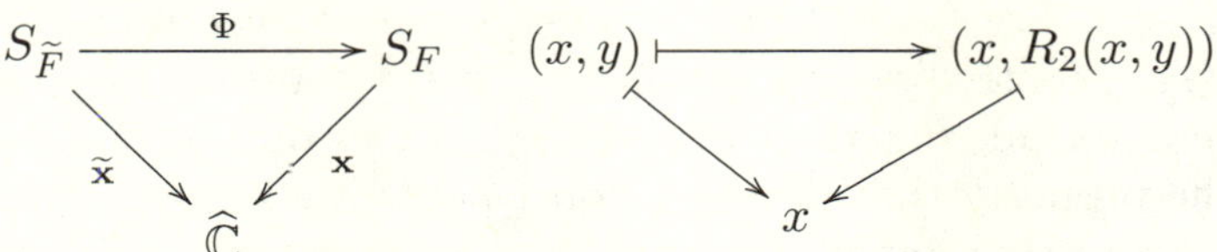

According to Definition 2.67, the morphisms $\pi : \widetilde{S} \longrightarrow S$ and $\widetilde{f} : \widetilde{S} \longrightarrow \mathbb{P}^1$ providing the normalization of $f : S \longrightarrow \mathbb{P}^1$ are given, up to equivalence, by $\Phi : S_{\widetilde{F}} \longrightarrow S_F$ and $\widetilde{\mathbf{x}} : S_{\widetilde{F}} \longrightarrow \widehat{\mathbb{C}}$ respectively.

2.9.1 The covering group of the normalization

We shall now describe the action of the monodromy group $\mathrm{Mon}(f)$ on the function field $\mathcal{M}(S_{\widetilde{F}}) = \mathbb{C}(\widetilde{\mathbf{x}}, \widetilde{\mathbf{y}})$.

Let $x^0 \in \widehat{\mathbb{C}}$ be a non-branching value of $\widetilde{\mathbf{x}}$ and hence of $\mathbf{x}$. We can assume that $x^0 \in \mathbf{x}(S_F^X)$ (see Chapter 1). Then the Implicit Function Theorem implies that there are functions $y_1, \ldots, y_d$ defined on a small disc D around x^0 satisfying $F(x, y_i(x)) \equiv 0$ on D, and such that the fibre $\mathbf{x}^{-1}(x^0)$ consists precisely of the points $(x^0, y_1^0), \ldots, (x^0, y_d^0) \in S_F$ where $y_i^0 = y_i(x^0)$.

Let us put $\widetilde{\mathbf{y}}_i = \alpha_1^{-1}(\widetilde{y}_i)$ and let $\widetilde{U}$ be an open set that maps isomorphically to D via $\widetilde{\mathbf{x}}$. From the identity

$$0 = F(\widetilde{x}, \widetilde{y}_i) = F(\alpha_1^{-1}(\widetilde{x}), \alpha_1^{-1}(\widetilde{y}_i)) = F(\widetilde{\mathbf{x}}, \widetilde{\mathbf{y}}_i)$$

we infer that the function

$$F(\widetilde{\mathbf{x}}, \widetilde{\mathbf{y}}_i) \circ \left(\widetilde{\mathbf{x}}|_{\widetilde{U}}\right)^{-1}(x) = F\left(x, \widetilde{\mathbf{y}}_i \circ \left(\widetilde{\mathbf{x}}|_{\widetilde{U}}\right)^{-1}(x)\right)$$

vanishes identically on D. Therefore, $\widetilde{\mathbf{y}}_i \circ \left(\widetilde{\mathbf{x}}|_{\widetilde{U}}\right)^{-1}$ agrees with one of the functions $y_1, \ldots, y_d$. Let us assume $\widetilde{\mathbf{y}}_i \circ \left(\widetilde{\mathbf{x}}|_{\widetilde{U}}\right)^{-1} = y_i$.

Since a meromorphic function is completely determined by its restriction to any open set we have an obvious bijective correspondence between the functions $\widetilde{\mathbf{y}}_i \in \mathcal{M}(S_{\widetilde{F}})$, the local functions $y_i : D \longrightarrow \mathbb{C}$ and the points $(x^0, y_i^0) \in \mathbf{x}^{-1}(x^0) \subset S_F$.

Before we proceed any further we recall (see e.g. [JS87]) that the *analytic continuation* of a meromorphic function ψ defined on a neighbourhood of some point $x^0 \in \mathbb{C}$ along a path $\gamma : [0, 1] \longrightarrow \mathbb{C}$ with $\gamma(0) = x^0$ is (if it exists) a finite sequence of meromorphic functions obtained by choosing a partition of the unit interval $0 = t_0 < t_1 < \cdots < t_n = 1$, a sequence of overlapping discs $D_0, \ldots, D_n$ centred at $x^0 = \gamma(t_0), \ldots, \gamma(t_n)$ covering the path γ and meromorphic functions $\psi_0 = \psi, \psi_1, \ldots, \psi_n$ on $D_0, D_1, \ldots, D_n$ with the property that $\psi_l = \psi_{l+1}$ on $D_l \cap D_{l+1}$.

The expression analytic continuation of ψ along γ will be also used here to refer to the final function $\psi_\gamma := \psi_n$.

We can now state the following lemma:

Lemma 2.68 *Keeping the notation as above, let $B \subset \widehat{\mathbb{C}}$ be the branching value set of* $\mathbf{x} : S_F \longrightarrow \widehat{\mathbb{C}}$ *and*

$$\pi_1(\widehat{\mathbb{C}} \smallsetminus B, x^0) \xrightarrow{\ M_{\mathbf{x}}\ } \Sigma_d$$

$$\gamma \longmapsto \sigma_\gamma^{-1}$$

the monodromy map as defined in Section 2.7. Then the analytic continuation of the local function y_i along γ is precisely $y_{\sigma_\gamma(i)}$.

Proof In order to avoid getting mixed up with subindices, let us set $\psi = y_i$. As $F(x, \psi(x)) = F(x, y_i(x)) = 0$ on $D_0 = D$ and ψ_1 agrees with $\psi_0 = \psi$ on $D_0 \cap D_1$ we see that $F(x, \psi_1(x)) = 0$ on D_1 as well, similarly $F(x, \psi_2(x)) = 0$ on D_2, etc. Clearly this process can be continued to reach a final function $\psi_n = \psi_\gamma$ defined in a neighbourhood of x^0 that must coincide with one of the local functions $y_1, \ldots, y_d$.

To see that ψ_γ is in fact the function $y_{\sigma_\gamma(i)}$ one only has to realize that the rule

$$\widetilde{\gamma}(t) = (\gamma(t), \psi_k(\gamma(t))) \quad \text{if } \gamma(t) \in D_k$$

defines a lift to S_F of the path γ whose initial point $\widetilde{\gamma}(0)$ is $(x^0, y_i(x^0)) = (x^0, y_i^0)$ and its endpoint is $\widetilde{\gamma}(1) = (x^0, \psi_\gamma(x^0))$. Then by the definition of the monodromy map we must have $\psi_\gamma(x^0) = y_{\sigma_\gamma(i)}^0$, that is $\psi_\gamma = y_{\sigma_\gamma(i)}$, as required. $\qquad\square$

We recall in passing that ψ_γ depends only on the homotopy class of γ, for if γ^t is a collection of loops with base point x^0 approaching γ as t tends to zero, then the corresponding functions ψ_{γ^t} will approach ψ_γ, and so all of them must equal the same local function y_j.

We can now define a map

$$\tau : \pi_1(\widehat{\mathbb{C}} \smallsetminus B, x^0) \longrightarrow \mathrm{Gal}(\mathcal{M}(S_{\widetilde{F}})/\mathbb{C}(\widetilde{\mathbf{x}}))$$

$$\gamma \longmapsto \tau_\gamma$$

where τ_γ is determined by

$$\widetilde{\mathbf{y}} = \sum_{i=1}^{d} a_i(\widetilde{\mathbf{x}})\widetilde{\mathbf{y}}_i \xmapsto{\ \tau_\gamma\ } \widetilde{\mathbf{y}}_\gamma := \sum_{i=1}^{d} a_i(\widetilde{\mathbf{x}})\widetilde{\mathbf{y}}_{\sigma_\gamma(i)} \qquad (2.29)$$

Of course, in order to show that τ_γ truly defines an element of $\mathrm{Gal}(\mathcal{M}(S_{\widetilde{F}})/\mathbb{C}(\widetilde{\mathbf{x}}))$ we must first prove that $\widetilde{\mathbf{y}}_\gamma$ is a root of $\widetilde{F}(\widetilde{\mathbf{x}}, Y) \in \mathbb{C}(\widetilde{\mathbf{x}})[Y]$.

Now

$$\widetilde{F}(\widetilde{\mathbf{x}}, \widetilde{\mathbf{y}}_\gamma) = 0$$
$$\Updownarrow$$
$$\widetilde{F}(\widetilde{\mathbf{x}}, \widetilde{\mathbf{y}}_\gamma)|_{\widetilde{U}} \equiv 0$$
$$\Updownarrow$$
$$\widetilde{F}(\widetilde{\mathbf{x}}, \widetilde{\mathbf{y}}_\gamma) \circ \left(\widetilde{\mathbf{x}}|_{\widetilde{U}}\right)^{-1} \equiv 0$$
$$\Updownarrow$$
$$\widetilde{F}\left(x, \widetilde{\mathbf{y}}_\gamma \circ \left(\widetilde{\mathbf{x}}|_{\widetilde{U}}\right)^{-1}(x)\right) = 0 \ \forall x \in D$$

and the latter holds because we know that $\widetilde{F}\left(x, \widetilde{\mathbf{y}} \circ \left(\widetilde{\mathbf{x}}|_{\widetilde{U}}\right)^{-1}(x)\right) \equiv 0$ on D, and by Lemma 2.68 the function

$$\widetilde{\mathbf{y}}_\gamma \circ \left(\widetilde{\mathbf{x}}|_{\widetilde{U}}\right)^{-1}(x) = \sum a_i(x) y_{\sigma_\gamma(i)}(x)$$

is clearly the analytic continuation of

$$\widetilde{\mathbf{y}} \circ \left(\widetilde{\mathbf{x}}|_{\widetilde{U}}\right)^{-1}(x) = \sum a_i(x) y_i(x)$$

along γ.

Proposition 2.69 *The automorphism τ_γ described by formula (2.29) is characterized by any of the two following properties:*

  (i) *For any $\widetilde{\psi} \in \mathcal{M}(S_{\widetilde{F}})$, the function $\tau_\gamma(\widetilde{\psi}) \circ \left(\widetilde{\mathbf{x}}|_{\widetilde{U}}\right)^{-1}$ is the analytic continuation of $\widetilde{\psi} \circ \left(\widetilde{\mathbf{x}}|_{\widetilde{U}}\right)^{-1}$ along γ.*

  (ii) *$\tau_\gamma(\widetilde{\mathbf{y}}_i) = \widetilde{\mathbf{y}}_{\sigma_\gamma(i)}$ for $i = 1, \ldots, d$.*

Proof (i) Write

$$\widetilde{\psi} = \sum_{k=1}^{d} b_k(\widetilde{\mathbf{x}}) \widetilde{\mathbf{y}}^k = \sum_{k=1}^{d} b_k(\widetilde{\mathbf{x}}) \left(\sum_{l=1}^{d} a_l(\widetilde{\mathbf{x}}) \widetilde{\mathbf{y}}_l\right)^k$$

Then

$$\widetilde{\psi} \circ \left(\widetilde{\mathbf{x}}|_{\widetilde{U}}\right)^{-1} = \sum_{k=1}^{d} b_k(\widetilde{\mathbf{x}}) \left(\sum_{l=1}^{d} a_l(\widetilde{\mathbf{x}}) \widetilde{\mathbf{y}}_l\right)^k \circ \left(\widetilde{\mathbf{x}}|_{\widetilde{U}}\right)^{-1}$$

that is for all $x \in D$ we have

$$\widetilde{\psi} \circ \left(\widetilde{\mathbf{x}}|_{\widetilde{U}}\right)^{-1}(x) = \sum_{k=1}^{d} b_k(x) \left(\sum_{l=1}^{d} a_l(x) y_l(x)\right)^k$$

for all $x \in D$.

We similarly have

$$\tau_\gamma(\widetilde{\psi}) \circ \left(\widetilde{\mathbf{x}}|_{\widetilde{U}}\right)^{-1}(x) = \sum_{k=1}^{d} b_k(x) \left(\sum_{l=1}^{d} a_l(x) y_{\sigma_\gamma(l)}(x)\right)^k$$

which is indeed the analytic continuation of $\widetilde{\psi} \circ \left(\widetilde{\mathbf{x}}|_{\widetilde{U}}\right)^{-1}(x)$.

Part (ii) is a consequence of part (i). $\qquad\square$

Corollary 2.70 *The map*

$$\widehat{\tau} : \pi_1(\widehat{\mathbb{C}} \smallsetminus B, x^0) \longrightarrow \mathrm{Gal}(\mathcal{M}(S_{\widetilde{F}})/\mathbb{C}(\widetilde{\mathbf{x}}))$$

$$\gamma \longmapsto \tau_\gamma^{-1}$$

is a group homomorphism that induces an injection of the monodromy group $\mathrm{Mon}(\mathbf{x})$ *in the Galois group* $\mathrm{Gal}(\mathcal{M}(S_{\widetilde{F}})/\mathbb{C}(\widetilde{\mathbf{x}}))$.

Proof The fact that $\widehat{\tau}$ is a homomorphism follows from the first characterization of τ_γ given in Proposition 2.69 along with the obvious observation that if $\gamma = \beta \circ \alpha$ is a composition of two paths then the analytic continuation of a function ψ along γ equals the analytic continuation of ψ_α along β.

On the other hand, by the second characterization in Proposition 2.69, the kernel of $\widehat{\tau}$ agrees with the kernel of the monodromy map

$$M_{\mathbf{x}} : \pi_1(\widehat{\mathbb{C}} \smallsetminus B, x^0) \longmapsto \Sigma_d$$

therefore we can write

$$\mathrm{Mon}(\mathbf{x}) = \frac{\pi_1(\widehat{\mathbb{C}} \smallsetminus B, x^0)}{\ker(M_{\mathbf{x}})} = \frac{\pi_1(\widehat{\mathbb{C}} \smallsetminus B, x^0)}{\ker(\tau)} \hookrightarrow \mathrm{Gal}(\mathcal{M}(S_{\widetilde{F}})/\mathbb{C}(\widetilde{\mathbf{x}}))$$

$\qquad\square$

In fact, the following stronger statement holds:

Theorem 2.71 $\mathrm{Mon}(\mathbf{x}) \simeq \mathrm{Gal}(\mathcal{M}(S_{\widetilde{F}})/\mathbb{C}(\widetilde{\mathbf{x}}))$

Proof By elementary Galois theory it is enough to show that

$$\mathcal{M}(S_{\widetilde{F}})^{\mathrm{Mon}(\mathbf{x})} = \mathbb{C}(\widetilde{\mathbf{x}}) \tag{2.30}$$

So, let us consider a function $\widetilde{\psi} \in \mathcal{M}(S_{\widetilde{F}})$ such that $\tau_\gamma(\widetilde{\psi}) = \widetilde{\psi}$, or equivalently such that $\widetilde{\psi} \circ (\widetilde{\mathbf{x}}|_{\widetilde{U}})^{-1} = \tau_\gamma(\widetilde{\psi}) \circ (\widetilde{\mathbf{x}}|_{\widetilde{U}})^{-1}$ for every $\gamma \in \pi_1(\widehat{\mathbb{C}} \smallsetminus B, x^0)$. We have to show that $\widetilde{\psi} = R(\widetilde{\mathbf{x}})$ for some rational function $R(X) \in \mathbb{C}(X)$.

By Proposition 2.69 this invariance property of $\widetilde{\psi}$ is equivalent to saying that the analytic continuation of the meromorphic function $\widetilde{\psi} \circ (\widetilde{\mathbf{x}}|_{\widetilde{U}})^{-1}$ along any loop $\gamma \in \pi_1(\widehat{\mathbb{C}} \smallsetminus B, x^0)$ agrees with itself. Using this fact we can construct a meromorphic function ψ on $\widehat{\mathbb{C}} \smallsetminus B$ via analytic continuation. More precisely, let x be a point in $\widehat{\mathbb{C}} \smallsetminus B$ and α a path inside $\widehat{\mathbb{C}} \smallsetminus B$ with initial point x^0 and final point x. If ψ_α denotes the analytic continuation of $\widetilde{\psi} \circ (\widetilde{\mathbf{x}}|_{\widetilde{U}})^{-1}$ along α then we set $\psi(x) = \psi_\alpha(x)$. The function ψ is well defined because if β is another such path and $\gamma = \beta^{-1} \circ \alpha$ then, clearly, ψ_γ can be obtained by performing analytic continuation of ψ_α along β^{-1}. Since $\psi_\gamma = \psi$ we conclude that $\psi_\beta(x)$ has to agree with $\psi_\alpha(x)$.

Clearly the identity $\psi \circ \widetilde{\mathbf{x}} = \widetilde{\psi}$ holds on $\widetilde{U}$, hence on $\widetilde{S} \smallsetminus \widetilde{\mathbf{x}}^{-1}(B)$. In view of the Casorati–Weierstrass Theorem, this relation implies that the points in B are not essential singularities of ψ. Therefore, ψ is a meromorphic function on the whole $\widehat{\mathbb{C}}$, that is $\psi(x) = R(x)$ for some rational function R. Now the identity $\psi \circ \widetilde{\mathbf{x}} = \widetilde{\psi}$ can be written as $\widetilde{\psi} = R(\widetilde{\mathbf{x}}) \in \mathbb{C}(\widetilde{\mathbf{x}})$ as required. $\qquad\square$

In view of this result we can rephrase our definition of normalization as follows:

Corollary 2.72 *The normalization of $f : S \longrightarrow \mathbb{P}^1$ is up to isomorphism the only Galois covering $\widetilde{f} : \widetilde{S} \longrightarrow \mathbb{P}^1$ of the form $\widetilde{f} = f \circ \pi$, for some morphism $\pi : \widetilde{S} \longrightarrow S$, whose degree equals $|\mathrm{Mon}(f)|$.*

Corollary 2.73 $\mathrm{Mon}(f) \simeq \mathrm{Aut}(\widetilde{S}, \widetilde{f})$

Proof The equivalent statement $\mathrm{Mon}(\mathbf{x}) \simeq \mathrm{Aut}(S_{\widetilde{F}}, \widetilde{\mathbf{x}})$ follows from Proposition 2.65 and Theorem 2.71. $\qquad\square$

We end this section by pointing out that this last result together with Corollary 2.59 allows the following Fuchsian group interpretation of normalization:

Proposition 2.74 *Let $f : S_1 \longrightarrow S$ be a covering of Riemann surfaces, $S_1 \smallsetminus f^{-1}(\{y_1,\ldots,y_n\}) \longrightarrow S \smallsetminus \{y_1,\ldots,y_n\}$ its associated unramified cover and $\pi : \mathbb{H}/\Gamma_1 \longrightarrow \Gamma$ its Fuchsian group representation. Then the normalization of $f : S_1 \longrightarrow S$ can be represented as (the compactification of) the following diagram:*

$$
\frac{\mathbb{H}}{\displaystyle\bigcap_{\gamma\in\Gamma}\gamma^{-1}\Gamma_1\gamma}
$$

$$
\frac{\mathbb{H}}{\Gamma_1} \longrightarrow \frac{\mathbb{H}}{\Gamma}
$$

Proof Clearly $\displaystyle\bigcap_{\gamma\in\Gamma}\gamma^{-1}\Gamma_1\gamma \lhd \Gamma$, hence

$$
\frac{\mathbb{H}}{\displaystyle\bigcap_{\gamma\in\Gamma}\gamma^{-1}\Gamma_1\gamma} \longrightarrow \frac{\mathbb{H}}{\Gamma}
$$

defines a Galois cover whose covering group is isomorphic to the quotient group

$$
\frac{\Gamma}{\displaystyle\bigcap_{\gamma\in\Gamma}\gamma^{-1}\Gamma_1\gamma}
$$

which by Corollary 2.59 is isomorphic to the monodromy group $\mathrm{Mon}(f)$. Now the result is a consequence of Corollaries 2.72 and 2.73. $\qquad\square$

Proposition 2.74 yields the following interesting observation:

Corollary 2.75 *The branching value set of a covering agrees with the branching value set of its normalization.*

Example 2.76 Let $F(X, Y) = Y^2 X - (Y - 1)^3$ and consider the covering

$$\mathbf{x} : S_F \longrightarrow \mathbb{P}^1$$
$$(x, y) \longmapsto x$$

The following statements hold:

(1) S_F has genus zero. It is sufficient to observe that

$$\mathcal{M}(S_F) = \mathbb{C}(\mathbf{x}, \mathbf{y}) = \mathbb{C}(\mathbf{y})$$

and, in fact, the map $\mathbf{y} : S_F \longrightarrow \widehat{\mathbb{C}}$ is an isomorphism with inverse

$$\varphi : \widehat{\mathbb{C}} \longrightarrow S_F$$
$$y \longmapsto \left(\frac{(y - 1)^3}{y^2}, y \right)$$

(2) $\mathbf{x} : S_F \longrightarrow \mathbb{P}^1$ is a function of degree 3 ramified at most over the values $0, -\frac{27}{4}$ and ∞. This is (see Theorem 1.86) because for a point $P = (x, y) \in \mathbb{C}^2$ the two conditions $F(P) = F_Y(P) = 0$ can simultaneously occur only for $P = (0, 1)$ or $P = \left(-\frac{27}{4}, -2 \right)$.

(3) $\mathrm{Mon}(\mathbf{x}) \simeq \Sigma_3$, hence $(S_F, \mathbf{x})$ is not a normal covering. To see this we first write down the Riemann–Hurwitz formula for this covering, namely

$$2 \cdot 0 - 2 = 3(2 \cdot 0 - 2) + b_0 + b_1 + b_\infty \Leftrightarrow b_0 + b_1 + b_\infty = 4$$

where $b_0, b_1, b_\infty \leq 2$ are the natural numbers accounting for the ramification over the values $0, -\frac{27}{4}$ and ∞ respectively.

Now we observe that the point $P = \left(-\frac{27}{4}, \frac{1}{4} \right) \in S_F$ is an unbranched point of $\mathbf{x}$, since $F_Y(P) \neq 0$. In fact

$$\mathbf{x}^{-1} \left(-\tfrac{27}{4} \right) = \left\{ \left(-\tfrac{27}{4}, -2 \right), \left(-\tfrac{27}{4}, \tfrac{1}{4} \right) \right\} = \varphi \left(\left\{ -2, \tfrac{1}{4} \right\} \right)$$

and we have $b_1 = 1$. On the other hand,

$$\mathbf{x}^{-1} (0) = \{ (0, 1) \} = \varphi(\{ 1 \})$$

consists of a single point of multiplicity 3, hence $b_0 = 2$ and therefore $b_\infty = 1$.

These computations imply that $\mathrm{Mon}(\mathbf{x}) \leq \Sigma_3$ contains a 2-cycle σ_1 corresponding to the value $-\frac{27}{4}$ and a 3-cycle σ_2

corresponding to the value 0. Thus $\mathrm{Mon}(\mathbf{x}) = \langle \sigma_1, \sigma_2 \rangle = \Sigma_3$ and, since $|\mathrm{Mon}(\mathbf{x})| = 6 > \deg(\mathbf{x}) = 3$, we conclude that $(S_F, \mathbf{x})$ is not a normal covering.

(4) The normalization of $(S_F, \mathbf{x})$ is the covering $\left(S_{\widetilde{F}}, \widetilde{\mathbf{x}}\right)$ where

$$\widetilde{F}(X, Y) = Y^2(1-Y)^2 X + (1-Y+Y^2)^3 = F(X, Y(1-Y))$$

Consider the commutative diagram

$$\begin{array}{ccc} S_{\widetilde{F}} & & \\ \pi \downarrow & \searrow^{\widetilde{\mathbf{x}}} & \\ S_F & \xrightarrow{\ \mathbf{x}\ } & \mathbb{P}^1 \end{array} \qquad\qquad (2.31)$$

where $\pi(x, y) = (x, y(1-y))$ and $\mathbf{x}, \widetilde{\mathbf{x}}$ stand for the obvious projections onto the first coordinate. Note that π is well defined because $F(\widetilde{\mathbf{x}}, \widetilde{\mathbf{y}}(1 - \widetilde{\mathbf{y}})) = \widetilde{F}(\widetilde{\mathbf{x}}, \widetilde{\mathbf{y}}) = 0$.

We now observe that $\mathrm{Aut}(S_{\widetilde{F}}, \widetilde{\mathbf{x}})$ is the group isomorphic to Σ_3 generated by the transformations $\tau_1(x, y) = (x, 1 - y)$ and $\tau_2(x, y) = (x, 1/(1 - y))$. To see this, it is enough to check that $\langle \tau_1, \tau_2 \rangle \leq \mathrm{Aut}(S_{\widetilde{F}}, \widetilde{\mathbf{x}})$ and note that the order of the group $\langle \tau_1, \tau_2 \rangle \simeq \Sigma_3$ coincides with the degree of the covering map $\widetilde{\mathbf{x}}$. This means that $\widetilde{\mathbf{x}} : S_{\widetilde{F}} \longrightarrow \mathbb{P}^1$ is a normal covering. Now our claim follows from Corollary 2.72.

In general it is not easy to find out the polynomial $\widetilde{F}(X, Y)$ representing the normalization of the covering $(S_F, \mathbf{x})$ defined by a given polynomial $F(X, Y)$. It may be illustrative to unveil how this example was constructed.

The classical j-function $j : \mathbb{P}^1 \longrightarrow \mathbb{P}^1$ defined by

$$j(t) = \frac{(1 - t + t^2)^3}{t^2(1 - t)^2}$$

(see Corollary 2.57) provides a well-known Galois cover with covering group $\mathrm{Aut}(\mathbb{P}^1, j) \simeq \Sigma_3$ generated by $\tau_1(t) = 1 - t$ and $\tau_2(t) = 1/(1 - t)$.

The subgroup $H = \langle \tau_1 \rangle$ is an order 2 non-normal subgroup of Σ_3, thus, by elementary Galois theory

$$\mathcal{M}(\mathbb{P}^1)^{\Sigma_3^*} = \mathbb{C}(j) \hookrightarrow \mathcal{M}(\mathbb{P}^1)^{H^*} = \mathbb{C}(t(1 - t))$$

is a non-normal field extension of degree 3 whose normalization is

$$\mathcal{M}(\mathbb{P}^1)^{\Sigma_3^*} = \mathbb{C}(j) \hookrightarrow \mathcal{M}(\mathbb{P}^1) = \mathbb{C}(t)$$

From the point of view of coverings this amounts to saying that $j : \mathbb{P}^1 \longrightarrow \mathbb{P}^1$ is the normalization of the covering

$$f : \mathbb{P}^1 \longrightarrow \mathbb{P}^1$$
$$u \longmapsto \frac{(1-u)^3}{u^2}$$

something that becomes obvious by considering the following commutative diagram:

$$\begin{array}{ccc} & \mathbb{P}^1 & \\ {\scriptstyle p}\downarrow & \searrow {\scriptstyle j} & \\ \mathbb{P}^1 & \xrightarrow{f} & \mathbb{P}^1 \end{array}$$

with $p(t) = t(1-t)$. But this diagram is equivalent to our diagram (2.31) above via the isomorphisms

$$\mathbb{P}^1 \longrightarrow S_{\widetilde{F}}$$
$$t \longmapsto (j(t), t)$$

and

$$\mathbb{P}^1 \longrightarrow S_F$$
$$u \longmapsto \left(\frac{(1-u)^3}{u^2}, u \right)$$

With this second chapter we conclude our account of the theory of Riemann surfaces, from the points of view of algebraic curves and Fuchsian groups. In the rest of the book we shall employ these ideas to introduce the reader to the Grothendieck–Belyi theory of dessins d'enfants and its connection to Riemann surfaces associated to algebraic curves with coefficients in a number field. For those who want to complement or go deeper into the subject of Riemann surfaces there are many excellent references avalaible. Among them the books by Jones–Singerman [JS87], Beardon [Bea84], Farkas–Kra [FK92], Siegel [Sie88], Jost [Jos02], Forster [For91], Miranda [Mir95], Buser [Bus92], Kirwan [Kir92] and Narasimhan [Nar92], just to mention a few.

3

Belyi's Theorem

We shall say that a Riemann surface S *is defined over a field* $K \subset \mathbb{C}$ (or that K is a *field of definition of S*) if $S \simeq S_F$ for some irreducible polynomial $F(X, Y) = \sum a_{ij} X^i Y^j$ with coefficients $a_{ij} \in K$. We recall from Chapter 1 that the statement $S \simeq S_F$ is the same as saying that its function field $\mathcal{M}(S)$ has generators f, h such that $F(f, h) = 0$.

For example, if S has genus 0 then $S \simeq \mathbb{P}^1$ and therefore it is defined over $\mathbb{Q}$ (take $F(X, Y) = Y$, Example 1.87). The Riemann surface S_1 corresponding to the curve $Y^2 = X^3 - \pi^3$ is obviously defined over the transcendental field extension $\mathbb{Q}(\pi)$. But it is also defined over $\mathbb{Q}$ since it is isomorphic to the Riemann surface $S_2 = \{y^2 = x^3 - 1\}$, an isomorphism being given by

$$S_1 \longrightarrow S_2$$

$$(x, y) \longmapsto \left(\frac{x}{\pi}, \frac{y}{\pi\sqrt{\pi}} \right)$$

The problem of deciding when S is defined over a *number field* (i.e. a finite extension of $\mathbb{Q}$) is particularly interesting. Since the number of coefficients involved in a polynomial $F(X, Y)$ is finite, the question is equivalent to asking when S is defined over $\overline{\mathbb{Q}}$, the field of algebraic numbers.

The main goal of this chapter is to prove the following result, due to Belyi ([Bel79]):

Theorem 3.1 (Belyi's Theorem) *Let S be a compact Riemann surface. The following statements are equivalent:*

(a) S is defined over $\overline{\mathbb{Q}}$.

169

(b) S admits a morphism $f : S \longrightarrow \mathbb{P}^1$ with at most three branching values.

A meromorphic function with less than four critical values is called *a Belyi function*. For such a function we will always assume the branching values to be contained in the set $\{0, 1, \infty\}$, something one can always achieve by composition with a Möbius transformation (Remark 1.28). Note that $n = 3$ is the lowest number of possible branching values for a morphism $f : S \longrightarrow \mathbb{P}^1$ with $S \neq \mathbb{P}^1$. Indeed using our knowledge of covering space theory, we see that:

- $n = 0$ means that f is an unramified covering, hence an isomorphism since $\mathbb{P}^1$ is simply connected.
- $n = 1$ means that the restriction $f : S \setminus f^{-1}(\infty) \to \widehat{\mathbb{C}} \setminus \{\infty\}$ is unramified, therefore an isomorphism since $\mathbb{C}$ is simply connected. It follows that $S \simeq \widehat{\mathbb{C}}$.
- $n = 2$ means that $f : S \setminus f^{-1}(\{0, \infty\}) \to \widehat{\mathbb{C}} \setminus \{0, \infty\}$ is unramified. By Example 1.71, this restriction must be isomorphic to

$$\mathbb{C} \setminus \{0\} \longrightarrow \mathbb{C} \setminus \{0\}$$
$$z \longmapsto z^k$$

for some k, so again $S \simeq \mathbb{P}^1$.

This is the reason why very often the term Belyi function is used as a synonym of *Belyi function with exactly three branching values*.

Example 3.2 Let K be a Fuchsian group contained in a triangle group $\Gamma = \Gamma_{n,m,l}$. The induced mapping

$$\mathbb{D}/K \longrightarrow \mathbb{D}/\Gamma$$
$$[z]_K \longmapsto [z]_\Gamma$$

is a Belyi function, since the quotient map $\mathbb{D} \longrightarrow \mathbb{D}/\Gamma$ has only three branching values.

Before going into the proof of Theorem 3.1 let us see how the so-called Belyi's algorithm works in a particular (but illustrative) example. Consider the curve

$$Y^2 = X(X - 1)(X - \lambda)$$

where $\lambda = \dfrac{m}{m+n}$ and $m, n \in \mathbb{N}$.

The corresponding Riemann surface S_λ is clearly defined over $\overline{\mathbb{Q}}$. According to Belyi's Theorem, one should be able to find a function $f : S_\lambda \longrightarrow \mathbb{P}^1$ with only three branching values. The first step is to recall that S_λ admits a degree two meromorphic function $\mathbf{x} : S_\lambda \to \mathbb{P}^1$ defined by $(x, y) \to x$ (see Example 1.32) which ramifies precisely over the values $0, 1, \lambda, \infty$. At this point Belyi's main idea comes in. He considers the following (Belyi) polynomial

$$P_{m,n}(x) = P_\lambda(x) = \frac{(m+n)^{m+n}}{m^m n^n} x^m (1-x)^n \qquad (3.1)$$

Proposition 3.3 *As a function $P_\lambda : \mathbb{P}^1 \to \mathbb{P}^1$, Belyi's polynomial satisfies the following properties:*

(i) *P_λ ramifies only at the points $x = 0, 1, \infty$ and λ.*
(ii) *$P_\lambda(0) = 0$, $P_\lambda(1) = 0$, $P_{m,n}(\infty) = \infty$ and $P_\lambda(\lambda) = 1$.*

Proof We only need to observe that the zeros of the derivative of P_λ are the solutions of $x^{m-1}(1-x)^{n-1}\left((m+n)x - m\right) = 0$. $\qquad \square$

We now see that the composition of the function $\mathbf{x} : S_\lambda \to \mathbb{P}^1$ with the polynomial $P_\lambda : \mathbb{P}^1 \longrightarrow \mathbb{P}^1$ produces a morphism

$$f = P_\lambda \circ \mathbf{x} : S_\lambda \to \mathbb{P}^1$$

that ramifies at the points $(0,0), (1,0), (\lambda, 0), \infty \in S$ with branching values $0, 0, 1, \infty \in \mathbb{P}^1$ respectively; that is, f is a Belyi function.

3.1 Proof of part (a) ⇒ (b) of Belyi's Theorem

This is actually the part of Theorem 3.1 proved by Belyi [Bel79]. The converse implication was already known to Grothendieck, see [Gro97].

We claim that for the proof of this part it is enough to show the existence of a morphism $f : S \longrightarrow \mathbb{P}^1$ ramified over a set of rational values $\{0, 1, \infty, \lambda_1, \ldots, \lambda_n\} \subset \mathbb{Q} \cup \{\infty\}$. To see this we observe that after composing with the Möbius transformations $T(x) = 1 - x$ and $M(x) = 1/x$ if necessary, we can assume that $0 < \lambda_1 < 1$. Therefore, λ_1 can be written in the desired form

$\lambda_1 = m/m + n$. Composing now f with the rational function P_{λ_1} we would get a morphism $P_{\lambda_1} \circ f$ with strictly less branching values, namely $\{0, 1, \infty, P_{\lambda_1}(\lambda_2), \ldots, P_{\lambda_1}(\lambda_n)\} \subset \mathbb{Q} \cup \{\infty\}$. From here the problem is solved inductively.

In order to show the existence of such a function $f : S \longrightarrow \mathbb{P}^1$, let us write S in the form $S = S_F$ with

$$F(X, Y) = p_0(X)Y^n + p_1(X)Y^{n-1} + \cdots + p_n(X) \in \overline{\mathbb{Q}}\,[X, Y]$$

and consider the morphism

$$S_F \xrightarrow{\ \mathbf{x}\ } \mathbb{P}^1$$
$$(x, y) \longmapsto x$$

Let us denote by $B_0 = \{\mu_1, \ldots, \mu_s\}$ the set of branching values of $\mathbf{x}$. According to Theorem 1.86 each μ_i is either a zero of $p_0(X)$ or the point $\infty \in \mathbb{P}^1$, or the first coordinate of a common zero of the polynomials $F, F_Y \in \overline{\mathbb{Q}}[X, Y]$. From Lemma 1.84 we deduce that B_0, is contained in $\overline{\mathbb{Q}} \cup \{\infty\}$. Now, if $B_0 \subset \mathbb{Q} \cup \{\infty\}$, we are done. If not, we start the following inductive argument.

Let $m_1(T) \in \mathbb{Q}[T]$ be the minimal polynomial of $\{\mu_1, \ldots, \mu_s\}$, i.e. the monic polynomial of lowest degree that vanishes at the points $\mu_1, \ldots, \mu_s$ (or at $\mu_1, \ldots, \mu_{s-1}$ if one of them, say μ_s, equals ∞). Equivalently, $m_1(T)$ is the product of the minimal polynomials of all algebraic numbers μ_i, avoiding repetition of factors. We denote by $\beta_1, \ldots, \beta_d$ the roots of $m_1'(T)$ and by $p(T)$ their minimal polynomial. By definition $\deg(p(T)) \leq \deg(m_1'(T))$.

We now make the general observation that if $g \circ f$ denotes the composition of two morphisms f and g then the following obvious identity between their branching values holds

$$\mathrm{Branch}(g \circ f) = \mathrm{Branch}(g) \cup g(\mathrm{Branch}(f))$$

This implies that the set of branching values of the composed morphism

$$S_F \xrightarrow{\ \mathbf{x}\ } \mathbb{P}^1 \xrightarrow{\ m_1\ } \mathbb{P}^1$$
$$(x, y) \longmapsto x \longmapsto m_1(x)$$

is precisely

$$B_1 = m_1\left(\{\text{roots of } m_1'\}\right) \cup \{0, \infty\}.$$

Again, if $B_1 \subset \mathbb{Q} \cup \{\infty\}$ we are done. If not, we denote by

$m_2(T) \in \mathbb{Q}[T]$ the minimal polynomial of the branch value set of m_1, that is $m_1(\{\text{roots of } m_1'\}) = \{m_1(\beta_1), \ldots, m_1(\beta_d)\}$. Clearly, $[\mathbb{Q}(m_1(\beta_i)) : \mathbb{Q}] \leq [\mathbb{Q}(\beta_i) : \mathbb{Q}]$, which means that the degree of the minimal polynomial of $m_1(\beta_i)$ is lower or equal to the degree of the minimal polynomial of β_i. Moreover, elementary Galois theory shows that two algebraic numbers β_i, β_j have the same minimal polynomial if and only if $\sigma(\beta_i) = \beta_j$ for some field embedding $\sigma : \mathbb{Q}(\beta_i) \to \overline{\mathbb{Q}}$. But in that case $\sigma(m(\beta_i)) = m(\beta_j)$ and so $m(\beta_i)$ and $m(\beta_j)$ also have the same minimal polynomial. Therefore,

$$\deg\left(m_2(T)\right) \leq \deg\left(p(T)\right) \leq \deg\left(m_1'(T)\right) < \deg\left(m_1(T)\right) \quad (3.2)$$

Next we ask ourselves for the set B_2 of branching values of the new morphism $m_2 \circ m_1 \circ \mathbf{x}$. The answer of course is

$$B_2 = m_2\left(\{\text{roots of } m_2'\}\right) \cup m_2(B_1).$$

By construction, $m_2(B_1) \subset \mathbb{Q} \cup \{\infty\}$; in fact $m_2(B_1)$ consists of the points $0, \infty$ and $m_2(0)$. Now if the whole set B_2 is contained in $\mathbb{Q} \cup \{\infty\}$ we have finished. If not, we continue the process denoting by $m_3(T) \in \mathbb{Q}[T]$ the minimal polynomial of $m_2(\{\text{roots of } m_2'\})$ and looking at the set B_3 of branching values of the morphism $m_3 \circ m_2 \circ m_1 \circ \mathbf{x}$, which is given by

$$B_3 = m_3\left(\{\text{roots of } m_3'\}\right) \cup m_3 \circ m_2(B_2)$$

etc.

This process ends when $B_k \subset \mathbb{Q} \cup \{\infty\}$, something that must happen after finitely many steps since by (3.2) we have the inequality $\deg\left(m_i(T)\right) \leq \deg\left(m_{i+1}(T)\right) - 1$.

Example 3.4 Let S be the Riemann surface

$$\left\{y^2 = x(x-1)(x-\sqrt{2})\right\} \cup \{\infty\}.$$

We can construct a Belyi function f on S by means of the following composition of maps:

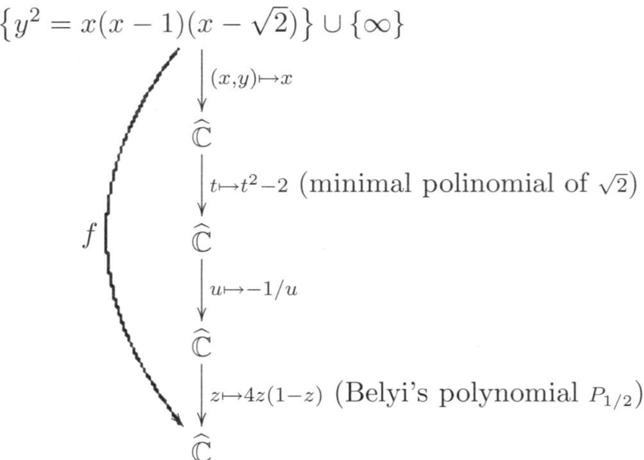

that is

$$f(x,y) = \frac{-4\left(x^2 - 1\right)}{\left(x^2 - 2\right)^2}$$

We can track back the ramification indices over the three ramification values $0, 1$, and ∞ from the bottom to the top in the previous diagram. We get

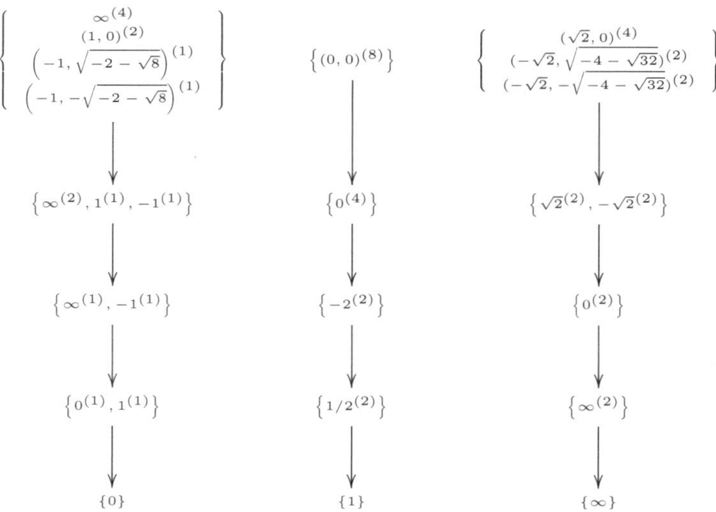

where the superindices denote branching orders.

3.1.1 Belyi's second proof of part (a) $\Rightarrow$ (b)

Some years after the proof described above, Belyi [Bel02] provided a shorter proof, which goes as follows.

First one argues as in the first proof in order to show the existence of a function $f : S \to \mathbb{P}^1$ ramified over a set of rational values $\lambda_1 < \lambda_2 < \cdots \lambda_n < \infty$. Moreover, applying a Möbius transformation of the form $z \mapsto Nz$ to clear out denominators, we can assume that these rational values are, in fact, integers. At this point one can avoid the iterated use of Belyi polynomials in the following way.

For $i = 1, \ldots, n$, set

$$y_i = \frac{1}{\prod_{j \neq i}(\lambda_i - \lambda_j)} \quad \text{and} \quad a_i = V y_i \in \mathbb{Z}$$

where $V := V(\lambda_1, \lambda_2, \cdots \lambda_n)$ stands for the Vandermonde determinant

$$V = \begin{vmatrix} 1 & \lambda_1 & \lambda_1^2 & \cdots & \lambda_1^{n-2} & \lambda_1^{n-1} \\ 1 & \lambda_2 & \lambda_2^2 & \cdots & \lambda_2^{n-2} & \lambda_2^{n-1} \\ 1 & \lambda_3 & \lambda_3^2 & \cdots & \lambda_3^{n-2} & \lambda_3^{n-1} \\ \cdots & \cdots & \cdots & \cdots & \cdots & \cdots \\ 1 & \lambda_{n-1} & \lambda_{n-1}^2 & \cdots & \lambda_{n-1}^{n-2} & \lambda_{n-1}^{n-1} \\ 1 & \lambda_n & \lambda_n^2 & \cdots & \lambda_n^{n-2} & \lambda_n^{n-1} \end{vmatrix} = \prod_{j>i}(\lambda_j - \lambda_i)$$

and consider the rational function

$$G(x) = \prod_{i=1}^{n}(x - \lambda_i)^{a_i} \in \mathbb{Q}(x)$$

Clearly $G(\{\lambda_1, \lambda_2, \ldots, \lambda_n, \infty\}) = \{0, \infty, G(\infty)\}$. To determine the ramification points of G one makes the computation

$$(logG(x))' = \frac{G'(x)}{G(x)} = \sum_{i=1}^{n} \frac{a_i}{x - \lambda_i} = \sum_{i=1}^{n} \frac{V y_i}{x - \lambda_i} = \frac{V}{\prod_{i=1}^{n}(x - \lambda_i)}$$

This computation shows that the set of ramification points of G is a subset of $\{\lambda_1, \lambda_2, \ldots, \lambda_n, \infty\}$ and so the branching values of the function $G \circ f : S \to \mathbb{P}^1$ are $0, \infty$ and $G(\infty)$. What remains to be seen is that $G(\infty) = 1$. This is the same as showing that $\sum a_i = \sum V y_i = 0$.

Clearly each term $V y_i$ agrees up to sign with the Vandermonde determinant $V_i := V(\lambda_1, \ldots, \lambda_{i-1}, \lambda_{i+1}, \ldots, \lambda_n)$ and since V and

V_i are positive numbers this sign coincides with the sign of y_i. Therefore, we can write

$$\sum a_i = V_n - V_{n-1} + \cdots + (-1)^n V_1$$

but the right-hand side term is the expression of the determinant

$$\begin{vmatrix} 1 & \lambda_1 & \lambda_1^2 & \cdots & \lambda_1^{n-2} & 1 \\ 1 & \lambda_2 & \lambda_2^2 & \cdots & \lambda_2^{n-2} & 1 \\ 1 & \lambda_3 & \lambda_3^2 & \cdots & \lambda_3^{n-2} & 1 \\ \cdots & \cdots & \cdots & \cdots & \cdots & \cdots \\ 1 & \lambda_{n-1} & \lambda_{n-1}^2 & \cdots & \lambda_{n-1}^{n-2} & 1 \\ 1 & \lambda_n & \lambda_n^2 & \cdots & \lambda_n^{n-2} & 1 \end{vmatrix}$$

when developed by the last column. Hence $\sum a_i = 0$ as required.

Our next goal is the proof of the (b)$\Rightarrow$(a) part of Belyi's Theorem. Since we are interested in understanding when a Riemann surface is defined over the field $K = \overline{\mathbb{Q}}$, we are naturally led to the study of concepts pertaining to Galois theory. But first of all we need a suitable algebraic description of morphisms between compact Riemann surfaces.

3.2 Algebraic characterization of morphisms

As we already saw in Remark 1.92, if two non-constant meromorphic functions f_1, f_2 of a Riemann surface S satisfy an identity

$$G(f_1, f_2) \equiv 0$$

for some irreducible polyomial $G(X, Y)$ then they define a morphism $f = (f_1, f_2) : S \longrightarrow S_G$. Conversely, any non-constant morphism $f : S \longrightarrow S_G$ is obviously determined by the pair of meromorphic functions (f_1, f_2) obtained by post-composition of f with the coordinate functions on S_G.

Suppose now that $S = S_F$, then $\mathcal{M}(S) = \mathbb{C}(\mathbf{x}, \mathbf{y})$ and so we can write

$$f_1 = R_1(\mathbf{x}, \mathbf{y}) = \frac{P_1(\mathbf{x}, \mathbf{y})}{Q_1(\mathbf{x}, \mathbf{y})}, \quad f_2 = R_2(\mathbf{x}, \mathbf{y}) = \frac{P_2(\mathbf{x}, \mathbf{y})}{Q_2(\mathbf{x}, \mathbf{y})}$$

for some polynomials $P_i, Q_i \in \mathbb{C}[X, Y]$ with $Q_i \notin (F)$, for otherwise the denominator would vanish identically.

This leads to the following algebraic description of morphisms of Riemann surfaces:

Proposition 3.5 *Defining a morphism $f : S_F \longrightarrow S_G$ is equivalent to specifying a pair of rational functions $f = (R_1, R_2)$, where*

$$R_i(X, Y) = \frac{P_i(X, Y)}{Q_i(X, Y)}$$

with $P_i, Q_i \in \mathbb{C}[X, Y]$ and $Q_i \notin (F)$, such that

$$Q_1^n Q_2^m G(R_1, R_2) = HF \tag{3.3}$$

where $n = \deg_X G$, $m = \deg_Y G$ and $H \in \mathbb{C}[X, Y]$.

Proof Clearing out denominators in the identity

$$G(R_1(x, y), R_2(x, y)) = 0$$

yields the relation

$$Q_1^n(x, y) Q_2^m(x, y) G(R_1(x, y), R_2(x, y)) = 0$$

which by the weak form of the Nullstelensatz (Lemma 1.84) is equivalent to the polynomial identity (3.3). $\qquad\square$

Next we address the question of when a morphism $f = (R_1, R_2)$ is an isomorphism. Of course this will occur if and only if there is an inverse morphism

$$h : S_G \to S_F.$$

This means first of all that there are two rational functions $W_i = U_i/V_i$ with $V_i \notin (G)$ satisfying a polynomial identity of the form

$$V_1^s V_2^t F(W_1, W_2) = T \cdot G \tag{3.4}$$

which, according to Proposition 3.5, is equivalent to the existence of a morphism $h = (W_1, W_2) : S_G \longrightarrow S_F$. Moreover, h must satisfy the following identity

$$
\begin{aligned}
h \circ f(x, y) &= h(R_1(x, y), R_2(x, y)) \\[2mm]
&= \left(\frac{U_1(R_1(x, y), R_2(x, y))}{V_1(R_1(x, y), R_2(x, y))}, \frac{U_2(R_1(x, y), R_2(x, y))}{V_2(R_1(x, y), R_2(x, y))} \right) \\[2mm]
&= (x, y)
\end{aligned}
$$

Arguing as above we see that this condition is satisfied if and only if the two following polynomial identities hold simultaneously

$$Q_1^d Q_2^k \left(U_1(R_1, R_2) - X V_1(R_1, R_2) \right) = H_1 F \qquad (3.5)$$

for some polynomial $H_1 \in \mathbb{C}[X, Y]$ and

$$Q_1^d Q_2^k \left(U_2(R_1, R_2) - Y V_2(R_1, R_2) \right) = H_2 F \qquad (3.6)$$

for some polynomial $H_2 \in \mathbb{C}[X, Y]$, where $d = \deg_X(U_i - X V_i)$ and $k = \deg_Y(U_i - Y V_i)$. Note that the identity $\mathrm{Id} = h \circ f$ alone already implies that f is injective, hence an isomorphism.

Example 3.6 In the case of the opening example of this chapter, that is the isomorphism

$$\{ y^2 = x^3 - \pi^3 \} \longrightarrow \{ y^2 = x^3 - 1 \}$$

$$(x, y) \longmapsto \left(\frac{x}{\pi}, \frac{y}{\pi \sqrt{\pi}} \right)$$

the polynomials ocurring in the identities (3.3) to (3.6) are as follows:

(3.3) $F = Y^2 - (X^3 - \pi^3)$, $G = Y^2 - (X^3 - 1)$; $P_1 = X$, $Q_1 = \pi$;
       $P_2 = Y$, $Q_2 = \pi \sqrt{\pi}$; $H = \pi^3$.
(3.4) $U_1 = \pi X$, $V_1 = 1$; $U_2 = \sqrt{\pi^3} Y$, $V_2 = 1$; $T = \pi^3$.
(3.5) $H_1 = 0$.
(3.6) $H_2 = 0$.

Example 3.7 For the isomorphism in Example 1.83

$$\left\{ y^2 = \prod_{i=1}^{2g+2} (x - a_i) \right\} \overset{F}{\longrightarrow} \left\{ y^2 = \prod_{j=1}^{2g+1} (x - b_j) \right\}$$

given by

$$F(x, y) = \left(\frac{1}{x - a_{2g+2}}, \frac{y}{(x - a_{2g+2})^{g+1} \sqrt{-\prod_{i=1}^{2g+1}(a_i - a_{2g+2})}} \right)$$

the polynomials involved in relation (3.3) are as follows:

$$P_1 = 1, \; P_2 = Y$$

$$Q_1 = (X - a_{2g+2})$$
$$Q_2 = (X - a_{2g+2})^{g+1}\sqrt{-\prod_{i=1}^{2g+1}(a_i - a_{2g+2})}$$
$$H = (X - a_{2g+2})^{2g+1}$$

According to the discussion made in this section, we have the following polynomial characterization of the concept of isomorphy between Riemann surfaces:

Theorem 3.8 *Two Riemann surfaces S_F and S_G are isomorphic if and only if there are polynomials P_i, Q_i, U_i, V_i, H_i ($i = 1, 2$) and H, T such that the identities (3.3) to (3.6) are satisfied. Moreover, in such case an isomorphism is provided by $\Phi = (R_1, R_2)$ where $R_i = P_i/Q_i$.*

Example 3.9 (Meromorphic functions) Let us now look at morphisms $f : S_F \longrightarrow S_G$ in the case in which $G(X, Y) = Y$ and so S_G is isomorphic to $\widehat{\mathbb{C}}$ via the obvious identification $(x, 0) \leftrightarrow x$ (see Example 1.37).

In this situation Proposition 3.5 tells us that f can be written in the form

$$f = \left(\frac{P_1}{Q_1}, \frac{P_2}{Q_2}\right)$$

where the polynomials P_i, Q_i satisfy the identity

$$Q_1^0 Q_2^1 \frac{P_2}{Q_2} = HF$$

hence

$$f = \left(\frac{P_1}{Q_1}, \frac{HF}{Q_2}\right) \equiv \left(\frac{P_1}{Q_1}, 0\right)$$

This way we rediscover the fact that every meromorphic function on S_F can be written as a rational function on the coordinate functions $\mathbf{x}$ and $\mathbf{y}$.

Remark 3.10 For later use we record here the fact that for three morphisms

$$f : S_F \longrightarrow S_G, \quad h : S_G \longrightarrow S_D, \quad \text{and} \quad u : S_F \longrightarrow S_D$$

with

$$f \;=\; (R_1, R_2) = (P_1/Q_1, P_2/Q_2)$$

$$h \;=\; (W_1, W_2) = (U_1/V_1, U_2/V_2)$$

$$u \;=\; (Z_1, Z_2) = (L_1/M_1, L_2/M_2)$$

the statement that the diagram

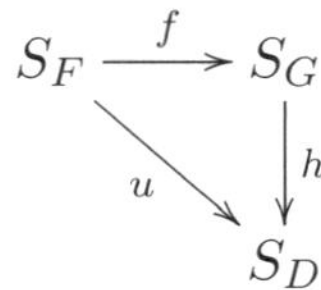

commutes is equivalent to the existence of two polynomial identities of the form

$$Q_1^d Q_2^k \left(U_1(R_1, R_2)M_1 - V_1(R_1, R_2)L_1 \right) \;=\; H_1^1 F$$

$$Q_1^{d'} Q_2^{k'} \left(U_2(R_1, R_2)M_2 - V_2(R_1, R_2)L_2 \right) \;=\; H_2^1 F$$

Note that when $u(X, Y) = (X, Y)$ these relations become (3.5) and (3.6).

3.3 Galois action

Let us denote by $\mathrm{Gal}(\mathbb{C}) = \mathrm{Gal}(\mathbb{C}/\mathbb{Q})$ the group of all field automorphisms of $\mathbb{C}$.

We point out that $\mathrm{Gal}(\mathbb{C})$ is a huge group. To realize this we first observe that if K is a subfield of $\mathbb{C}$ and $\alpha \in \mathbb{C}$, then any monomorphism $\sigma : K \hookrightarrow \mathbb{C}$ admits an extension monomorphism $\overline{\sigma} : K(\alpha) \hookrightarrow \mathbb{C}$. Let us recall how $\overline{\sigma}$ can be constructed:

(1) If α is algebraic over K, then elementary Galois theory shows that for each root β of the minimal polynomial of α over K there is an extension $\overline{\sigma} : K(\alpha) \hookrightarrow \mathbb{C}$ characterized by sending α to $\sigma(\beta)$.

(2) If α is transcendental over K, then by definition there is an isomorphism $K(X) \simeq K(\alpha)$ determined by sending X to α. Therefore for each $\beta \in \mathbb{C}$ transcendental over K there is an isomorphism $\overline{\sigma} : K(\alpha) \to K(\beta) \subseteq \mathbb{C}$ characterized by the property $\overline{\sigma}(\alpha) = \sigma(\beta)$.

Now the extension property (1) (resp. (1) + (2) together) implies that any monomorphism $\sigma : K \hookrightarrow \mathbb{C}$ extends to an automorphism of the algebraic closure $\overline{K}$ of K (resp. to an automorphism of $\mathbb{C}$). This is because if $K \subset L$ is an algebraic field extension (resp. a field extension) which is maximal among the extensions enjoying the property that $\sigma : K \hookrightarrow \mathbb{C}$ extends to L, then necessarily $L = \overline{K}$ (resp. $L = \mathbb{C}$), for if there were an element $\alpha \in \overline{L} \setminus K$ (resp. $\mathbb{C} \setminus K$) then by (1) (resp. (1)+(2)) $\sigma : L \longrightarrow \mathbb{C}$ could still be extended to $\overline{\sigma} : L(\alpha) \hookrightarrow \mathbb{C}$ contradicting the maximality of L. We remind the reader that the rigorous proof of this fact requires the use of Zorn's Lemma to guarantee that such a maximal field extension $K \subset L$ exists (see [Lan84], Chapter VII §2).

Definition 3.11 For given $\sigma \in \mathrm{Gal}(\mathbb{C})$ and $a \in \mathbb{C}$, we shall write a^σ instead of $\sigma(a)$. Accordingly we shall employ the following notation:

(i) If $P = \sum a_{ij} X^i Y^j \in \mathbb{C}[X,Y]$ is a polynomial, we will write $P^\sigma = \sum a_{ij}^\sigma X^i Y^j$. Also if $R(X,Y) = P(X,Y)/Q(X,Y)$ is a rational function we will put $R^\sigma = P^\sigma/Q^\sigma$. This way σ induces automorphisms of the polynomial ring $\mathbb{C}[X,Y]$ and the function field $\mathbb{C}(X,Y)$ respectively.

(ii) If $S \simeq S_F$ we set $S^\sigma \simeq S_{F^\sigma}$.

(iii) If $\Psi : S_F \longrightarrow S_G$ is a morphism given by $\Psi = (R_1, R_2)$ we define $\Psi^\sigma : S_{F^\sigma} \longrightarrow S_{G^\sigma}$ to be given by $\Psi^\sigma = (R_1^\sigma, R_2^\sigma)$.

In particular, for a meromorphic function

$$f = R(\mathbf{x}, \mathbf{y}) = P(\mathbf{x}, \mathbf{y})/Q(\mathbf{x}, \mathbf{y}) \in \mathcal{M}(S_F)$$

we put

$$f^\sigma = R^\sigma(\mathbf{x}, \mathbf{y}) = P^\sigma(\mathbf{x}, \mathbf{y})/Q^\sigma(\mathbf{x}, \mathbf{y}) \in \mathcal{M}(S_{F^\sigma}).$$

(iv) For an equivalence class $(S, f) \equiv (S_F, R(\mathbf{x}, \mathbf{y}))$ of ramified covers of the projective line we set $(S, f)^\sigma = (S_{F^\sigma}, R^\sigma(\mathbf{x}, \mathbf{y}))$.

Before going any further we check that this Galois action is well defined.

Lemma 3.12 (a) *Let* $Q(\mathbf{x}, \mathbf{y}) \in \mathcal{M}(S_F)$. *Then* $Q \not\equiv 0$ *implies* $Q^\sigma(\mathbf{x}, \mathbf{y}) \not\equiv 0$.

(b) *If*

$$\frac{P_1(\mathbf{x},\mathbf{y})}{Q_1(\mathbf{x},\mathbf{y})} = \frac{P_2(\mathbf{x},\mathbf{y})}{Q_2(\mathbf{x},\mathbf{y})} \in \mathcal{M}(S_F)$$

then

$$\frac{P_1^\sigma(\mathbf{x},\mathbf{y})}{Q_1^\sigma(\mathbf{x},\mathbf{y})} = \frac{P_2^\sigma(\mathbf{x},\mathbf{y})}{Q_2^\sigma(\mathbf{x},\mathbf{y})} \in \mathcal{M}(S_F^\sigma)$$

(c) *If* $\Psi : S_F \longrightarrow S_G$ *is a morphism (resp. an isomorphism) that transforms* $R_2(\mathbf{x},\mathbf{y}) \in \mathcal{M}(S_G)$ *into* $R_1(\mathbf{x},\mathbf{y}) \in \mathcal{M}(S_F)$ *then* $\Psi^\sigma : S_{F^\sigma} \longrightarrow S_{G^\sigma}$ *is a morphism (resp. an isomorphism) that transforms* $R_2^\sigma(\mathbf{x},\mathbf{y}) \in \mathcal{M}(S_{G^\sigma})$ *into* $R_1^\sigma(\mathbf{x},\mathbf{y}) \in \mathcal{M}(S_{F^\sigma})$

Proof By the Nullstellensatz (see Lemma 1.84) $Q(\mathbf{x},\mathbf{y}) \equiv 0$ if and only if $Q(X,Y) = H(X,Y)F(X,Y)$ for some $H \in \mathbb{C}[X,Y]$. But this is so if and only if $Q^\sigma(X,Y) = H^\sigma(X,Y)F^\sigma(X,Y)$, which is again equivalent to the condition $Q^\sigma(\mathbf{x},\mathbf{y}) = 0 \in \mathcal{M}(S_{F^\sigma})$.

Similarly, the function $P_1(x,y)Q_2(x,y) - Q_1(x,y)P_2(x,y)$ vanishes identically on S_F if and only if there is a polynomial relation of the form

$$P_1(X,Y)Q_2(X,Y) - Q_1(X,Y)P_2(X,Y) = H(X,Y)F(X,Y)$$

As above, applying σ to this identity yields the proof of (b).

For the proof of part (c) one only needs to apply σ to the polynomial identities (3.3) to (3.6) in Section 3.2, e.g. (3.3) becomes

$$(Q_1^\sigma)^n (Q_2^\sigma)^m G^\sigma(R_1^\sigma, R_2^\sigma) = H^\sigma F^\sigma$$

which means that $\Psi^\sigma = (R_1^\sigma, R_2^\sigma)$ is a morphism from S_F^σ to S_G^σ. Finally, the proof of the statement about the action of Ψ and Ψ^σ in the corresponding function fields follows from a similar action of σ on the identity $R_2 \circ \Psi = R_1$, something allowed by Remark 3.10. $\qquad\square$

Example 3.13 When $\sigma \in \mathrm{Gal}(\mathbb{C})$ is the complex conjugation $\sigma(z) = \bar{z}$ it is costumary to write $F^\sigma = \bar{F}$ and $S^\sigma = \bar{S}$. Fortunately this notation will not lead to ambiguity with the notation used in Example 1.22 to denote the complex conjugate Riemann surface. This is because, as was observed there, $S_{\bar{F}}$ is isomorphic to the complex conjugate of S_F (by means of the isomorphism $(x,y) \longmapsto (\bar{x},\bar{y})$).

Recall now that the compact Riemann surface S_F was constructed out of a non-compact Riemann surface S_F^X by adding finitely many points in a rather cumbersome way (see Theorem 1.86). If $P = (a,b)$ is a point in $S_F^X \subset \mathbb{C}^2$ it is clear how to define its Galois transform: of course, one writes $P^\sigma = (a^\sigma, b^\sigma)$. It is a trivial matter to check that this rule defines a bijection between S_F^X and $S_{F^\sigma}^X$ for one obviously has

$$F(a,b) = 0 \quad \Rightarrow \quad F^\sigma(a^\sigma, b^\sigma) = 0^\sigma = 0$$
$$F_y(a,b) = 0 \quad \Rightarrow \quad F_y^\sigma(a^\sigma, b^\sigma) = 0$$

and so on.

But in order to extend the action of $\mathrm{Gal}(\mathbb{C})$ to all points of S_F, we need a purely algebraic interpretation of the remaining points. This leads us to the algebraic concept of valuation, which we introduce next.

3.4 Points and valuations

From Section 1.3.1 we recall that the term *function field* (in one variable) refers to a field $\mathcal{M}$ which is $\mathbb{C}$-isomorphic to a finite extension of the field $\mathbb{C}(X)$ of rational functions. This is the same as saying that $\mathcal{M} \simeq \mathcal{M}(S)$ for some compact Riemann surface S for (see Corollary 1.93) if $\mathcal{M} = \mathbb{C}(X)(y)$ is an algebraic extension of $\mathbb{C}(X)$ and $F(X,Y) \in \mathbb{C}[X,Y]$ is the minimal polynomial of y over the field $\mathbb{C}(X)$ in the variable Y, then $\mathcal{M} \simeq \mathcal{M}(S_F)$.

Definition 3.14 Let $\mathcal{M}$ be a function field, and $\mathcal{M}^* = \mathcal{M} \smallsetminus \{0\}$ its multiplicative group. A *discrete valuation* (from now on simply a *valuation*) of $\mathcal{M}$ is a map

$$v : \mathcal{M}^* \longrightarrow \mathbb{Z}$$

satisfying the following properties:

 (i) $v(\varphi\psi) = v(\varphi) + v(\psi)$, i.e. v is a group homomorphism.
 (ii) $v(\varphi \pm \psi) \geq \min(v(\varphi), v(\psi))$, the equality holding whenever $v(\varphi) \neq v(\psi)$.
 (iii) $v(\varphi) = 0$ if $\varphi \in \mathbb{C}^*$.
 (iv) v is non-trivial, i.e. there is $\varphi \in \mathcal{M}$ such that $v(\varphi) \neq 0$.

Usually one extends v to the whole $\mathcal{M}$ by setting $v(0) = \infty$.

The following properties hold:

- $A_v = \{\varphi \in \mathcal{M} : v(\varphi) \geq 0\}$ is a subring of $\mathcal{M}$.
- $\varphi \in A_v$ is a unit $\Leftrightarrow v(\varphi) = 0$.
- The set of all non-units $M_v = \{\varphi \in \mathcal{M} : v(\varphi) > 0\}$ forms an ideal of A_v.
- A_v is therefore a *local ring* whose unique maximal ideal is M_v.
- Let $v(\mathcal{M}^*) = (m_v) = m_v \mathbb{Z}$ with $m_v > 0$. Then $M_v = (\varphi) = \varphi A_v$ if and only if $v(\varphi) = m_v$.

These facts are readily checked. For instance if $\psi\varphi = 1$ then $v(\psi) + v(\varphi) = 0$ and therefore $v(\psi) = v(\varphi) = 0$ if both $\psi, \varphi \in A_v$. Conversely, this also shows that if $\varphi \in A_v$ and $v(\varphi) = 0$ then necessarily $v(\psi) = 0$, hence $\psi \in A_v$ and it is the inverse of φ, etc. The reason why M_v is a maximal ideal is that it could only grow by adding on unit elements, thus becoming the whole ring A_v. This also explains the fact that it is the only maximal ideal. Finally, $M_v = (\varphi)$ because if $\psi \in M_v$ then $\psi = \dfrac{\psi}{\varphi}\varphi$ with $\dfrac{\psi}{\varphi} \in A_v$.

One says that A_v is a *local ring with maximal ideal* M_v *and uniformizing parameter* φ.

It is quite often convenient to assume that v is surjective, i.e. that $m_v = 1$. This can be achieved simply by performing the *normalization* $v^*(\varphi) = \dfrac{v(\varphi)}{m_v}$. Obviously this normalization does not alter the local ring, nor its maximal ideal. We will say that two valuations are equivalent if their normalizations coincide.

The following result explains why we are interested in valuations:

Proposition 3.15 *Every point P of a compact Riemann surface S defines a valuation on the field $\mathcal{M}(S)$ by means of the formula*
$$v_P(\varphi) = \operatorname{ord}_P(\varphi).$$

Proof This is an obvious consequence of the definition of $\operatorname{ord}_P(\varphi)$. We shall only point out here that the true reason why these valuations are never trivial is the existence of meromorphic functions with a given pole or zero. $\qquad\square$

Example 3.16 Let S be the hyperelliptic curve
$$S = \{y^2 = (x - a_1) \cdots (x - a_{2g+1})\}$$

$P = (a, b) \in S$ and $v = v_P$. Then clearly $\mathbf{x}, \mathbf{y} \in A_v$ and $\mathbf{x} - a$, $\mathbf{y} - b \in M_v$.

Moreover, if $a \neq a_i$ then $v(\mathbf{x} - a) = 1$ and so $M_v = (\mathbf{x} - a)A_v$ and $\mathbf{x} - a$ is a uniformizing parameter. However, if $P = (a_i, 0)$ we have $v(\mathbf{x} - a_i) = 2$ (see Example 1.32). The computation of $v(\mathbf{y})$ can also be done indirectly as follows

$$
\begin{aligned}
2v(\mathbf{y}) &= v(\mathbf{y}^2) = v\left(\prod(\mathbf{x} - a_j)\right) \\
&= v(\mathbf{x} - a_i) + v\left(\prod_{j \neq i}(\mathbf{x} - a_j)\right) \\
&= v(\mathbf{x} - a_i) = 2
\end{aligned}
$$

At the remaining point $P = \infty$ one has $\mathbf{x}, \mathbf{y} \notin A_v$, in fact $1/\mathbf{x}, 1/\mathbf{y} \in A_v$. Moreover, since $v(\mathbf{x}) = -2$ and $v(\mathbf{y}) = -(2g + 1)$ (see again Example 1.32), it follows that $\phi = \mathbf{x}^g/\mathbf{y}$ is a uniformizing parameter in this case.

Our next main goal is to show that the valuations v_P just defined account for all normalized valuations of the field $\mathcal{M}(S)$ (this will be the content of Theorem 3.23). In other words, the assignment $P \to v_P$ produces a bijection between the set of points of S and the set of equivalence classes of valuations on its function field.

Once we do that we will be in position to extend the action of $\mathrm{Gal}(\mathbb{C})$ to all points of $P \in S$ by means of the rule

$$
v_{P^\sigma}(\varphi^\sigma) = v_P(\varphi)
$$

Before we address the general case we shall check the result when $S = \mathbb{P}^1$.

Proposition 3.17 *Every valuation of the field $\mathcal{M}(\mathbb{P}^1) = \mathbb{C}(x)$ is, up to equivalence, of the form v_z for some point $P \in \mathbb{P}^1 = \widehat{\mathbb{C}}$.*

Proof Let v be a valuation of $\mathbb{P}^1$. Any element $f \in \mathcal{M}(\mathbb{P}^1)$ is of the form

$$
f = c\frac{\prod(x - a_i)^{n_i}}{\prod(x - b_j)^{m_j}}
$$

hence $v(f) = \sum n_i v(x - a_i) - \sum m_j v(x - b_j)$. Therefore, v is completely determined by its value on the degree one functions $(x - a)$. One of the following two possibilities occurs:

(i) There is $a \in \mathbb{C}$ such that $v(x-a) = -k$, where k is a positive integer. We claim that in this case $v = kv_\infty$. This can be seen as follows. If $b \neq a$ we can write

$$
\begin{aligned}
-k &= v(x-a) \\
&= v(x - b + (b - a)) \geq \min(v(x-b), v(b-a)) \\
&= \min(v(x-b), 0)
\end{aligned}
$$

with equality if $v(x-b) \neq 0$. But this must be indeed the case since otherwise we would get $-k \geq 0$. That is, we have

$$
v(x-b) = -k = kv_\infty(x-b)
$$

for every $b \in \mathbb{C}$.

(ii) For every $b \in \mathbb{C}$, $v(x-b) \geq 0$. In this case, since v is not allowed to be trivial there must be some $a \in \mathbb{C}$ such that $v(x-a) = k > 0$ and then what we have is $v = kv_a$. To see this it is enough to observe that if $b \neq a$ then $v(x-b) = 0$ for otherwise we would obviously get

$$
k = v(x-a) = v(x - b + (b - a)) = \min(v(x-b), 0) = 0
$$

which is a contradiction. $\qquad\square$

The next step towards proving the equivalence between points and valuations is to show that just as meromorphic functions separate points (Corollary 2.12 and Proposition 2.16), so they also separate valuations (Proposition 3.19).

Lemma 3.18 *Let v_1, v_2 be (normalized) valuations of $\mathcal{M}$. Then $v_1 = v_2$ if and only if $A_{v_1} = A_{v_2}$.*

Proof If $A_{v_1} = A_{v_2}$ then $M_{v_1} = M_{v_2} = (\varphi)$. In other words, $v_1(\varphi) = v_2(\varphi) = 1$. Now for an arbitrary element $\psi \in \mathcal{M}^*$ we have

$$v_1(\psi) = n$$
$$\Updownarrow$$
$$v_1(\varphi^n/\psi) = 0$$
$$\Updownarrow$$
$$\varphi^n/\psi \text{ is a unit of } A_{v_1}$$
$$\Updownarrow$$
$$\varphi^n/\psi \text{ is a unit of } A_{v_2}$$
$$\Updownarrow$$
$$v_2(\varphi^n/\psi) = 0$$
$$\Updownarrow$$
$$v_2(\psi) = n$$

$\square$

Proposition 3.19 *For any pair of distinct valuations v_1, v_2 of $\mathcal{M}$ there exists an element $\varphi \in \mathcal{M}$ such that $v_1(\varphi) \geq 0$ and $v_2(\varphi) < 0$.*

Proof Neglecting the statement would mean that $A_{v_1} \subset A_{v_2}$. By Lemma 3.18 all we need to prove is that this in turn implies that $A_{v_1} = A_{v_2}$. We shall do it in three steps:

(1) $M_{v_2} \subset M_{v_1}$.
(2) $M_{v_1} \subset M_{v_2}$.
(3) The units of A_{v_2} lie in A_{v_1}.

(1) If $y \in M_{v_2} \backslash M_{v_1}$ then

$$\left\{ \begin{array}{l} v_2(y) > 0 \\ v_1(y) \leq 0 \end{array} \right. \Rightarrow \left\{ \begin{array}{l} v_2(1/y) < 0 \\ v_1(1/y) \geq 0 \end{array} \right. \Rightarrow \left\{ \begin{array}{l} 1/y \notin A_{v_2} \\ 1/y \in A_{v_1} \end{array} \right.$$

This contradiction implies that $M_{v_2} \subset M_{v_1}$.

(2) Since we are assuming that $A_{v_1} \subset A_{v_2}$ it is enough to show that a uniformizing parameter φ_1 of A_{v_1} lies in M_{v_2}, for then we would have $M_{v_1} = \varphi_1 A_{v_1} \subset \varphi_1 A_{v_2} \subset M_{v_2}$.

So, suppose that $\varphi_1 \in A_{v_2} \smallsetminus M_{v_2}$, that is φ_1 is a unit of A_{v_2}. Since by (1) $M_{v_2} \subset M_{v_1}$, any element $y \in M_{v_2} \smallsetminus \{0\}$ could be written in the form $y = u\varphi_1^k$ where u is a unit of A_{v_1}, hence of A_{v_2}. This way we would conclude that the non-zero elements of M_{v_2} are units of A_{v_2}, which is a contradiction.

(3) If u is a unit of A_{v_2} which is not in A_{v_1}, i.e. such that $v_1(u) < 0$, then $v_1(u^{-1}) > 0$ and therefore $u^{-1} \in M_{v_1} = M_{v_2}$. Contradiction.

$\square$

Remark 3.20 Note that when $\mathcal{M}$ is the function field of a Riemann surface S and $v_1 = v_P$, $v_2 = v_Q$ for two distinct points $P, Q \in S$, Proposition 3.19 only means that there are meromorphic functions in S that separate P and Q (Corollary 2.12 and Proposition 2.16).

Corollary 3.21 *Let* $v_1, v_2, \ldots, v_n$ *be* $n \geq 2$ *distinct valuations of a function field* $\mathcal{M}$. *Then:*

(1) *There exists an element* $y \in \mathcal{M}$ *such that* $v_1(y) > 0$ *and* $v_k(y) < 0$, *for* $k = 2, \ldots, n$.

(2) *There exists an element* $y \in \mathcal{M}$ *such that* $v_1(y) = 0$ *and* $v_k(y)$ *is arbitrary large for* $k = 2, \ldots, n$.

Proof (1) We argue by induction. Let us consider first the case $n = 2$. By Proposition 3.19 above there exists y_1 (resp. y_2) such that $v_1(y_1) \geq 0$ and $v_2(y_1) < 0$ (resp. $v_2(y_2) \geq 0$ and $v_1(y_2) < 0$). Therefore $v_1(y_1/y_2) > 0$ and $v_2(y_1/y_2) < 0$ and so the first step in the induction argument is settled.

Let now $x \in \mathcal{M}$ be an element such that $v_1(x) > 0$ whereas $v_2(x) < 0, \ldots, v_{n-1}(x) < 0$. By the previous step applied to v_1 and v_n there exists an element $z \in \mathcal{M}$ such that $v_1(z) > 0$ and $v_n(z) < 0$. If we denote $y = x + z^r$ then for r sufficiently large we get $v_n(y) = v_n(x + z^r) = \min(v_n(x), v_n(z^r)) = r v_n(z) < 0$. Arguing in a similar way one finds that $v_1(x + z^r) = v_1(x) > 0$ whereas for $k = 2, \ldots, n-1$, one has $v_k(x + z^r) = \min(v_k(x), r v_k(z)) < 0$, for suitable r.

(2) Let y be the element obtained in part (1) and set $z = 1 + y^d$. Then $v_1(z^{-1}) = -v_1(z) = v_1(1) = 0$. On the other hand, for $k \geq 2$ one has

$$v_k(z^{-1}) = -v_k(z) = -v_k(1 + y^d) = -v_k(y^d) = d(-v_k(y))$$

a positive integer which grows with d. $\qquad\square$

Let $\mathcal{M}_1 \subset \mathcal{M}_2$ be a finite extension of function fields. Clearly a valuation v on $\mathcal{M}_2$ induces by restriction a valuation $v = v_{|\mathcal{M}_1}$ on the field $\mathcal{M}_1$. Note that the latter will not, in general, be a normalized valuation even if v is. In fact for a normalized valuation v on $\mathcal{M}_2$ one has $v(\mathcal{M}_1^*) = e\mathbb{Z}$ where $e = [v(\mathcal{M}_2^*) : v(\mathcal{M}_1^*)]$, the index of the subgroup $v(\mathcal{M}_1^*)$ in $v(\mathcal{M}_2^*)$.

Theorem 3.22 *Let $\mathcal{M}_1 \subset \mathcal{M}_2$ be a finite extension of function fields and let $v_1, v_2, \ldots, v_n$ be distinct normalized valuations on $\mathcal{M}_2$ that restrict to a same valuation v on $\mathcal{M}_1$, up to equivalence. Let e_k be the index of $v_k(\mathcal{M}_1)$ in $v_k(\mathcal{M}_2) = \mathbb{Z}$. Then*

$$\sum e_k \leq [\mathcal{M}_2 : \mathcal{M}_1]$$

Proof Let $\pi_1, \ldots, \pi_n \in \mathcal{M}_2$ be uniformizing parameters of the valuations $v_1, \ldots, v_n$ respectively. By Corollary 3.21 we know that there exist elements $y_1, y_2, \ldots, y_n \in \mathcal{M}_2$ such that $v_k(y_k) = 0$ and $v_j(y_k) \geq N_{jk}$ for $j \neq k$, where N_{jk} is an arbitrary large positive integer to be determined below.

The proof would be done if we could show that the following $\sum e_k$ elements

$$\begin{aligned}
&y_1\pi_1, y_1\pi_1^2, \ldots, y_1\pi_1^{e_1} \\
&y_2\pi_2, y_2\pi_2^2, \ldots, y_2\pi_2^{e_2} \\
&\qquad\qquad \vdots \\
&y_n\pi_n, y_n\pi_n^2, \ldots, y_n\pi_n^{e_n}
\end{aligned}$$

of $\mathcal{M}_2$ are linearly independent over $\mathcal{M}_1$.

Suppose that we could arrange a linear identity

$$\sum c_{kj}(y_k\pi_k^j) = 0$$

with coefficients $c_{kj} \in \mathcal{M}_1$ not all equal to zero.

Let us look at the coefficient $c_{rs} \in \mathcal{M}_1$ at which the valuation v, hence each of the valuations v_i, reaches a minimal value. Dividing our linear combination by c_{rs} we can assume that one of its coefficients, say c_{11}, equals 1 and that the rest of them satisfy $v(c_{kj}) \geq 0$.

By construction, we have

$$\begin{aligned}
v_1(y_1\pi_1) &= 1 \\
v_1(c_{12}y_1\pi_1^2) &= v(c_{12}) + 2 \\
&\;\;\vdots \\
v_1(c_{1e_1}y_1\pi_1^{e_1}) &= v(c_{1e_1}) + e_1
\end{aligned}$$

As by hypothesis $v(c_{1j}) \in e_1\mathbb{Z}$ we find that these values are pairwise distinct and therefore $v_1\left(\sum c_{1j}y_1\pi_1^j\right) = v_1(y_1\pi_1) = 1$.

Also by construction we see that for $k \geq 2$ one has

$$v_1\left(c_{kj}y_k\pi_k^j\right) = v_1\left(c_{kj}\right) + v_1\left(y_k\right) + jv_1\left(\pi_k\right) \geq N_{1k} + jv_1\left(\pi_k\right) \geq 2$$

for a suitable choice of N_{jk} (hence of the elements y_k).

Finally, applying v_1 to the identity

$$\sum c_{1j}y_1\pi_1^j = -\sum_{k \geq 2} c_{kj}y_k\pi_k^j,$$

we get

$$
\begin{aligned}
1 &= v_1\left(\sum c_{1j}y_1\pi_1^j\right) \\
&= v_1\left(\sum_{k \geq 2} c_{kj}y_k\pi_k^j\right) \\
&\geq \min_{k \geq 2} v_1\left(c_{kj}y_k\pi_k^j\right) \\
&\geq 2
\end{aligned}
$$

which is a contradiction. $\qquad\square$

Let us now consider the particular case in which the algebraic extension $\mathcal{M}_1 \subset \mathcal{M}_2$ comes from a morphism $f : S \to \mathbb{P}^1$.

That is, let us consider the $\mathbb{C}$-algebra embedding given by

$$
\begin{aligned}
f^* : \quad \mathcal{M}_1 = \mathcal{M}\left(\mathbb{P}^1\right) &\hookrightarrow \quad \mathcal{M}(S) = \mathcal{M}_2 \\
R(x) &\to \quad f^*(R(x)) := R \circ f
\end{aligned}
$$

As it was shown in Section 1.3.1 this is a finite field extension whose degree is precisely $\deg(f)$.

Choose $a \in \mathbb{P}^1$ and suppose that its preimages $P_1, \ldots, P_r$ have branching orders $m_1, \ldots, m_r$. Then the value of the valuations v_{P_i} at an element $R(x) \in \mathcal{M}_1$ viewed as an element of $\mathcal{M}_2$ via this embedding is

$$
\begin{aligned}
v_{P_i}(R(x)) &:= v_{P_i}(f^*(R(x))) \\
&= v_{P_i}(R \circ f) \\
&= \mathrm{ord}_{P_i}(f) \cdot \mathrm{ord}_a(R(x)) \\
&= m_i \cdot \mathrm{ord}_a(R(x))
\end{aligned}
$$

We infer that $v_{P_i|\mathcal{M}_1} = m_i v_a$.

We emphasize the two following facts arising from this discussion:

(i) The restriction of any of the valuations v_{P_i} on $\mathcal{M}(S)$ produces a valuation on $\mathcal{M}\left(\mathbb{P}^1\right)$ which is equivalent to v_a.

(ii) The index $e_i = [v_{P_i}(\mathcal{M}_2^*) : v_{P_i}(\mathcal{M}_1^*)]$ coincides with the branching order $m_i = \mathrm{ord}_{P_i}(f)$.

Therefore, Theorem 3.22 applied to this particular field extension yields

$$\sum m_i + \sum \left(\begin{array}{c} \text{indices } e_j \text{ coming from other} \\ \text{possible extensions of } v_a \end{array} \right) \leq \deg(f)$$

On the other hand, by the degree formula (Proposition 1.57), $\sum m_i$ already equals $\deg(f)$. This means that there are no other extensions of v_a apart from the valuations v_{P_i}.

In other words we have almost proved the result announced at the beginning of this section:

Theorem 3.23 *For any compact Riemann surface S, the rule*

$$P \in S \longrightarrow v_P := \mathrm{ord}_P$$

establishes a one-to-one correspondence between points of S and valuations on $\mathcal{M}(S)$.

Proof The injectivity is simply a manifestation of the fact that functions separate points (Theorem 1.90), hence only surjectivity remains to be shown.

Let v be any valuation on $\mathcal{M}(S)$. Fix a non-constant morphism $f : S \to \mathbb{P}^1$. Via this morphism v restricts to a valuation on $\mathcal{M}(\mathbb{P}^1)$ which is necessarily of the form v_a (Proposition 3.17). By what has gone above, $v = v_{P_i}$ for some $P_i \in S$ such that $f(P_i) = a$. The proof is complete. $\qquad\qquad\square$

3.4.1 Galois action on points

As mentioned in the last paragraph of Section 3.3, for any element $\sigma \in \mathrm{Gal}(\mathbb{C})$ the correspondence $P = (a, b) \to P^\sigma = (a^\sigma, b^\sigma)$ provides a bijection between S_F^X and $S_{F^\sigma}^X$. A nice application of Theorem 3.23 is that it allows us to extend the action of $\mathrm{Gal}(\mathbb{C})$ at all points of the Riemann surface $S = S_F$.

Definition 3.24 (1) Given a valuation v on $\mathcal{M}(S)$ we define a valuation v^σ on $\mathcal{M}(S^\sigma)$ by the formula

$$v^\sigma = v \circ \sigma^{-1}$$

that is

$$v^\sigma(\psi^\sigma) = v(\psi), \text{ for every } \psi \in \mathcal{M}(S)^*$$

(2) Accordingly, for a point $P \in S^\sigma$ we define P^σ to be the only point of S^σ such that $v_{P^\sigma} = (v_P)^\sigma$.

Proposition 3.25 (1) *For any $\sigma \in \mathrm{Gal}(\mathbb{C})$ the correspondence $P \to P^\sigma$ defines a bijection between S and S^σ.*

(2) *On points $P \in S_F^X$ the definition above agrees with the obvious one.*

(3) *In particular, $a^\sigma = a$ for every $a \in \mathbb{Q} \cup \{\infty\} \subset \mathbb{P}^1$ and every $\sigma \in \mathrm{Gal}(\mathbb{C})$.*

Proof (1) The inverse is obviously given by $Q \to Q^{\sigma^{-1}}$.

(2) Let us start by considering the field extension $\mathbb{C}(\mathbf{x}) \subset \mathbb{C}(\mathbf{x}, \mathbf{y})$ which is nothing but the embedding of $\mathcal{M}(\mathbb{P}^1)$ in $\mathcal{M}(S_F)$ induced by the x coordinate morphism $\mathbf{x} : S_F \to \mathbb{P}^1$.

Let $P = (a, b) \in S_F^X$. Then for the function $\mathbf{x} - a^\sigma \in \mathcal{M}(S_{F^\sigma})$ we have

$$(v_P)^\sigma(\mathbf{x} - a^\sigma) = (v_P)(\mathbf{x} - a) = 1$$

since P is a simple zero of the function $\mathbf{x} - a$. This shows that the restriction of $(v_P)^\sigma$ to the subfield $\mathbb{C}(\mathbf{x})$ agrees with v_{a^σ}. This means (see Theorem 3.23 and its proof) that the valuation $(v_P)^\sigma$ on $\mathcal{M}(S_{F^\sigma})$ is one of the $n = \deg_Y(F)$ valuations corresponding to the points of the form $(a^\sigma, y) \in S_{F^\sigma}^X$, which are the preimages of a^σ via the function $\mathbf{x} : S_{F^\sigma} \to \mathbb{P}^1$.

Among all of them there is only one taking a positive value on the function $\mathbf{y} - b^\sigma$ (as obviously does the valuation $(v_P)^\sigma$). This is precisely the one corresponding to the point (a^σ, b^σ). The proof is done.

(3) Note that part (2) includes the case of points $a \in \mathbb{P}^1 \setminus \{\infty\}$; simply take $F(X, Y) = Y$ so that $S_F^X \simeq \mathbb{P}^1 \setminus \{\infty\}$ and identify a to $(a, 0)$. Finally, if in part (1) we let P be the point $\infty \in \mathbb{P}^1$, the corresponding identity is

$$(v_\infty)^\sigma(\mathbf{x} - a) = (v_\infty)(\mathbf{x} - a^{\sigma^{-1}}) = -1 = v_\infty(\mathbf{x} - a)$$

for any $a \in \mathbb{C}$. $\qquad\square$

Example 3.26 Let $S = \left\{ y^2 = \prod_{i=1}^{2g+1}(x - a_i) \right\}$ and $\sigma \in \mathrm{Gal}(\mathbb{C})$.

If $P = (a, b) \in S$ then $P^\sigma = (a^\sigma, b^\sigma) \in S^\sigma$, whereas if $P = \infty \in S$ then $\sigma(P) = \infty \in S^\sigma$, for otherwise σ would not be a bijection.

In the case of a hyperelliptic curve $S = \left\{ y^2 = \prod_{i=1}^{2g+2}(x - a_i) \right\}$, the two points at infinity ∞_1, ∞_2 added for the compactification can be either preserved or interchanged by each particular $\sigma \in$ Gal($\mathbb{C}$). The same phenomenon can be seen in the next example.

Example 3.27 In Example 1.10 and Example 1.41 we considered the Fermat curves

$$S = \left\{ (x, y) \in \mathbb{C}^2 : x^d + y^d = 1 \right\}$$

Using the parametrizations around the points at infinity given in Example 1.10, we see that the meromorphic function $\mathbf{y}/\mathbf{x}$ takes the value ξ_d^j (where $\xi_d = e^{2\pi i/d}$) at the point $P_j = \infty_j$ ($j = 1, \ldots, d$). Moreover $\pi_j = \mathbf{y}/\mathbf{x} - \xi_d^j$ is a uniformizing parameter for P_j.

Let now $\sigma \in$ Gal($\mathbb{C}$), then its action on $\mathbb{Q}(\xi_d)$ is determined by $\sigma(\xi_d) = \xi_d^k$ for some exponent k coprime to d. It is rather simple to show that $P_1^\sigma = P_k$, or equivalently that $v_{P_1^\sigma} = v_{P_k}$, for on the one hand we know that P_1^σ must be one of the points at infinity, since the restriction of the action of σ to S_F^X is always bijective, and on the other the computation

$$v_{P_1^\sigma}(\pi_k) = v_{P_1^\sigma}(\pi_1^\sigma) = v_{P_1}(\pi_1) = 1$$

implies that indeed $v_{P_1^\sigma} = v_{P_k}$.

3.5 Elementary invariants of the action of Gal($\mathbb{C}$)

We make the obvious observation that the bijection $S \leftrightarrow S^\sigma$ performed by a Galois element $\sigma \in \mathbb{C}$ is never holomorphic (not even continuous with the exception of the case in which σ is the complex conjugation). Therefore, S and S^σ will in general be non-isomorphic. Likewise there is not an a priori reason why these two Riemann surfaces should be even homeomorphic. Nevertheless, the action $(S, f)^\sigma = (S^\sigma, f^\sigma)$ defined at the beginning of this section (see Definition 3.11) preserves the genus as well as some other interesting properties.

Theorem 3.28 *The action of* Gal($\mathbb{C}$) *on pairs* (S, f) *enjoys the following properties:*

(1) $\deg(f^\sigma) = \deg(f)$.

(2) $(f(P))^\sigma = f^\sigma(P^\sigma)$.

(3) $\mathrm{ord}_{P^\sigma}(f^\sigma) = \mathrm{ord}_P(f)$.

(4) $a \in \widehat{\mathbb{C}}$ *is a branching value of* f *if and only if* a^σ *is a branching value of* f^σ.

(5) *The genus of* S^σ *is the same as the genus of* S, *i.e.* S *and* S^σ *are homeomorphic.*

(6) *The rule*

$$\mathrm{Aut}(S, f) \longrightarrow \mathrm{Aut}(S^\sigma, f^\sigma)$$
$$h \longmapsto h^\sigma$$

is a group isomorphism.

(7) *The monodromy group* $\mathrm{Mon}(f)$ *of the covering* (S, f) *is isomorphic to the monodromy group* $\mathrm{Mon}(f^\sigma)$ *of the covering* (S^σ, f^σ).

Proof (1) Since σ is a field isomorphism we clearly have

$$\deg(f) = [\mathcal{M}(S) : \mathcal{M}(f)] = [\mathcal{M}(S^\sigma) : \mathcal{M}(f^\sigma)] = \deg(f^\sigma)$$

(2), (3) Recall that in the discussion leading to Theorem 3.23 we have seen that $f(P) = a$ means

$$v_P(R \circ f) = m_P(f) \cdot \mathrm{ord}_a(R)$$

for every rational function $R(x)$. Thus, showing that $f^\sigma(P^\sigma) = a^\sigma$ is tantamount to showing that

$$v_{P^\sigma}(R^\sigma \circ f^\sigma) = m_{P^\sigma}(f^\sigma) \cdot \mathrm{ord}_{a^\sigma}(R^\sigma)$$

for every rational function $R^\sigma(x)$.

Now, for any rational function $R(x)$ and any meromorphic function $f = P(x,y)/Q(x,y)$ one clearly has $(R \circ f)^\sigma = (R^\sigma \circ f^\sigma)$, therefore

$$\begin{aligned}
v_{P^\sigma}(R^\sigma \circ f^\sigma) &= v_{P^\sigma}(R \circ f)^\sigma = v_P(R \circ f) \\
&= m_P(f) \cdot \mathrm{ord}_a(R) \\
&= m_{P^\sigma}(f^\sigma) \cdot \mathrm{ord}_{a^\sigma}(R^\sigma)
\end{aligned}$$

since by the definition of P^σ one has

$$\mathrm{ord}_{P^\sigma}(f^\sigma) = v_{P^\sigma}(f^\sigma) = v_P(f) = \mathrm{ord}_P(f)$$

and similarly

$$\mathrm{ord}_{a^\sigma}(R^\sigma) = v_{a^\sigma}(R^\sigma) = v_a(R) = \mathrm{ord}_a(R)$$

(4) This is an obvious consequence of (2) and (3).

(5) Parts (1) and (3) imply that the Riemann–Hurwitz formula for the function $f^\sigma : S^\sigma \longrightarrow \mathbb{P}^1$ (which gives the genus of S^σ) is exactly the same as the Riemann–Hurwitz formula for the function $f : S \longrightarrow \mathbb{P}^1$ (which gives the genus of S).

(6) This is because for a given $h \in \mathrm{Aut}(S, f)$ one clearly has $f^\sigma = (f \circ h)^\sigma = f^\sigma \circ h^\sigma$.

(7) In view of Definition 2.67, applying σ to the diagram

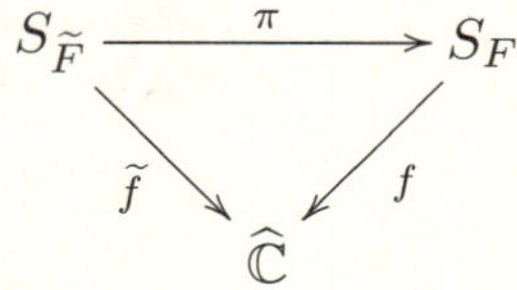

that defines the normalization of $f : S \longrightarrow \widehat{\mathbb{C}}$ gives a new commutative diagram

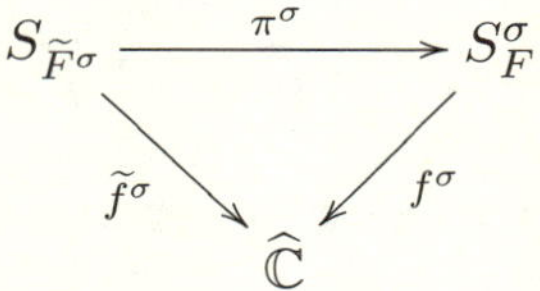

that defines the normalization of f^σ. Now apply (6) together with Corollary 2.73. $\square$

3.6 A criterion for definability over $\overline{\mathbb{Q}}$

We now present what is going to be our main tool to decide whether or not a compact Riemann surface S is defined over $\overline{\mathbb{Q}}$. Clearly if S is defined over a number field, that is if $S = S_F$ for a polynomial F with coefficients in some finite Galois extension K of $\mathbb{Q}$, then the family $\{F^\sigma\}_{\sigma \in \mathrm{Gal}(\mathbb{C})}$ consists of only finitely many (actually, at most $[K : \mathbb{Q}]$) different polynomials. In particular, the family $\{S^\sigma\}_{\sigma \in \mathrm{Gal}(\mathbb{C})}$ contains only finitely many Riemann surfaces, hence finitely many isomorphism classes of Riemann surfaces. It turns out that the latter is also a sufficient condition for S to be defined over $\overline{\mathbb{Q}}$.

Criterion 3.29 ([GD06]) *For a compact Riemann surface S the following conditions are equivalent:*

(i) S *is defined over* $\overline{\mathbb{Q}}$.

(ii) *The family* $\{S^\sigma\}_{\sigma\in\mathrm{Gal}(\mathbb{C})}$ *contains only finitely many isomorphism classes of Riemann surfaces.*

The proof of this criterion will be the object of Section 3.7. For the time being let us just point out that for the Riemann surface $S = S_F$, with $F = Y^2 - (X^3 - \pi^3)$, introduced at the beginning of this chapter, the family of Galois transforms S_{F^σ}, $\sigma \in \mathrm{Gal}(\mathbb{C})$, certainly satisfies the condition (ii) of the above criterion. In fact, Galois conjugation yields only one isomorphism class of Riemann surfaces, the isomorphism between S_F and S_{F^σ} being given by

$$(x, y) \longrightarrow \left(\frac{\sigma(\pi)}{\pi} x, y \sqrt{\left(\frac{\sigma(\pi)}{\pi} \right)^3} \right)$$

We now show how our criterion readily brings the proof of Belyi's Theorem to a quick end. We note that in the original proof Belyi used for this part a more powerful criterion due to A. Weil [Wei56], whose statement and proof we find more difficult to understand.

3.6.1 Proof of part (b) $\Rightarrow$ (a) of Belyi's Theorem

With the above criterion at one's disposal the proof is almost done. If $f : S \longrightarrow \mathbb{P}^1$ is a morphism of degree d whose only branching values are $0, 1, \infty$, then for any $\sigma \in \mathrm{Gal}(\mathbb{C})$ the conjugated morphism $f^\sigma : S^\sigma \longrightarrow \mathbb{P}^1$ is by Theorem 3.28 a morphism of the same degree d having the same branching values $\sigma(0) = 0, \sigma(1) = 1, \sigma(\infty) = \infty$. In particular, this family $\{f^\sigma\}$ of morphisms gives rise to only finitely many different monodromy homomorphisms $M_{f^\sigma} : \pi_1(\mathbb{P}^1 \smallsetminus \{0, 1, \infty\}) \longrightarrow \Sigma_d$ (recall that the fundamental group of the sphere minus three points is a free group with two generators, see Example 2.33). Therefore, Theorem 2.61 implies now that among all the Riemann surfaces $S^\sigma, \sigma \in \mathrm{Gal}(\mathbb{C})$, there are only finitely many equivalence classes. Finally, apply Criterion 3.29.

3.7 Proof of the criterion for definibility over $\overline{\mathbb{Q}}$

In order to prove Criterion 3.29 we need to introduce a new operation on polynomials which is in a way similar and yet radically different to the Galois action.

Example 3.30 As a first glimpse to what is coming ahead let us now go back again to our running example $S = S_F$, where $F = Y^2 - (X^3 - \pi^3)$. The question is why the fact that $S_F \simeq S_{F^\sigma}$ for $\sigma \in \mathrm{Gal}(\mathbb{C})$ should imply that S_F is defined over $\overline{\mathbb{Q}}$. The idea is as follows. Choose $\sigma \in \mathrm{Gal}(\mathbb{C})$ such that $\sigma(\pi) = e$. Then $F^\sigma = Y^2 - (X^3 - e^3)$ and the isomorphism $S_F \simeq S_{F^\sigma}$ is given by

$$\Psi : S_F \longrightarrow S_{F^\sigma}$$

$$(x, y) \longmapsto \left(\frac{e}{\pi}x, \sqrt{\left(\frac{e}{\pi}\right)^3} y \right)$$

Now consider the following $\overline{\mathbb{Q}}$-algebra homomorphism

$$\mathbf{s} : \overline{\mathbb{Q}}\left[\sqrt{\pi}, \sqrt{e}\right] \longrightarrow \mathbb{C} \tag{3.7}$$

$$\sqrt{\pi} \longmapsto \sqrt{\pi}$$

$$\sqrt{e} \longmapsto 1$$

We see that if we formally apply $\mathbf{s}$ to the coefficients defining the isomorphism $\Psi : S_F \longrightarrow S_{F^\sigma}$ in the same way as we have applied σ so far, we get a morphism between the Riemann surfaces of the curves $F^{\mathbf{s}} = F$ and $(F^\sigma)^{\mathbf{s}} = F_1 = Y^2 - (X^3 - 1)$, which is given by

$$\Psi_1 : S_F \longrightarrow S_{F_1}$$

$$(x, y) \longmapsto \left(\frac{1}{\pi}x, \frac{1}{\sqrt{\pi^3}} y \right)$$

Furthermore, the same process can be applied to the inverse morphism

$$\Phi = (\Psi)^{-1} : S_{F^\sigma} \longrightarrow S_F$$

$$(x, y) \longmapsto \left(\frac{\pi}{e}x, \sqrt{\left(\frac{\pi}{e}\right)^3} y \right)$$

to obtain the morphism

$$\Phi_1 : S_{F_1} \longrightarrow S_F$$

$$(x, y) \longmapsto \left(\pi x, \sqrt{\pi^3} y\right)$$

which is the inverse of Ψ_1. We conclude that S_F is isomorphic to the Riemann surface of a curve F_1 which has coefficients in a number field.

Now suppose that in part (c) of Lemma 3.12 instead of a field isomorphism $\sigma \in \mathrm{Gal}\,(\mathbb{C})$ we only have a ring homomorphism $\mathbf{s} : A \longrightarrow \mathbb{C}$ from a ring A containing the coefficients of the polynomials $F, G, P_1, P_2, Q_1, Q_2, H$ to the field $\mathbb{C}$ (as in Example 3.30). Then $\mathbf{s} : A \longrightarrow \mathbb{C}$ induces a ring homomorphism $\mathbf{s} : A[X, Y] \to \mathbb{C}[X, Y]$ and it still makes sense to apply $\mathbf{s}$ to the identity (3.3) that defines a morphism $f : S_F \longrightarrow S_G$ to obtain the identity $(Q_1^{\mathbf{s}})^n (Q_2^{\mathbf{s}})^m G^{\mathbf{s}}(R_1^{\mathbf{s}}, R_2^{\mathbf{s}}) = H^{\mathbf{s}} F^{\mathbf{s}}$. This in turn defines a morphism $f^{\mathbf{s}} = (R_1^{\mathbf{s}}, R_2^{\mathbf{s}})$ between the Riemann surface of the polynomial $F^{\mathbf{s}}$ (which may agree with F) and that of the polynomial $G^{\mathbf{s}}$ (which may now be defined over $\overline{\mathbb{Q}}$). This is exactly what we did in Example 3.30, and this is what we intend to do in the general case.

3.7.1 *Specialization of transcencendental coefficients*

We now recall some basic facts of the theory of transcendental field extensions that can be found, for example, in [Lan02].

Let k be a subfield of $\mathbb{C}$. A finite set of complex numbers $\{\pi_1, \ldots, \pi_d\}$ is said to be *algebraically independent over k* if the *evaluation* map

$$k[X_1, \ldots, X_d] \to \mathbb{C}$$

$$a(X_1, \ldots, X_d) \to a(\pi_1, \ldots, \pi_d)$$

is injective, i.e. if it induces an isomorphism between $k[X_1, \ldots, X_d]$ and its image, usually denoted by $k[\pi_1, \ldots, \pi_d]$. Because of this, an arbitrary d-tuple $(q_1, \ldots, q_d)$ of complex numbers induces a well defined k-algebra homomorphism $\mathbf{s} : k[\pi_1, \ldots, \pi_d] \to \mathbb{C}$ by the rule that sends $a(\pi_1, \ldots, \pi_d)$ to $a(q_1, \ldots, q_d)$.

By a *specialization* of $(\pi_1, \ldots, \pi_d)$ we shall only mean a d-tuple $(q_1, \ldots, q_d) \in \mathbb{C}^d$. The non-negative real number

$$\max_i |\pi_i - q_i|$$

will be called the *distance* of the specialization.

Note that saying that a singleton $\{\pi\} \subset \mathbb{C}$ is an algebraically independent subset over k only means that π is transcendental over k. Similarly, two elements $\pi_1, \pi_2 \in \mathbb{C}$ will be algebraically independent over k if for any non-zero polynomial $p(X, Y) = \sum a_{ij} X^i Y^j$ with coefficients in k one has

$$p(\pi_1, \pi_2) = \sum a_{ij} \pi_1^i \pi_2^j = \sum (\sum a_{ij} \pi_1^i) \pi_2^j = \sum (\sum a_{ij} \pi_2^j) \pi_1^i \neq 0$$

that is if π_2 is transcendental over $k(\pi_1)$ or, equivalently, π_1 is transcendental over $k(\pi_2)$. More generally, we see that a finite set of elements $\pi_1, \ldots, \pi_d \in \mathbb{C}$ is algebraically independent over a subfield k if and only if each π_i is transcendental over the field $k(\pi_1, \ldots, \pi_{i-1})$.

A field extension K of k is called *purely transcendental* if it is generated over k by a set of algebraically independent elements over k. Note that one cannot expect all transcendental field extensions to be purely transcendental. This can be already realized by considering fields of meromorphic functions for, as we well know by now, the only function field isomorphic to the purely transcendental extension $\mathbb{C}(x)$ is $\mathcal{M}(\mathbb{P}^1)$.

In the case $k = \mathbb{Q}$ one simply speaks of algebraically independent sets of complex numbers without reference to the base field $k = \mathbb{Q}$. We observe that there are algebraically independent sets of arbitrary large cardinality. This is because if there were a maximal algebraically independent set with only finitely many elements, say $\{\pi_1, \ldots, \pi_d\}$, then the field $\mathbb{C}$ would have to be an algebraic extension of the field $\mathbb{Q}(\pi_1, \ldots, \pi_d)$. Since the latter is obviously a countable set, so would be $\mathbb{C}$.

3.7.2 Infinitesimal specializations

Throughout this section $(\pi_1, \ldots, \pi_d; u)$ will denote a $(n + 1)$-tuple of complex numbers such that $\pi_1, \ldots, \pi_d$ are algebraically independent and u is algebraic over the field $\mathbb{Q}(\pi_1, \ldots, \pi_d)$.

Let $(q_1, \ldots, q_d)$ be a specialization of $(\pi_1, \ldots, \pi_d)$ and let

$$\mathbf{s} : \mathbb{Q}[\pi_1, \ldots, \pi_d] \to \mathbb{C}$$

be the $\mathbb{Q}$-algebra homomorphism determined by $\mathbf{s}(\pi_i) = q_i$. We would like to understand the extent to which $\mathbf{s}$ can be extended to the ring $\mathbb{Q}[\pi_1, \ldots, \pi_d, u]$. In order to do that we introduce the following notation, which imitates the one we have used for the Galois action, namely

$$a^{\mathbf{s}} = \mathbf{s}(a) = a(q_1, \ldots, q_d)$$

for an element $a \in \mathbb{Q}[\pi_1, \ldots, \pi_d]$, and

$$q^{\mathbf{s}}(X) = \sum a_l^{\mathbf{s}} X^l$$

for a polynomial $q(X) = \sum a_l X^l \in \mathbb{Q}[\pi_1, \ldots, \pi_d][X]$.

We first observe that we cannot choose the image of u, say $\mathbf{s}(u) = b$, arbitrarily. To explain why this is so, let us consider the minimal polynomial of u over $\mathbb{Q}(\pi_1, \ldots, \pi_d)$

$$m_u(X) = \sum a_j X^j = a_0 + a_1 X + \cdots + a_n X^n \qquad (3.8)$$

chosen so that the coefficients a_l lie in the ring $\mathbb{Q}[\pi_1, \ldots, \pi_d]$ and are coprime (this is achieved by the usual clearing out denominators procedure).

Then we can apply $\mathbf{s}$ to relation (3.8) to find

$$0 = \mathbf{s}(0) = \mathbf{s}(m_u(u)) = m_u^{\mathbf{s}}(\mathbf{s}(u))$$

In other words, $\mathbf{s}(u)$ must be a root of the polynomial $m_u^{\mathbf{s}}(X)$. We claim that the converse is true.

Lemma 3.31 *Notation being as above, let $b \in \mathbb{C}$ be any root of the polynomial $m_u^{\mathbf{s}}(X)$. Then the assignment $\pi_i \to q_i$ and $u \to b$ extends $\mathbf{s}$ to a $\mathbb{Q}$-algebra homomorphism*

$$\mathbf{s}_b : \mathbb{Q}[\pi_1, \ldots, \pi_d, u] \to \mathbb{C}$$

Proof For any element $x = \sum a_i u^i, a_i \in \mathbb{Q}[\pi_1, \ldots, \pi_d]$ we set $\mathbf{s}_b(x) = \sum a_i^{\mathbf{s}} b^i$. By definition $\mathbf{s}_b$ preserves sums and products, thus to conclude that $\mathbf{s}_b$ is a $\mathbb{Q}$-algebra homomorphism one only needs to check that $\mathbf{s}_b(x)$ does not depend of the chosen representation of x.

Suppose that $x = \sum c_i u^i, c_i \in \mathbb{Q}[\pi_1, \ldots, \pi_d]$ is another way of writing the element x. Then we have $\sum(a_i - c_i)u^i = 0$, which implies that $\sum(a_i - c_i)X^i = p(X)m_u(X)$ for certain polynomial $p(X) \in \mathbb{Q}(\pi_1, \ldots, \pi_d)[X]$. As the coefficients of $m_u(X)$ are coprime the Gauss lemma implies that $p(X)$ lies in $\mathbb{Q}[\pi_1, \ldots, \pi_d][X]$. Hence we can write

$$0 = p^{\mathbf{s}}(b)m_u^{\mathbf{s}}(b) = \sum a_i^{\mathbf{s}}b^i - \sum c_i^{\mathbf{s}}b^i$$

and therefore $\sum a_i^{\mathbf{s}}b^i = \sum c_i^{\mathbf{s}}b^i$ as required. $\square$

Lemma 3.32 *Let $u = u_1, u_2, \ldots, u_n \in \mathbb{C}$ be the roots of $m_u(X)$ and let $\delta = \min_{k,l}|u_k - u_l|$. There is a real number $\epsilon(u) > 0$ such that if $\mathbf{s} : \mathbb{Q}[\pi_1, \ldots, \pi_d] \to \mathbb{C}$ is the homomorphism determined by an arbitrary specialization of distance less than $\epsilon(u)$, then the polynomial $m_u^{\mathbf{s}}(X)$ possesses a unique root $u_{\mathbf{s}}$ with the property that $|u - u_{\mathbf{s}}| < \delta$.*

Proof Write $m_u(X)$ in the form $m_u(X) = a_n \prod(X - u_i)$ and let $(q_1, \ldots, q_n)$ be a specialization of $(\pi_1, \ldots, \pi_n)$ of distance ϵ. As ϵ gets small, the coefficients of $m_u^{\mathbf{s}}(X)$ get close to those of $m_u(X)$.

This obvious observation has two implications. One is that in that case we would have $a_n^{\mathbf{s}} \neq 0$ and so it makes sense to write $m_u^{\mathbf{s}}(X) = a_n^{\mathbf{s}} \prod_{i=1}^{n}(X - \alpha_i)$. The other one is that for each root u_j of $m_u(X)$ the difference

$$|m_u(u_j) - m_u^{\mathbf{s}}(u_j)| = |m_u^{\mathbf{s}}(u_j)| = |a_n^{\mathbf{s}}| \prod_i |u_j - \alpha_i|$$

will become arbitrarily small, and therefore so must do at least one of the factors, say $|u_j - \alpha_j|$.

This means that if ϵ is sufficiently small then for each root u_j of $m_u(X)$ there is at least one root of $m_u^{\mathbf{s}}(X)$, say α_j, such that $|u_j - \alpha_j| < \delta/2$. From here the result follows by simply taking $u_{\mathbf{s}} = \alpha_1$. Note that if we also had $|u_1 - \alpha_2| < \delta/2$ then we could write

$$|u_1 - u_2| \leq |u_1 - \alpha_2| + |u_2 - \alpha_2| < 2 \cdot \delta/2 = \delta$$

which is a contradiction. $\square$

Definition 3.33 An *infinitesimal specialization*, of $(\pi_1, \ldots, \pi_d; u)$ is nothing but a specialization of $(\pi_1, \ldots, \pi_d)$ of distance $\epsilon < \epsilon(u)$

for some $\epsilon(u)$ as in Lemma 3.32. By the *homomorphism associated to an infinitesimal specialization* of $(\pi_1, \ldots, \pi_d; u)$ we shall refer to the $\mathbb{Q}$-algebra homomorphism $\mathbf{s} : \mathbb{Q}[\pi_1, \ldots, \pi_d, u] \longrightarrow \mathbb{C}$ determined by $\mathbf{s}(\pi_i) = q_i$ and $\mathbf{s}(u) = u_{\mathbf{s}}$.

3.7.3 End of the proof

We are now ready to finish the proof of Criterion 3.29.

Let $\Sigma_1 := \{\pi_1, \ldots, \pi_d\}$ be a maximal set of algebraically independent coefficients of $F = F(X, Y)$. By the Primitive Element Theorem the field generated by all coefficients of F is of the form

$$K_1 = \mathbb{Q}(\pi_1, \ldots, \pi_d, v)$$

with v algebraic over $\mathbb{Q}(\pi_1, \ldots, \pi_d)$ and with minimal polynomial $m_v(T) \in \mathbb{Q}(\pi_1, \ldots, \pi_d)[T]$ chosen as in (3.8). Therefore, for any $\sigma \in \mathrm{Gal}(\mathbb{C})$ the field $K_2 := \sigma(K_1) = \mathbb{Q}(\sigma(\pi_1), \ldots, \sigma(\pi_d), \sigma(v))$ will be the field generated by the coefficients of F^σ.

Now consider Galois elements $\sigma \in \mathrm{Gal}\,(\mathbb{C})$ such that

$$\Sigma_2 := \{\pi_1, \ldots, \pi_d, \pi_{d+1} := \sigma(\pi_1), \ldots, \pi_{2d} := \sigma(\pi_d)\}$$

is a set of algebraically independent elements. By the finiteness condition in Criterion 3.29 there must be plenty of pairs of elements $\beta, \tau \in \mathrm{Gal}\,(\mathbb{C})$ for which there is an isomorphism $\Phi : S_{F^\tau} \to S_{F^\beta}$, hence an isomorphism $\Psi = \Phi^{\tau^{-1}} : S_F \to S_{F^\sigma}$ with $\sigma = \tau^{-1} \circ \beta$.

Next we enlarge the set Σ_2 by adding a number of coefficients of the polynomials $P_i, Q_i, U_i, V_i, T, H_i, H$ intervening in the identities (3.3) to (3.6) that express the fact that Ψ is an isomorphism (see Theorem 3.8). More precisely, we adjoin a maximal possible collection of such coefficients $\{\pi_{2d+1}, \ldots, \pi_n\}$ with the condition that the set $\Sigma_3 = \{\pi_1, \ldots, \pi_n\}$ is still an algebraically independent set. Then the field K_3 generated by K_1, K_2 and the whole set of coefficients of these polynomials will be of the form

$$K_3 = \mathbb{Q}(\pi_1, \ldots, \pi_n, u)$$

where the element u is algebraic over the pure transcendental extension $\mathbb{Q}(\pi_1, \ldots, \pi_n)$.

For $j = d + 1, \ldots, n$ let $q_j \in \mathbb{Q}(\sqrt{-1})$ be such that the n-tuple $(\pi_1, \ldots, \pi_d, q_{d+1}, \ldots, q_n)$ is an infinitesimal specialization of

$(\pi_1, \ldots, \pi_n; u)$, and let $\mathbf{s} : \mathbb{Q}[\pi_1, \ldots, \pi_n, u] \longrightarrow \mathbb{C}$ denote the associated homomorphism. As usual, let us denote by $\mathbb{Q}(\pi_1, \ldots, \pi_n, u)$ the field of fractions of $\mathbb{Q}[\pi_1, \ldots, \pi_n, u]$. Remember that an element $z \in \mathbb{Q}(\pi_1, \ldots, \pi_n, u)$ can be written in the form

$$z = \frac{A(\pi_1, \ldots, \pi_n, u)}{B(\pi_1, \ldots, \pi_n, u)} \tag{3.9}$$

where A and B are polynomials in $n+1$ variables with coefficients in $\mathbb{Q}$.

Let $\mathbb{Q}[\pi_1, \ldots \pi_n, u]_{\mathbf{s}}$ denote the subring consisting of the elements z as in (3.9) whose denominators $B = B(\pi_1, \ldots, \pi_n, u)$ do not lie in the kernel of $\mathbf{s}$, i.e. those $z \in \mathbb{Q}(\pi_1, \ldots, \pi_n, u)$ admitting an expression of the form (3.9) such that

$$\mathbf{s}(B) = B(\pi_1, \ldots, \pi_d, q_{d+1}, \ldots, q_n, u_{\mathbf{s}}) \neq 0$$

Clearly $\mathbf{s} : \mathbb{Q}[\pi_1, \ldots, \pi_n, u] \longrightarrow \mathbb{C}$ extends to a unique homomorphism $\mathbf{s} : \mathbb{Q}[\pi_1, \ldots, \pi_n, u]_{\mathbf{s}} \longrightarrow \mathbb{C}$, and hence to a homomorphism $\mathbf{s} : \mathbb{Q}[\pi_1, \ldots, \pi_n, u]_{\mathbf{s}}[X, Y] \longrightarrow \mathbb{C}[X, Y]$, simply by writting $\mathbf{s}(A/B) = \mathbf{s}(A)/\mathbf{s}(B)$.

Now if the elements q_i are chosen sufficiently close to the elements π_i (i.e. if the distance of our specialization becomes small), then all the elements of the finite set consisting of the coefficients of the polynomials $P_i, Q_i, U_i, V_i, T, H_i, H$ along with the element $v \in K_1$ lie in $\mathbb{Q}[\pi_1, \ldots, \pi_n, u]_{\mathbf{s}}$. In these circumstances we will be in position to apply $\mathbf{s}$ to the polynomial identities (3.3) to (3.6). For instance, applying $\mathbf{s}$ to the first relation

$$Q_1^n Q_2^m G(R_1, R_2) = HF$$

which defines the morphism $\Psi = (R_1, R_2)$, gives the identity

$$(Q_1^n)^{\mathbf{s}} (Q_2^m)^{\mathbf{s}} G^{\mathbf{s}}(R_1^{\mathbf{s}}, R_2^{\mathbf{s}}) = H^{\mathbf{s}} F^{\mathbf{s}}$$

which defines a morphism $\Psi^{\mathbf{s}} = (R_1^{\mathbf{s}}, R_2^{\mathbf{s}}) : S_{F^{\mathbf{s}}} \to S_{(F^\sigma)^{\mathbf{s}}}$. Doing the same with the rest of the identities we end up with a collection of polynomial identities which define an isomorphism

$$\Psi^{\mathbf{s}} =: S_{F^{\mathbf{s}}} \to S_{(F^\sigma)^{\mathbf{s}}}$$

Now, by construction, the coefficients of the polynomial $(F^\sigma)^{\mathbf{s}}$ lie in the field generated by certain Gaussian numbers $q_j \in \mathbb{Q}(\sqrt{-1})$ together with $\mathbf{s}(\sigma(v))$, which must be a root of the polynomial

$(m_v^\sigma)^{\mathbf{s}}(X) \in \mathbb{Q}(\sqrt{-1})[X]$ and therefore an algebraic number. We thus conclude that $(F^\sigma)^{\mathbf{s}}$ has coefficients in a number field.

Finally, it remains to be seen that $F^{\mathbf{s}} = F$ for which, by our construction of $\mathbf{s}$, it is enough to see that $\mathbf{s}(v) = v$. Clearly $\mathbf{s}(v)$ must also be a root of $m_v(X) = m_v^{\mathbf{s}}(X)$. Moreover, a simple look at the corresponding expression (3.9) for v shows that if the distance of our specialization gets small then $\mathbf{s}(v)$ is as close to the root v as wanted, hence it must agree with v.

3.8 The field of definition of Belyi functions

The theory of specialization employed to prove Belyi's Theorem can also be used to show that Belyi functions are defined over $\overline{\mathbb{Q}}$ as well.

Proposition 3.34 *Belyi functions are defined over* $\overline{\mathbb{Q}}$.

Proof Let (S, f) be a Belyi pair for which we have $S = S_F$ and $f(x,y) = \dfrac{P_1(x,y)}{Q_1(x,y)}$. We have to show that there exists an irreducible polynomial G with coefficients in a number field, an isomorphism $\Phi = \left(\dfrac{V_1}{W_1}, \dfrac{V_2}{W_2} \right) : S_F \to S_G$ and a function $h \in \mathcal{M}(S_G)$ also defined over $\overline{\mathbb{Q}}$ such that the following diagram commutes:

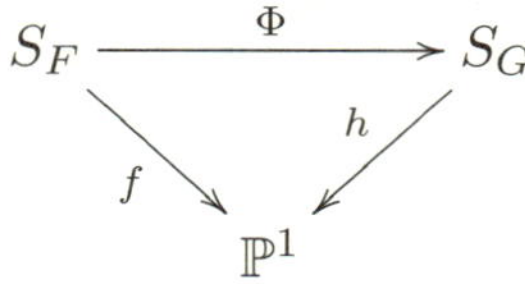

To prove this one merely has to mimic what we have done in the previous sections. Let $K_1 = \mathbb{Q}(\pi_1, \ldots, \pi_d, v)$ be the field generated by the coefficients of F, P_1 and Q_1 and consider $\sigma \in \mathrm{Gal}(\mathbb{C})$ such that $\pi_1, \ldots, \pi_d, \pi_{d+1} = \sigma(\pi_1), \ldots, \pi_{2d} = \sigma(\pi_d)$ are algebraically independent. Clearly for each such σ the field generated by the coefficients of F^σ and f^σ will be $K_2 = \mathbb{Q}(\pi_{d+1}, \ldots, \pi_{2d}, \sigma(v))$. Since all coverings (S_{F^σ}, f^σ), $\sigma \in \mathrm{Gal}(\mathbb{C})$ have the same degree and branching values $\sigma(0) = 0$, $\sigma(1) = 1$, $\sigma(\infty) = \infty$ (Theorem 3.28) there can only be finitely many equivalence classes of such pairs (Proposition 2.63). It follows that there must be some σ such that the ramified coverings (S_F, f) and (S_{F^σ}, f^σ) are equivalent, that is there must be an isomorphism $\Phi = (\frac{V_1}{W_1}, \frac{V_2}{W_2}) : S_F \to S_{F^\sigma}$ such that the identity $f^\sigma \circ \Phi = f$ holds. Arguing as in Section 3.2 we see that this identity is equivalent to a polynomial relation of the form

$$W_1^k W_2^{rd}(P_1^\sigma(V_1/W_1, V_2/W_2)Q_1 - Q_1^\sigma(V_1/W_1, V_2/W_2)P_1 = HF$$

(3.10)

together with the polynomial identities corresponding to (3.3) to (3.6) which express the fact that Φ has an inverse (see Section 3.2).

Let now $K_3 = \mathbb{Q}(\pi_1, \ldots, \pi_{2d}, \ldots, \pi_r, u)$ be the field generated by the fields K_1, K_2 and the coefficients of all polynomials occurring in identity (3.10) as well as in (3.3) to (3.6). For $j = d+1, \ldots, 2d$ choose algebraic numbers $q_j \in \mathbb{Q}(\sqrt{-1})$ such that

$$(\pi_1, \ldots, \pi_d, q_{d+1}, \ldots, q_{2d}, \pi_{2d+1}, \ldots, \pi_r)$$

is an infinitesimal specialization of $(\pi_1, \ldots, \pi_r; u)$ and consider the homomorphism $\mathbf{s} : \mathbb{Q}[\pi_i, e_j, u] \to \mathbb{C}$ associated to it. As in the previous section, if the distance of this specialization is sufficiently small it will make sense to apply $\mathbf{s}$ to the identities (3.10) and (3.3) to (3.6). This way we will obtain a covering $(S_G, h) = ((S_{F^\sigma})^{\mathbf{s}}, (f^\sigma)^{\mathbf{s}})$, which solves our problem. $\qquad\square$

This concludes our proof of Belyi's Theorem. For different approaches consult Köck [Köc04] and Wolfart [Wol97]. An extension of Belyi's theorem to the case of complex surfaces (holomorphic manifolds of complex dimension 2) can be found in [GD08], see also [GGD10].

4

Dessins d'enfants

4.1 Definition and first examples

Definition 4.1 A *dessin d'enfant*, or simply a *dessin*, is a pair $(X, \mathcal{D})$ where X is an oriented compact topological surface, and $\mathcal{D} \subset X$ is a finite graph such that:

    (i) $\mathcal{D}$ is connected.

    (ii) $\mathcal{D}$ is bicoloured, i.e. the vertices have been given either white or black colour and vertices connected by an edge have different colours.

    (iii) $X \smallsetminus \mathcal{D}$ is the union of finitely many topological discs, which we call *faces* of $\mathcal{D}$.

The *genus* of $(X, \mathcal{D})$ is simply the genus of the topological surface X.

We consider two dessins $(X_1, \mathcal{D}_1)$ and $(X_2, \mathcal{D}_2)$ *equivalent* when there exists an orientation-preserving homeomorphism from X_1 to X_2 whose restriction to $\mathcal{D}_1$ induces an isomorphism between the coloured graphs $\mathcal{D}_1$ and $\mathcal{D}_2$.

Remark 4.2 In fact condition (i) is a consequence of condition (iii) as it is fairly obvious that any path in X connecting two given points of $\mathcal{D}$ is homotopic to a path supported on the boundary of the faces encountered along the way.

Remark 4.3 Some authors remove condition (ii) with the understanding that to any (single-coloured) graph satisfying conditions (i) and (iii), a dessin is associated by placing a new vertex in the middle of each edge. This process produces only dessins where all

207

the white vertices have degree 2, a restriction that looks rather unnatural from the point of view of bicoloured graphs. These graphs are classically known as *maps* (see [JS78] and the references given there), and the associated dessins are the ones originally introduced by Grothendieck [Gro97]. Sometimes these sort of dessins are referred to as *clean dessins d'enfants*. We will come back to this later (see Remark 4.27).

For example, the tetrahedron can be naturally considered as a graph embedded in the topological sphere. Being a complete graph in four vertices, it cannot be made into a bicoloured graph. Hence the tetrahedron does not fall into the class of graphs introduced in Definition 4.1, despite satisfying the most restrictive condition (iii). Nevertheless, the method of placing extra vertices at the edge midpoints produces the dessin at the left-hand side of Figure 4.1.

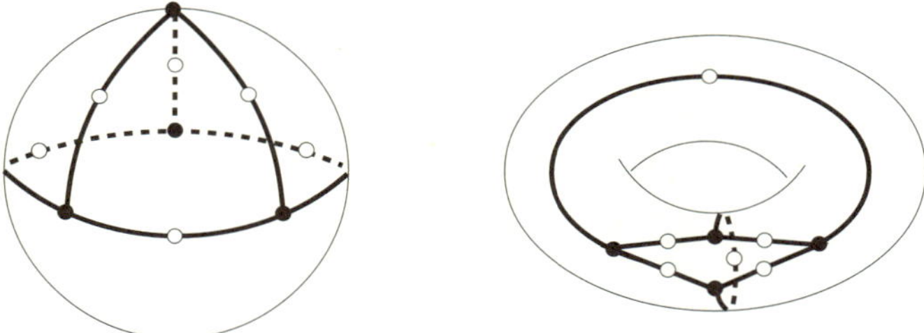

Fig. 4.1. Two dessins with the same underlying abstract graph.

Note that a dessin is more than a mere abstract graph, as it comes equipped with a certain embedding in a given topological surface. For example, the abstract graphs underlying both dessins in Figure 4.1 coincide, but the two dessins are different (they have different genus). Nevertheless, when the topological surface underlying $(X, \mathcal{D})$ is clear from the context, we will denote the dessin simply by $\mathcal{D}$.

Example 4.4 Any tree contained in the sphere is obviously a dessin with only one face (note that trees can always be bicoloured).

The graph given by the vertices, edges and faces of a cube is a dessin with six faces, twelve edges, four white vertices and four black ones. The boundary of every face consists of four edges, and all vertices are incident with three edges (we say that all the vertices of the cube have *degree* 3).

Spherical polyhedra (all platonic solids, for instance) can also be seen as dessins, sometimes (as in the case of the tetrahedron) after placing extra vertices to make them bicoloured.

Example 4.5 The boundary of the polygon in Figure 1.11 is a dessin embedded in a surface of genus three. It has one face, seven edges, one white vertex and one black vertex.

Example 4.6 The boundary of the hyperbolic octogon in Figure 2.4 gives rise to a dessin of genus 2 if we place black vertices at the vertices of the octogon and white vertices at the midpoints of the edges of the octogon.

4.1.1 *The permutation representation pair of a dessin*

Suppose $(X, \mathcal{D})$ has N edges, and label them with integer numbers from 1 to N. Consider the two permutations $\sigma_0, \sigma_1 \in \Sigma_N$ defined as follows.

Draw small (topological) circles around each of the white vertices and set $\sigma_0(i) = j$ if j is the edge that follows i under a positive rotation (cf. Section 1.2.1). Similarly, σ_1 is obtained by applying the same construction to the black vertices (Figure 4.2).

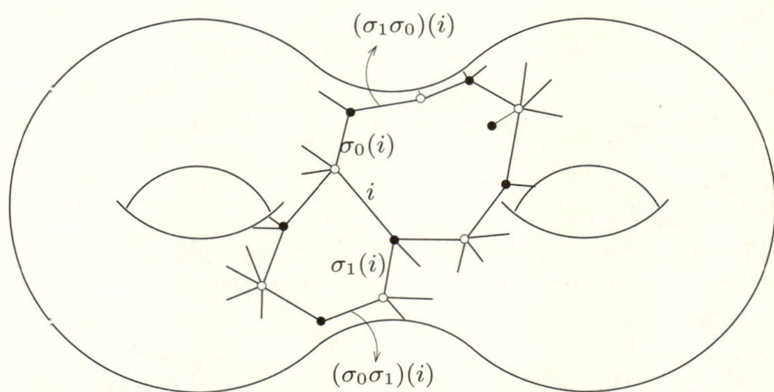

Fig. 4.2. The permutations σ_0 and σ_1.

Definition 4.7 We call (σ_0, σ_1) the *permutation representation pair* of the dessin.

Example 4.8 For the dessin in Figure 1.11 mentioned in Example 4.5, replacing the letters a to g by the corresponding numbers 1 to 7 in the labels of Figure 1.11 we get the permutation representation pair $(\sigma_0, \sigma_1) = ((1, 4, 3, 6, 5, 7, 2), (1, 5, 4, 7, 3, 2, 6))$.

Given a dessin $(X, \mathcal{D})$ with permutation representation pair (σ_0, σ_1), the cycles of σ_0 are clearly in one-to-one correspondence with the white vertices of $\mathcal{D}$, the length of each cycle being the degree of the corresponding vertex. The same statement holds true replacing σ_0 with σ_1 and white vertices with black vertices.

The cycles of $\sigma_1\sigma_0$ (or, equivalently, those of $\sigma_0\sigma_1$) are in one-to-one correspondence to the faces of $\mathcal{D}$. More precisely, the sequence

$$i, \sigma_1\sigma_0(i), (\sigma_1\sigma_0)^2(i), \ldots, (\sigma_1\sigma_0)^{m-1}(i), (\sigma_1\sigma_0)^m(i) = i$$

produced by an m-cycle of $\sigma_1\sigma_0$ enumerates in counterclockwise direction one half of the $2m$ edges of a face containing the edge i (see Figure 4.2). Similarly, the sequence

$$i, \sigma_0\sigma_1(i), (\sigma_0\sigma_1)^2(i), \ldots, (\sigma_0\sigma_1)^{r-1}(i), (\sigma_0\sigma_1)^r(i) = i$$

corresponding to a cycle of $\sigma_0\sigma_1$ of length r enumerates in clockwise direction one half of the $2r$ edges of another face containing the edge i (this face may agree with the previous one if the edge i meets the same face at both sides). Note that $\mathcal{D}$ being a bicoloured graph, a face must be bounded by an even number of edges, with the same edge counted twice if it meets the same face at both sides.

Remark 4.9 The connectedness of $\mathcal{D}$ ensures that the group generated by the two permutations σ_0, σ_1 is a transitive subgroup of permutations of Σ_N.

The genus of the surface X where $\mathcal{D}$ is embedded is also encoded in the permutation representation pair. This is a consequence of the following result:

Proposition 4.10 *Let g denote the genus of X. Then the following formula holds*

$$2 - 2g = (\#\{cycles\ of\ \sigma_0\} + \#\{cycles\ of\ \sigma_1\}) - N \\ + \#\{cycles\ of\ \sigma_1\sigma_0\}$$

Of course this is nothing but the Euler–Poincaré characteristic of X corresponding to the polygonal decomposition induced by the dessin. Although this is not a triangulation in the standard sense (see Section 1.2.1), it is known that the usual formula (vertices)−(edges)+(faces) works for more general polygonal decompositions. In any case, later on (Proposition 4.18) we shall give a proof of Proposition 4.10 as an application of the Riemann–Hurwitz formula.

Example 4.11 Let $(\sigma_0, \sigma_1) = ((1,4,3,6,5,7,2),(1,5,4,7,3,2,6))$ as in Example 4.8. As $\sigma_1\sigma_0 = (1,7,6,4,2,5,3)$, we have in this case $2 - 2g = 1 + 1 + 1 - 7 = -4$, therefore $g = 3$ as we already found in Example 1.51.

Remark 4.12 Let the edges of a given dessin be labelled in a certain way, and let (σ_0, σ_1) be the corresponding permutation representation pair. If relabelling the edges in a different way the new permutation representation pair is (σ_0', σ_1'), there is a permutation in Σ_N conjugating σ_0 to σ_0' and σ_1 to σ_1'. The permutation representation pair is therefore defined only up to conjugation in Σ_N.

Proposition 4.13 *Let σ_0, σ_1 be two permutations in Σ_N such that $\langle \sigma_0, \sigma_1 \rangle$ is a transitive group. There exists a dessin d'enfant $(X, \mathcal{D})$ such that its permutation representation pair is precisely (σ_0, σ_1).*

Proof The result follows from a sort of cutting and pasting argument similar to the ones in Section 1.2.1. The genus of the resulting surface will of course agree with the one predicted in Proposition 4.10.

We start by computing $\sigma_1\sigma_0$. If its decomposition in disjoint cycles is $\sigma_1\sigma_0 = \tau_1 \cdots \tau_k$, where τ_j has order n_j and $\sum_{j=1}^{k} n_j = N$, we create k (for the moment, disjoint) faces bounded by $2n_1, \ldots, 2n_k$

edges respectively. These will account for the faces of the dessin after glueing them together in the right way.

After assigning white and black colours to the vertices of each face, we label half of the edges as prescribed by the cycles of $\sigma_1\sigma_0$ and then we use σ_0 to label the remaining edges. It suffices now to glue together these pieces along edges with the same label in order to form a connected body. Note that the transitivity of $\langle\sigma_0,\sigma_1\rangle$ ensures that no face can remain disconnected from the rest. The boundary of this final object consists of a certain number of edges identified in pairs, and therefore it represents a compact surface. This surface is orientable since the double colouring of vertices ensures that no Möbius band can be created after the side-identifications. $\qquad\square$

The main feature of the permutation representation pair is the fact that it characterizes the dessin completely. More precisely, if two dessins have conjugate permutation representation pairs (σ_0,σ_1) and $(\tau\sigma_0\tau^{-1},\tau\sigma_0\tau^{-1})$, then they are equivalent. This will be a direct consequence of the results in Section 4.3.1.

Example 4.14 Let $\sigma_0 = (1,5,4)(2,6,3)$ and $\sigma_1 = (1,2)(3,4)(5,6)$. The corresponding dessin $\mathcal{D}$ has six edges, two white vertices of degree three (since σ_0 has two cycles of length three), and three black vertices of degree two. Moreover, $\sigma_1 \cdot \sigma_0 = (1,6,4,2,5,3)$, hence $\mathcal{D}$ has only one face. As for the genus of the surface where $\mathcal{D}$ is embedded, Proposition 4.10 gives

$$2 - 2g = 3 + 2 + 1 - 6 = 0,$$

thus $\mathcal{D}$ is embedded in a topological torus. The right-hand side of Figure 4.3 shows a model of $\mathcal{D}$ constructed by the method described in the proof of Proposition 4.13. The dessin on the left-hand side of the same figure, depicted in a topological torus in Euclidean 3-space, has the same monodromy and therefore it is equivalent to (or, simply, it agrees with) $\mathcal{D}$.

4.2 From dessins d'enfants to Belyi pairs

The main reason why we are interested in the theory of dessins d'enfants is because of its connection with Belyi functions. Giving the details of this connection will be the goal of this section.

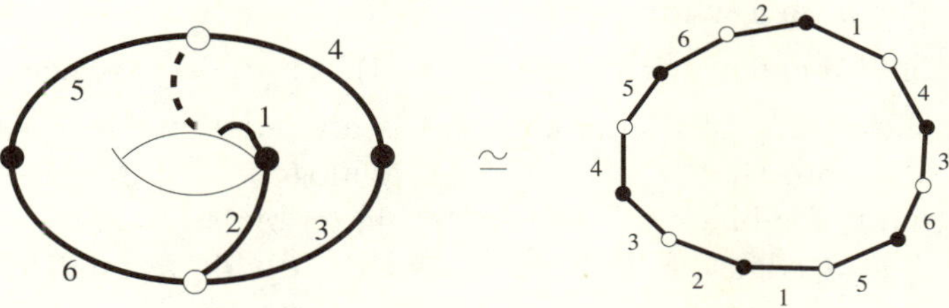

Fig. 4.3. A dessin in a topological torus.

4.2.1 The triangle decomposition associated to a dessin

For a given dessin $(X, \mathcal{D})$ we shall construct a triangle decomposition $\mathcal{T} = \mathcal{T}(\mathcal{D})$ of X associated to $\mathcal{D}$. By the term *triangle decomposition* of X we mean a collection of triangles that cover the whole X, and such that the intersection of two triangles consists of a union of edges or vertices. Note that triangle decompositions need not be triangulations in the usual sense as triangles are allowed to meet at more than one edge (see for instance Figures 4.4 and 4.5).

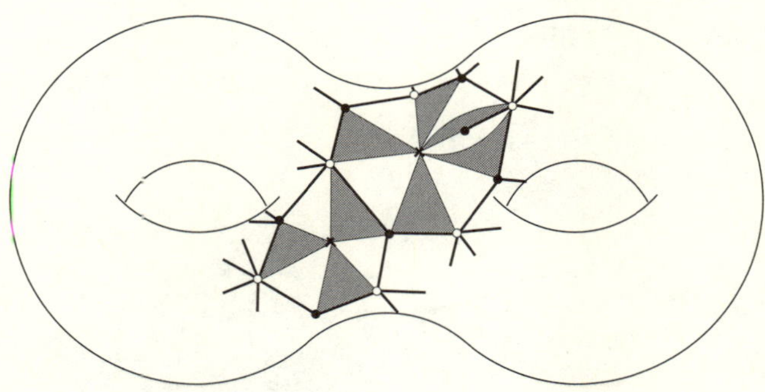

Fig. 4.4. The triangle decomposition $\mathcal{T}(\mathcal{D})$.

While reading through the description of $\mathcal{T}(\mathcal{D})$, it will be very convenient to keep an eye on Figure 4.4, where the triangle decomposition associated to the dessin depicted in Figure 4.2 is shown.

(1) We choose a *centre* in each of the faces of which $X \smallsetminus \mathcal{D}$ consists of. In our pictures, the centres will be marked with the symbol $\times$.

(2) Let v be a white vertex of $\mathcal{D}$. Recall that the set of edges of $\mathcal{D}$ incident with v are $\left\{j, \sigma_0(j), \ldots, \sigma_0^{d-1}(j)\right\}$, where d is the degree of v and $j = j(v)$ is any edge incident with v. We note that $\sigma_0^d(j) = j$. An analogous statement can be made for black vertices replacing σ_0 by σ_1.

(3) The d edges of $\mathcal{D}$ incident with v divide a small disc D around v into d open disjoint sectors, say $D \smallsetminus \mathcal{D} = A_1^v \cup \cdots \cup A_d^v$.

(4) Draw a (topological) segment γ_i^v that starts at v, runs first inside A_i^v, and ends at the centre of the face of $\mathcal{D}$ that contains A_i^v. During the process, γ_i^v is not allowed to meet any edge except at v itself.

(5) This construction associates to each edge $j = [v, w]$ of $\mathcal{D}$ two triangles T_j^1 and T_j^2 of which j is a common edge and whose remaining two edges are the obvious segments in the collection $\{\gamma_i^v\}$ and $\{\gamma_k^w\}$ respectively. This way, a triangle decomposition $\mathcal{T} = \left\{T_j^\delta\right\}$ of X is provided where $j = 1, \ldots, N = \#\{\text{edges of } \mathcal{D}\}$ and $\delta = 1, 2$. Note that, as long as it may happen $\sigma_0(j) = j$, two given triangles of $\mathcal{T}$ might have two (see Figure 4.4), or even three common edges (see Figure 4.5).

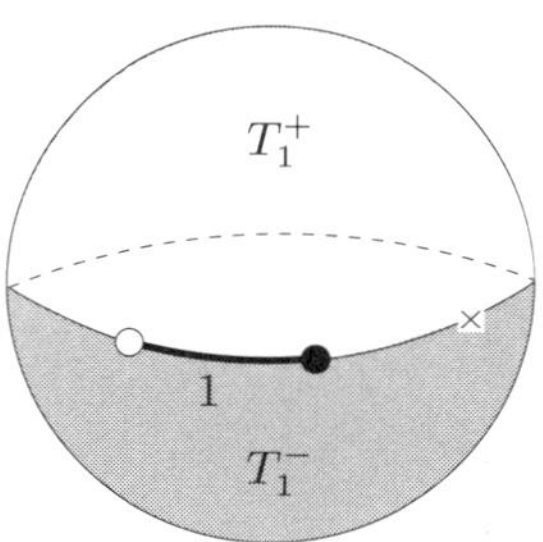

Fig. 4.5. A triangle decomposition for the most simple dessin on the sphere.

(6) The orientation of X enables us to classify the triangles T_j^δ into two types '+' and '−' (or *white* and *black*). We denote T_j^δ as T_j^+ if the circuit $\circ \rightarrow \bullet \rightarrow \times \rightarrow \circ$ follows the positive orientation of ∂T_j^δ (cf. Section 1.2.1), and as T_j^- otherwise. By construction, adjacent triangles belong

to different types; this shows in particular that each face of $\mathcal{D}$ contains exactly the same number of triangles of both types.

We now summarize what we have so far achieved in this section:

Summary 4.15 Starting from $(X, \mathcal{D})$ we have obtained a triangle decomposition $\mathcal{T} = \mathcal{T}(\mathcal{D})$ of X satisfying the following properties:

(i) Each triangle contains one vertex of each type $\circ$, $\bullet$ and $\times$.

(ii) Each triangle is completely determined by the following data: its type (or colour), and the only edge that $\mathcal{D}$ has in common with it.

(iii) Each edge j of $\mathcal{D}$ belongs to *two triangles* T_j^+ and T_j^-. When j belongs to only one face of $\mathcal{D}$, then T_j^+ and T_j^- belong to the same face of $\mathcal{D}$ (see Figure 4.4). On the other hand, if two faces meet at j then one of them contains T_j^+ and the other one contains T_j^-.

(iv) Two adjacent triangles are of different type ($+$ or $-$).

(v) Each face of $\mathcal{D}$ is decomposed into a reunion of an even number of triangles, half of them of each type.

4.2.2 The Belyi function associated to a dessin

Choose a triangle T_j^+ of type $+$, and a homeomorphism f_j^+ from the triangle T_j^+ to $\overline{\mathbb{H}}^+ := \mathbb{H} \cup \mathbb{R} \cup \{\infty\}$ (the closure of the upper halfplane) satisfying the following condition

$$f_j^+ : \begin{cases} \partial T_j^+ & \longrightarrow & \mathbb{R} \cup \{\infty\} \\ \circ & \longrightarrow & 0 \\ \bullet & \longrightarrow & 1 \\ \times & \longrightarrow & \infty \end{cases} \tag{4.1}$$

Next, take the triangle T_j^- adjacent to T_j^+ along the edge labelled j, and map T_j^- to the closure of the lower halfplane $\overline{\mathbb{H}}^-$ by a homeomorphism $f_j^- : T_j^- \longrightarrow \overline{\mathbb{H}}^-$ that coincides with f_j^+ in the intersection $T_j^+ \cap T_j^-$ and verifies also (4.1).

The existence of such a homeomorphism would follow from the

following two statements. Firstly, every homeomorphism from an edge of a triangle T to a segment in $\partial\overline{\mathbb{H}} = \mathbb{R} \cup \{\infty\}$ extends to a homeomorphism from the whole ∂T to $\partial\overline{\mathbb{H}}$, and secondly every homeomorphism $\partial T \longrightarrow \partial\overline{\mathbb{H}}$ extends to a homeomorphism $T \longrightarrow \overline{\mathbb{H}}$. Now, the first statement is equivalent to saying that every homeomorphism from an arc $c \subset \mathbb{S}^1$ to an arc $c' \subset \mathbb{S}^1$ extends to a homeomorphism that maps the complement of c to the complement of c', which is clear. The second part is equivalent to saying that every homeomorphism $\theta \longrightarrow f(\theta)$ from $\mathbb{S}^1$ to itself extends to a homeomorphism $\overline{\mathbb{D}} \longrightarrow \overline{\mathbb{D}}$, which, in terms of polar coordinates, can be explicitly done by the formula $(r, \theta) \to (r, f(\theta))$.

More generally, this shows that for two triangles T_1, T_2 whose boundaries meet we can define homeomorphisms f_1, f_2 to the upper and lower half-planes such that f_1 and f_2 satisfy (4.1) and agree on $\partial T_1 \cap \partial T_2$. When this occurs we say that f_1 and f_2 can be *glued together*.

We can glue together the collection of homeomorphisms $f_j^{\pm}$ to construct a continuous function $f_{\mathcal{T}(\mathcal{D})} : X \longrightarrow \widehat{\mathbb{C}}$ whose restriction $f_{\mathcal{T}(\mathcal{D})} : X^* = X \smallsetminus f_{\mathcal{T}(\mathcal{D})}^{-1} \{0, 1, \infty\} \to \widehat{\mathbb{C}} \smallsetminus \{0, 1, \infty\}$ is a topological covering. This allows us to provide X^* with the only Riemann surface structure that makes $f_{\mathcal{T}(\mathcal{D})}$ holomorphic. Once we do that, we can use Lemma 1.80 to convert X into a Riemann surface denoted $S_{\mathcal{T}(\mathcal{D})}$. Clearly $f_{\mathcal{T}(\mathcal{D})}$ becomes a morphism from $S_{\mathcal{T}(\mathcal{D})}$ to $\widehat{\mathbb{C}}$.

Suppose that in the above conditions we change the collection of homeomorphisms $f_j^{\pm} : T_j^{\pm} \longrightarrow \overline{\mathbb{H}}^{\pm}$ by a new collection $h_j^{\pm} : L_j^{\pm} \longrightarrow \overline{\mathbb{H}}^{\pm}$, where the $L_j^{\pm}$ are triangles of a new triangle decomposition $\mathcal{L}$ associated to $\mathcal{D}$ which may have been modified by a new choice of the segments γ_i^v or the centres of the faces. Let $h : X \longrightarrow \widehat{\mathbb{C}}$ be the function resulting after glueing together the new local homeomorphisms in the same way as above. Then, for every triangle $T_j^{\pm}$ we have an orientation-preserving homeomorphism $F_j^{\pm} : T_j^{\pm} \longrightarrow L_j^{\pm}$ defined by $F_j^{\pm} = \left(h_j^{\pm} \right)^{-1} \circ f_j^{\pm}$. By construction, $F_i^{\pm}$ and $F_j^{\mp}$ coincide in the intersections $T_i^{\pm} \cap T_j^{\mp}$ and, therefore, they can be glued together to construct a homeomorphism $F : X \longrightarrow X$. In particular $F : S_{\mathcal{T}(\mathcal{D})} \longrightarrow S_{\mathcal{L}(\mathcal{D})}$ is an isomorphism of Riemann surfaces that preserves each edge of $\mathcal{D}$

and makes the diagram

$$S_{\mathcal{T}} \xrightarrow{\ F\ } S_{\mathcal{L}} \tag{4.2}$$

$$\left. \begin{array}{c} f \downarrow \quad \swarrow h \\ \widehat{\mathbb{C}} \end{array} \right.$$

commutative. That is, F is an isomorphism of the ramified coverings $\left(S_{\mathcal{T}(\mathcal{D})}, f_{\mathcal{T}(\mathcal{D})}\right)$ and $\left(S_{\mathcal{L}(\mathcal{D})}, f_{\mathcal{L}(\mathcal{D})}\right)$. In other words, modulo equivalence of coverings, the pair $\left(S_{\mathcal{T}(\mathcal{D})}, f_{\mathcal{T}(\mathcal{D})}\right)$ depends only on the dessin but not on the particular choice of the associated triangle decomposition, nor on the choice of the collection of local homeomorphisms $f_j^{\pm} : T_j^{\pm} \longrightarrow \overline{\mathbb{H}}^{\pm}$.

Therefore from now on we shall write $(S_{\mathcal{D}}, f_{\mathcal{D}})$ instead of writing $\left(S_{\mathcal{T}(\mathcal{D})}, f_{\mathcal{T}(\mathcal{D})}\right)$.

Summary 4.16 (Properties of the function $f_{\mathcal{D}}$) From the above construction one can deduce the following properties of $f_{\mathcal{D}}$:

(i) $f_{\mathcal{D}}$ ramifies only at the vertices of $\mathcal{T}(\mathcal{D})$ (i.e. at the points $\bullet, \circ$ and $\times$), since $f_{\mathcal{D}}$ is a local homeomorphism away from them.

(ii) In particular, $f_{\mathcal{D}}$ has no other branch value apart from $0, 1, \infty$. Therefore $f_{\mathcal{D}}$ is a Belyi function.

(iii) $\deg\left(f_{\mathcal{D}}\right)$ agrees with the number of edges of $\mathcal{D}$, as can be seen by simply counting the number of preimages of the value $1/2 \in [0, 1]$).

(iv) The multiplicity of $f_{\mathcal{D}}$ at a vertex v of $\mathcal{D}$ is half the number of triangles surrounding v, as each pair of adjacent triangles cover a complete neighbourhood of $f_{\mathcal{D}}(v)$, hence it coincides with the degree of the vertex (count the number of preimages of some value in a neighbourhood of v).

(v) The multiplicity of $f_{\mathcal{D}}$ at the centre $\times$ of a face of $\mathcal{D}$ equals half the number of triangles around it, hence it agrees with half the number of edges of that face (for this statement to be correct, an edge that belongs at both sides to the same face must be counted twice). This is a consequence of part (iii) in Summary 4.15.

(vi) By construction, $f_{\mathcal{D}}^{-1}\left([0, 1]\right) = \mathcal{D}$.

Remark 4.17 The arguments above show also that for a given orientation-preserving homeomorphism $F : X \longrightarrow X$, the coverings $(S_{\mathcal{D}}, f_{\mathcal{D}})$ and $\left(S_{F(\mathcal{D})}, f_{F(\mathcal{D})}\right)$ are equivalent via the isomorphism $F : S_{\mathcal{D}} \longrightarrow S_{F(\mathcal{D})}$. To see this one only has to observe that $\mathcal{T}(F(\mathcal{D}))$ can be chosen to be $F(\mathcal{T}(\mathcal{D}))$.

We can now provide the promised proof of Proposition 4.10.

Proposition 4.18 *For a given dessin* $(X, \mathcal{D})$ *with* v *vertices,* e *edges and* f *faces, Euler formula* $\chi(S) = v - e + f$ *holds.*

Proof The Riemann–Hurwitz formula for $f_{\mathcal{D}}$ reads

$$-\chi(S) = \deg(f_{\mathcal{D}})(-2) + \sum (m_P(f_{\mathcal{D}}) - 1) = -2e + b_0 + b_1 + b_\infty$$

where $b_0 = \sum_{f_{\mathcal{D}}(P)=0} (m_P(f_{\mathcal{D}}) - 1)$, etc.

But $b_0 = e - v_0$ and $b_1 = e - v_1$, where v_0 and v_1 is the number of white and black vertices respectively (every edge contains one white and one black vertex).

On the other hand $2 \sum_{f_{\mathcal{D}}(P)=\infty} m_P(f_{\mathcal{D}}) = 2e$ and therefore $b_\infty = e - f$. $\qquad\qquad\square$

Definition 4.19 (1) By the term *Belyi pair* we will refer to a pair (S, f) in which S is a compact Riemann surface and f is a Belyi function.

(2) Two Belyi pairs will be considered *equivalent* when they are equivalent as ramified coverings.

(3) The pair $(S_{\mathcal{D}}, f_{\mathcal{D}})$ constructed above shall be referred as the *Belyi pair associated to the dessin* $\mathcal{D}$.

We have then the following result:

Proposition 4.20 *The rule*

$$\{Dessins\} \longrightarrow \{Belyi\ pairs\}$$
$$(X, \mathcal{D}) \longmapsto (S_{\mathcal{D}}, f_{\mathcal{D}})$$

sends equivalent dessins to equivalent Belyi pairs, and therefore it induces a well-defined map

$$\{Equiv.\ classes\ of\ dessins\} \longrightarrow \{Equiv.\ classes\ of\ Belyi\ pairs\}$$

Example 4.21 We shall determine the Belyi pair corresponding to the dessin $\mathcal{D}$ of Example 4.14 (see Figure 4.3).

First of all, let us represent $\mathcal{D}$ in the equivalent topological way depicted on top of Figure 4.6. In other words, assume that the only face of $\mathcal{D}$ is a regular Euclidean hexagon H of area $\sqrt{3}/2$. Proposition 4.13 tells us that the topological surface where $\mathcal{D}$ is embedded is constructed by identifying the edges of H as the labels indicate. Let the side pairings be induced by the obvious translations in the complex plane, in such a way that H is a fundamental domain of the action of the group Λ generated by these translations (the images of H by the elements of this group form a tessellation of $\mathbb{C}$, as indicated in the middle part of Figure 4.6).

Now, one can check directly that the quadrilateral at the bottom of Figure 4.6 is also a fundamental domain for Λ. This in turn shows that Λ is generated by the two translations identifying opposite sides. More precisely $\Lambda = \mathbb{Z} \oplus \xi_6 \mathbb{Z}$ where $\xi_6 = e^{2\pi i/6}$.

It is obvious from Figure 4.6 that the map

$$\mathbb{C}/\Lambda \longrightarrow \mathbb{C}/\Lambda$$

$$[z] \longmapsto [\xi_6 z]$$

induced by the rotation of order 6 around the origin is an automorphism. This is the same as saying that $\Lambda = \xi_6 \Lambda$ (cf. proof of Proposition 2.54), something that can be deduced directly from the identity $\xi_6^2 - \xi_6 + 1 = 0$.

The Weierstrass $\wp$ function corresponding to this lattice verifies the identity

$$\wp(z) = \xi_6^2 \wp(\xi_6 z) \tag{4.3}$$

since clearly

$$
\begin{aligned}
\wp(\xi_6 z) &= \frac{1}{\xi_6^2 z^2} + \sum_{0 \neq \omega \in \Lambda} \left(\frac{1}{(\xi_6 z - \omega)^2} - \frac{1}{\omega^2} \right) \\
&= \frac{1}{\xi_6^2 z^2} + \sum_{0 \neq \omega \in \Lambda} \left(\frac{1}{(\xi_6 z - \xi_6 \omega)^2} - \frac{1}{(\xi_6 \omega)^2} \right) \\
&= \frac{1}{(\xi_6^2)} \wp(z)
\end{aligned}
$$

the second identity being a consequence of the fact that $\xi_6 \Lambda = \Lambda$.

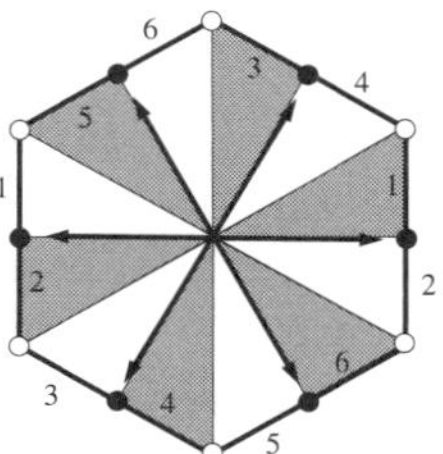

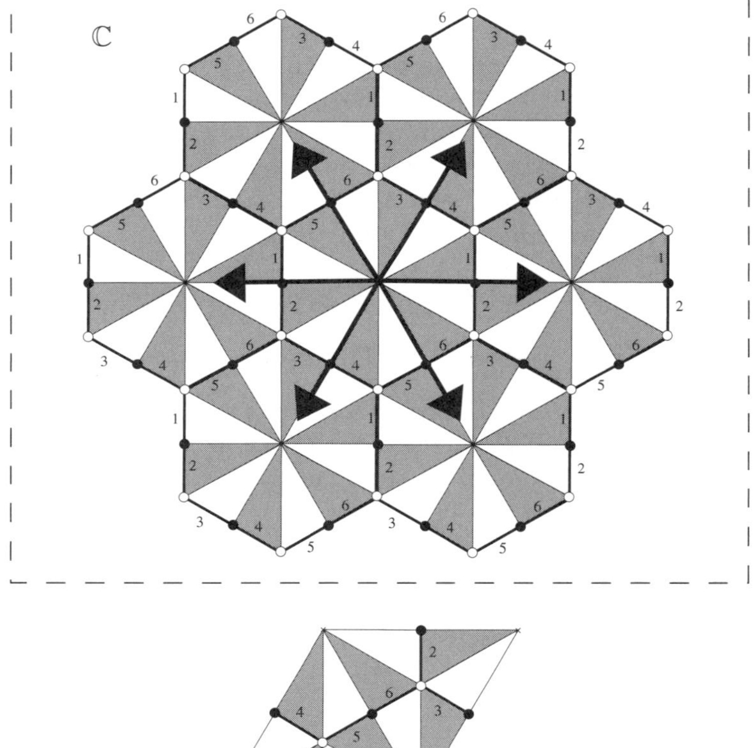

Fig. 4.6. An equivalent topological representation of the dessin in Figure 4.3. The arrows of the figure in the middle represent translations that generate a group Λ. Both the hexagon at the top and the quadrilateral at the bottom are fundamental domains of Λ, and the action of Λ identifies opposite sides.

Accordingly, for the derivative we have

$$\wp'(\xi_6 z) = \frac{1}{(\xi_6^3)}\wp'(z) = -\wp'(z) \tag{4.4}$$

The bicoloured graph $\mathcal{D}$ at the bottom of Figure 4.6 is for the moment a mere topological object embedded in the Riemann surface $\mathbb{C}/\Lambda$. If we could find a meromorphic function f on $\mathbb{C}/\Lambda$ such that $\mathcal{D} = f^{-1}([0,1])$ (with white and black vertices mapping to 0 and 1 respectively), it would follow that $S_{\mathcal{D}} = \mathbb{C}/\Lambda$ and $f_{\mathcal{D}} = f$.

In fact, up to a multiplicative constant, the function we are looking for is $\wp^3$. In order to check this, we notice some properties of this (degree 6) function:

- $\wp^3(\xi_6 z) = \wp^3(z)$.

  This is a consequence of identity (4.3).
- $\wp^3(\xi_6^k \mathbb{R}) \subset \mathbb{R}$ for $k = 0, 1, 2$.

  To see this note first that given $x \in \mathbb{R}$ we have

$$\begin{aligned}
\overline{\wp(x)} &= \frac{1}{x^2} + \sum_{0 \neq \omega \in \Lambda}\left(\frac{1}{(x-\overline{\omega})^2} - \frac{1}{\overline{\omega}^2}\right) \\
&= \frac{1}{x^2} + \sum_{0 \neq \omega \in \Lambda}\left(\frac{1}{(x-\omega)^2} - \frac{1}{\omega^2}\right) \\
&= \wp(x)
\end{aligned}$$

the second identity being a consequence of the fact that $\overline{\Lambda} = \Lambda$. The statement for $k = 1$ or 2 follows from this one by means of identity (4.3).

- $\wp^3(\xi_{12}^{-1}\mathbb{R}) = \wp^3(\xi_{12}\mathbb{R}) = \wp^3(i\mathbb{R}) \subset \mathbb{R}$.

  This is again a combination of identity (4.3) with the fact that for $x \in \mathbb{R}$ we have

$$\begin{aligned}
\overline{\wp(ix)} &= -\frac{1}{x^2} + \sum_{0 \neq \omega \in \Lambda}\left(\frac{1}{(-ix-\overline{\omega})^2} - \frac{1}{\overline{\omega}^2}\right) \\
&= -\frac{1}{x^2} + \sum_{0 \neq \omega \in \Lambda}\left(\frac{1}{(ix+\overline{\omega})^2} - \frac{1}{\overline{\omega}^2}\right) \\
&= -\frac{1}{x^2} + \sum_{0 \neq \omega \in \Lambda}\left(\frac{1}{(ix-\omega)^2} - \frac{1}{\omega^2}\right) \\
&= \wp(ix)
\end{aligned}$$

since $-\overline{\Lambda} = \Lambda$.

- $\wp^3\left(\{\mathrm{Re}(z)=\tfrac{1}{2}\}\right)\subset\mathbb{R}$ and $\wp^3\left(\{\mathrm{Re}(z)=1\}\right)\subset\mathbb{R}$.

  If $z=\xi_6+ix$ with $x\in\mathbb{R}$ we have $\wp^3(z)=\wp^3(ix)\in\mathbb{R}$. The same argument applies to $\{\mathrm{Re}(z)=1\}$ taking $z=1+ix$.

- Up to a multiplicative constant, $\wp^3$ is a Belyi function.

  Recall first that $\wp$, and therefore $\wp^3$, has a single pole located at $z=[0]$. Now, evaluating at $\xi_6 z$ the algebraic relation (cf. Section 2.2.1)

$$\wp'^2(z)=4\wp^3(z)-g_2\wp(z)-g_3$$

and using identities (4.3) and (4.4), we find that $g_2=0$ and therefore $g_3\neq 0$. In particular $\dfrac{g_3}{4}$ is the image under $\wp^3$ of any of the three finite ramification points of $\wp$, which are located at $[1/2]$, $[\xi_6/2]$ and $[(1+\xi_6)/2]$. Note that $g_3/4$ is a positive real number.

  From the identity $(\wp^3)'=3\wp'(\wp)^2$ it follows that besides the pole, $\wp^3$ has two other ramification points located at the two zeros of $\wp$. Hence $\wp^3$ has only three ramification values.

- The zeros of $\wp^3$ are the points $[z_i]$ where z_1,z_2 are the white vertices in Figure 4.6.

  An elementary computation shows that $z_1=\frac{1}{\sqrt{3}}\xi_{12}=\frac{1}{2}+\frac{\sqrt{3}}{6}i$ and $z_2=\frac{2}{\sqrt{3}}\xi_{12}=1+\frac{\sqrt{3}}{3}i$. We have

$$\begin{aligned}
\wp(z_1) &= \xi_6^2\wp(\xi_6 z_1)=\xi_6^2\wp(\tfrac{\sqrt{3}}{3}i)\\
&= \xi_6^2\wp\left((\tfrac{1}{2}+\tfrac{\sqrt{3}}{2}i)-(\tfrac{1}{2}+\tfrac{\sqrt{3}}{6}i)\right)\\
&= \xi_6^2\wp(\xi_6-z_1)=\xi_6^2\wp(z_1)
\end{aligned}$$

and, similarly,

$$\begin{aligned}
\wp(z_2) &= \xi_6^2\wp(\xi_6 z_2)=\xi_6^2\wp(\tfrac{2\sqrt{3}}{3}i)\\
&= \xi_6^2\wp\left(2(\tfrac{1}{2}+\tfrac{\sqrt{3}}{2}i)-(1+\tfrac{\sqrt{3}}{3}i)\right)=\xi_6^2\wp(2\xi_6-z_2)
\end{aligned}$$

The above properties show that white triangles (resp. black triangles) are mapped bijectively to $\overline{\mathbb{H}}^+$ (resp. $\overline{\mathbb{H}}^-$) by $\wp^3$.

  Finally, normalizing $\wp^3$ we see that the Belyi pair corresponding to $\mathcal{D}$ is

$$(S_{\mathcal{D}},f_{\mathcal{D}})=\left(\frac{\mathbb{C}}{\mathbb{Z}\oplus\xi_6\mathbb{Z}},\frac{4}{g_3}\wp^3\right)$$

which is equivalent to the Belyi pair

$$\left(\left\{y^2 = x^3 - 1\right\}, \mathbf{x}^3\right)$$

as an obvious consequence of the isomorphism (2.15), once we know that $g_2 = 0$.

4.3 From Belyi pairs to dessins

Suppose that (S, f) is a Belyi pair, and set $\mathcal{D}_f = f^{-1}([0, 1])$. We consider $\mathcal{D}_f$ as a bicoloured graph embedded in S whose set of white (resp. black) vertices is $f^{-1}(0)$ (resp. $f^{-1}(1)$).

Proposition 4.22 (i) *$\mathcal{D}_f$ is a dessin d'enfant.*

(ii) *Each of the sets $f^{-1}([-\infty, 0])$, $f^{-1}([0, 1])$ and $f^{-1}([1, \infty])$ is a union of topological segments. All of them together are the complete set of edges of a triangle decomposition $\mathcal{T}(\mathcal{D}_f)$.*

(iii) *$f = f_{\mathcal{D}_f}$.*

Proof We first observe that Theorem 1.74, applied to our Belyi function $f : S \longrightarrow \widehat{\mathbb{C}}$, has the following immediate consequences:

(a) Each point $p \in f^{-1}(\{0, 1\})$ admits a parametric disc D_p on which f is of the form $z \longmapsto z^{m_p}$. Hence, for ε sufficiently small, $\Gamma_p^\varepsilon := f^{-1}([0, \varepsilon]) \cap D_p$ is a collection of m_p small non-intersecting segments emanating from p.

(b) $f^{-1}\left(\widehat{\mathbb{C}} \smallsetminus [0, 1]\right) = S \smallsetminus \mathcal{D}_f$ is a disjoint union of holomorphic discs $D_1, \ldots, D_k$, one for each point $q_i \in f^{-1}(\infty)$, which serves as the centre of D_i. On each D_i, f is of the form $z \longmapsto z^{m_{q_i}}$.

(c) f induces a smooth covering map

$$f : S \smallsetminus f^{-1}(\{0, 1, \infty\}) \longrightarrow \widehat{\mathbb{C}} \smallsetminus \{0, 1, \infty\}.$$

Now we prove the three statements.

(i) $\mathcal{D}_f$ is certainly a bicoloured graph. Its edges are the various lifts of the segment $[0, 1] \subset \widehat{\mathbb{C}}$. Thus $\mathcal{D}$ is a dessin by (b).

(ii), (iii) Consider the diagram

$$
\begin{array}{ccc}
 & & S \smallsetminus f^{-1}(\{0, 1, \infty\}) \\
 & \nearrow^{\widetilde{i_k}} & \downarrow^{f} \\
\overline{\mathbb{H}}^+ \smallsetminus \{0, 1, \infty\} \xrightarrow{\ \ i\ \ } & & \widehat{\mathbb{C}} \smallsetminus \{0, 1, \infty\}
\end{array}
$$

where $\widetilde{i_k}$ is a lift of the inclusion $\overline{\mathbb{H}}^+ \smallsetminus \{0,1,\infty\} \overset{i}{\hookrightarrow} \widehat{\mathbb{C}} \smallsetminus \{0,1,\infty\}$. Remember that this can be done because $\overline{\mathbb{H}}^+ \smallsetminus \{0,1,\infty\}$ is simply connected, and that the number of possible lifts agrees with the degree of f, each one being determined by the choice of the image of, say, $\varepsilon \in (0,1)$. The commutativity of the diagram implies that $\widetilde{i_k}$ is injective. Moreover, $\widetilde{i_k}$ extends continuously to $\overline{\mathbb{H}}^+$. In order to define $\widetilde{i_k}(0)$ we use again property (a) and take $\widetilde{i_k}(0)$ to be the point $p_l \in f^{-1}(0)$ determined by the condition $\widetilde{i_k}(\varepsilon) \in D_{p_l}$. The extension $\widetilde{i_k} : \overline{\mathbb{H}}^+ \longrightarrow S$ is, by construction, still continuous and bijective. It is even a homeomorphism, as $\overline{\mathbb{H}}^+$ is compact. Therefore, $T_k^+ = \widetilde{i_k}\left(\overline{\mathbb{H}}^+\right)$ can be seen as a triangle with edges $\widetilde{i_k}([-\infty,0])$, $\widetilde{i_k}([0,1])$ and $\widetilde{i_k}([1,\infty])$, where $k = 1,\ldots,\deg(f)$ and $\infty = -\infty$. Similarly, $T_k^- = \widetilde{i_k}\left(\overline{\mathbb{H}}^-\right)$. All these triangles together give a triangle decomposition $\mathcal{T}(\mathcal{D}_f)$ of S associated to $\mathcal{D}$ and, by construction, $f = f_{\mathcal{T}(\mathcal{D})}$. $\qquad\square$

We have then defined a correspondence which goes in the opposite direction of that of Proposition 4.20, namely:

Proposition 4.23 *The rule*

$$\{Belyi\ pairs\} \longrightarrow \{Dessins\}$$
$$(S,f) \longmapsto (S,\mathcal{D}_f)$$

sends equivalent Belyi pairs to equivalent dessins, therefore it induces a well-defined map

$$\{Equiv.\ classes\ of\ Belyi\ pairs\} \longrightarrow \{Equiv.\ classes\ of\ dessins\}$$

Proof If two Belyi pairs (S_1,f_1) and (S_2,f_2) are equivalent through an isomorphism $\tau : S_1 \longrightarrow S_2$, then the homeomorphism induced by τ on the underlying topological surfaces S_1 and S_2 gives an equivalence of the corresponding dessins $\mathcal{D}_1 = f_1^{-1}([0,1])$ and $\mathcal{D}_2 = f_2^{-1}([0,1]) = \tau^{-1}(\mathcal{D}_1)$. $\qquad\square$

Example 4.24 We find now the dessin corresponding to the Belyi pair (C, f) in Example 3.4, that is

$$C = \left\{ y^2 = x(x-1)(x-\sqrt{2}) \right\} \xrightarrow{\ \ f\ \ } \widehat{\mathbb{C}}$$

$$(x, y) \longmapsto \frac{-4\left(x^2 - 1\right)}{\left(x^2 - 2\right)^2}$$

To find $f^{-1}([0,1])$ it is helpful to regard f as a composition of simpler maps and then compute the inverse image step by step.

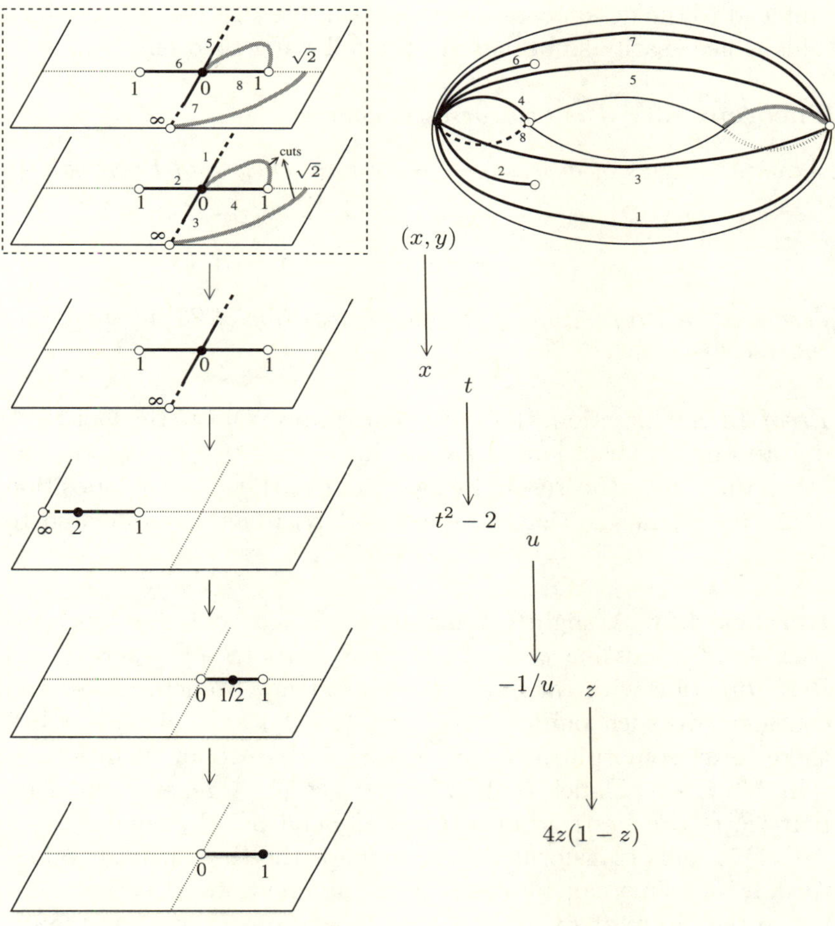

Fig. 4.7. $\mathcal{D} = f^{-1}([0,1])$; f as in Example 3.4.

The result is shown in Figure 4.7, where the topological torus underlying the algebraic curve C is constructed by glueing two copies of the complex plane along cuts from 0 to 1 and from $\sqrt{2}$ to ∞ (see Remark 2.62).

An analogous computation with the Belyi pair

$$C = \{y^2 = x^3 - 1\} \xrightarrow{\ f\ } \widehat{\mathbb{C}}$$
$$(x, y) \longmapsto x^3$$

will lead to the dessin considered in Example 4.21 (see Figure 4.3). This is necessarily so because of the following theorem:

Theorem 4.25 *The two correspondences*

$$\{Equiv.\ classes\ of\ dessins\} \longrightarrow \{Equiv.\ classes\ of\ Belyi\ pairs\}$$
$$(X, \mathcal{D}) \longmapsto (S_\mathcal{D}, f_\mathcal{D})$$
$$(S, \mathcal{D}_f) \longleftarrow (S, f)$$

described in Proposition 4.20 and Proposition 4.23 are mutually inverse.

Proof In one direction, the statement simply reflects the fact that, due to our construction of $f_\mathcal{D}$, we have $f_\mathcal{D}^{-1}([0, 1]) = \mathcal{D}$. In the other direction, the result follows from part (iii) of Proposition 4.22, since it means that f serves as Belyi function associated to $\mathcal{D}_f$. $\qquad\square$

Remark 4.26 A slightly different approach to Belyi functions consists of admitting *generalized* Belyi pairs (S, f), where f is a Belyi function with (at most) three arbitrary branch values, and consider two such pairs (S_1, f_1) and (S_2, f_2) as equivalent when there is an isomorphism $F : S_1 \longrightarrow S_2$ and a Möbius transformation $M : \widehat{\mathbb{C}} \longrightarrow \widehat{\mathbb{C}}$ such that $M \circ f_1 = f_2 \circ F$. A generalized Belyi pair (S, f) can be transformed into a Belyi pair by composing f with a Möbius transformation that maps the three branch values to $0, 1, \infty$. This can be done in six different ways, corresponding to the group of six Möbius transformations permuting the set $\{0, 1, \infty\}$. On the topological setting, these six transformations correspond to a permutation of the labels $\circ, \bullet, \times$ in the triangle decomposition $\mathcal{T}(\mathcal{D})$, which in turn corresponds to transposing the colours of the vertices of $\mathcal{D}$, or replacing $\mathcal{D}$ by its dual graph, or a combination of both operations. These constructions produce dessins that here we do not consider equivalent to $\mathcal{D}$.

Remark 4.27 As we mentioned already at the beginning of this chapter, Grothendieck considered only what some people afterwards called *clean dessins*. These are the dessins in which all white (or black) vertices have degree 2. This is not an important restriction, since to an arbitrary dessin we can always associate a clean one in a very simple manner.

Let $(X, \mathcal{D})$ be an arbitrary dessin whose associated Belyi pair is (S, f). The combinatorial process of giving all vertices of $\mathcal{D}$ black colour and placing extra white vertices in the middle of the edges produces a clean dessin $(X, \mathcal{D}^c)$ whose associated Belyi pair is clearly $(S, 1 + 4f(f - 1))$. The function

$$f_c = 1 + 4f(f - 1) = c \circ f, \quad \text{where } c(x) = 1 + 4x(x - 1)$$

has a ramification index equal to 2 at all points in the fibre above 0 (such Belyi functions are called *clean*), reflecting the fact that all white vertices of $\mathcal{D}^c$ have degree 2. Similarly, the function

$$C \circ f, \quad \text{where } C(x) = 1 - c(x) = -4x(x - 1)$$

is a Belyi function on S with a ramification index equal to 2 at all points in the fibre above 1. The corresponding dessin $\mathcal{D}_{C \circ f}$ is obtained by interchanging the colours of the vertices of $\mathcal{D}^c$.

Note, in particular, that $\mathcal{D}$ and $\mathcal{D}^c$ determine the same complex structure S on X, therefore all Belyi surfaces admit clean Belyi functions.

Example 4.28 In fact the dessin in Example 4.21 is the clean dessin associated to the one depicted in Figure 4.8.

The corresponding Belyi function $h : \mathbb{C}/\mathbb{Z} \oplus \xi_6 \mathbb{Z} \longrightarrow \widehat{\mathbb{C}}$ such that $f_{\mathcal{D}} = 4\wp^3/g_3 = C \circ h$ is

$$h := \frac{1 + \wp'/\sqrt{-g_3}}{2}$$

This is easily checked:

$$1 - (2h - 1)^2 = 1 - (\wp'/\sqrt{-g_3})^2 = 1 + (\wp')^2/g_3 = 4\wp^3/g_3$$

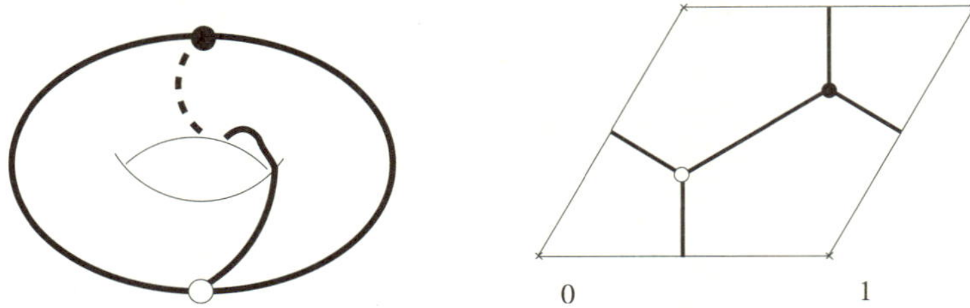

Fig. 4.8. Another dessin inducing the Riemann surface of Example 4.21.

Note that h is indeed a Belyi function. To see it one takes derivatives in the identity $\wp'^2(z) = 4\wp^3(z) - g_3$ to get $\wp'' = 6\wp^2$. This means that $\wp'$ ramifies only at the pole located at $[0]$ and at the two zeros $[z_1], [z_2]$ of $\wp$, i.e. the white vertices in Figure 4.6.

4.3.1 The monodromy of a Belyi pair

When $f : S \longrightarrow \widehat{\mathbb{C}}$ is a Belyi function, say with degree d, its monodromy can be described in a particularly simple way. In this case $\pi_1\left(\widehat{\mathbb{C}} \setminus \{0, 1, \infty\}, y\right)$ is a rank 2 free group (Theorem 2.34) generated by two loops γ_0 and γ_1 based at $y = 1/2$, say, turning counterclockwise once around the points 0 and 1 respectively (Figure 2.20).

The monodromy homomorphism

$$M_f : \pi_1\left(\widehat{\mathbb{C}} \setminus \{0, 1, \infty\}, y\right) \longrightarrow \Sigma_d$$

is therefore determined by the two permutations $M_f(\gamma_0) = \sigma_{\gamma_0}^{-1}$ and $M_f(\gamma_1) = \sigma_{\gamma_1}^{-1}$. Note that if (σ_0, σ_1) is the permutation representation pair of $\mathcal{D}_f$ then $\sigma_{\gamma_0} = \sigma_0$. This is because if x_j is the point in $f^{-1}(y)$ which lies in the edge j of $\mathcal{D}_f$ then the lift of γ_0 with initial point x_j ends at $x_{\sigma_0(j)}$ since f is of the form $z \mapsto z^n$ in a neighbourhood of a point in $f^{-1}(0)$ (a white point of $\mathcal{D}$), see Figure 4.9. A similar argument shows that $\sigma_{\gamma_1} = \sigma_1$.

Therefore, we have the following proposition:

Proposition 4.29 *The permutation representation pair of a dessin and the monodromy of the corresponding Belyi pair are determined by each other.*

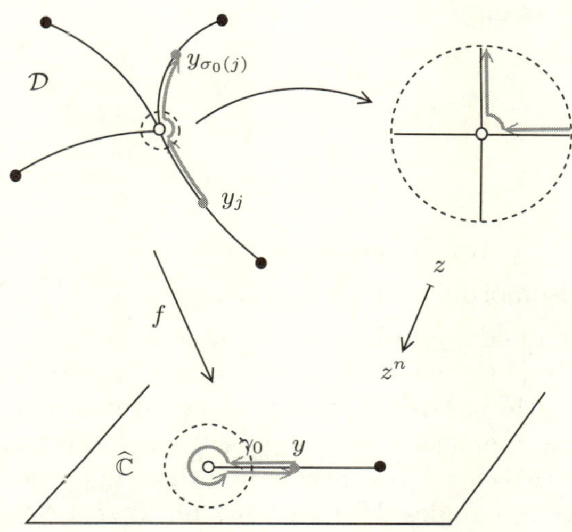

Fig. 4.9. A lift of γ_0.

If (S, f) is the Belyi pair corresponding to a dessin d'enfant $(X, \mathcal{D})$ with permutation representation pair (σ_0, σ_1), we will refer to the permutation group generated by σ_0 and σ_1 as the *monodromy group of the dessin* (or monodromy group of the Belyi pair), and we will denote it by $\mathrm{Mon}(\mathcal{D})$ or $\mathrm{Mon}(f)$.

4.4 Fuchsian group description of Belyi pairs

Suppose (S, f) is a Belyi pair corresponding to a dessin $\mathcal{D}$ with permutation representation pair (σ_0, σ_1). The restriction of f to the Riemann surface S punctured at the vertices and face centres of $\mathcal{D}$ is an unramified covering

$$f_0 : S \smallsetminus f^{-1}\{0, 1, \infty\} \longrightarrow \widehat{\mathbb{C}} \smallsetminus \{0, 1, \infty\}$$

that can be described in terms of Fuchsian groups by the arguments in Section 2.7.1.

Recall that the Fuchsian group uniformizing $\widehat{\mathbb{C}} \smallsetminus \{0, 1, \infty\}$ is, by Theorem 2.34, the triangle group $\Gamma_{\infty, \infty, \infty} = \Gamma(2)$. Therefore, f_0 is isomorphic to the covering induced by an inclusion $K_0 < \Gamma(2)$, where $K_0 = M_j^{-1}(I(1))$ is the preimage of the stabilizer of 1 under the monodromy map

$$\Gamma(2) \xrightarrow{\ M_f\ } \Sigma_N$$
$$\gamma_0 \longmapsto \sigma_0^{-1}$$
$$\gamma_1 \longmapsto \sigma_1^{-1}$$

Here γ_0 and γ_1 correspond to the generating loops around 0 and 1 under the isomorphism $\Gamma(2) \simeq \pi_1\left(\widehat{\mathbb{C}} \setminus \{0, 1, \infty\}\right)$. Notice that these were denoted $\widetilde{x}_2$ and $\widetilde{x}_3$ in Example 2.33.

Let now n, m, l be the order of σ_0, σ_1 and $\sigma_0\sigma_1$ respectively, and let $\Gamma := \Gamma_{n,m,l}$ be a triangle group of signature (n, m, l). Let us assume for the moment that $\frac{1}{n} + \frac{1}{m} + \frac{1}{l} < 1$. We recall (Section 2.4.3) that in this case Γ is generated by three hyperbolic rotations x_1, x_2, x_3 through angles $2\pi/n$, $2\pi/m$ and $2\pi/l$ around the three vertices v_1, v_2, v_3 of the triangle T (see Figure 2.13) with angles π/n, π/m and π/l.

There is a commutative diagram

$$
\begin{array}{ccc}
\Gamma(2) & & \\
\Big\downarrow{\scriptstyle\rho} & \searrow^{M_f} & \\
\Gamma & \xrightarrow{\ M_\Gamma\ } & \Sigma_N
\end{array}
$$

where the epimorphism $\rho : \Gamma(2) \longrightarrow \Gamma$ is defined by sending γ_0, γ_1 to x_1, x_2, and M_Γ is (necessarily) defined by $M_\Gamma(x_1) = \sigma_0^{-1}$ and $M_\Gamma(x_2) = \sigma_1^{-1}$. We recall that while it is completely clear that ρ defines a homomorphism (since $\Gamma(2)$ is a free group), the reason why M_Γ is a well-defined homomorphism is that it preserves the defining relations of $\Gamma = \Gamma_{n,m,l}$, namely

$$M_\Gamma(x_1)^n = M_\Gamma(x_2)^m = M_\Gamma(x_3)^l = M_\Gamma(x_1x_2x_3) = 1$$

Notice that once ρ has been fixed M_f and M_Γ are determined by each other.

If we take now $K = M_\Gamma^{-1}(I(1))$ then ρ induces a bijection $\Gamma(2)\backslash K_0 \simeq \Gamma\backslash K$.

Recall that the inclusion $K < \Gamma$ defines a Belyi function

$$\mathbb{D}/K \xrightarrow{\ \widehat{f}\ } \mathbb{D}/\Gamma$$
$$[z]_K \longmapsto [z]_\Gamma$$

on the compact Riemann surface $\mathbb{D}/K$. We identify $\mathbb{D}/\Gamma$ with $\widehat{\mathbb{C}}$ in such a way that $[v_1]_\Gamma$, $[v_2]_\Gamma$ and $[v_3]_\Gamma$ correspond to $0, 1$ and ∞ respectively.

As we have identified $\widehat{\mathbb{C}} \setminus \{0, 1, \infty\}$ with $\mathbb{D}/\Gamma \setminus \{[v_1]_\Gamma, [v_2]_\Gamma, [v_3]_\Gamma\}$, the loops γ_0 and γ_1 can be represented in a fundamental domain Q of Γ. For instance, we can think of γ_0 as the projection to $\mathbb{D}/\Gamma$ of a piecewise geodesic path C_0 in $\mathbb{D}$ connecting first a chosen point $z \neq v_i$ in Q (whose projection will serve as base point of the fundamental group) with a point z' near v_1, followed by a path joining z' to $x_1(z')$ and from here travelling geodesically to $x_1(z)$, see Figure 4.10.

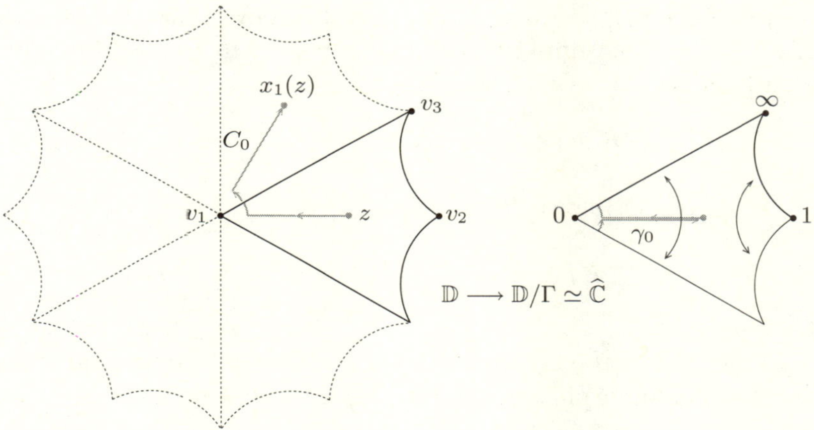

Fig. 4.10. A generator of Γ corresponds to a loop of $\pi_1\left(\widehat{\mathbb{C}} \setminus \{0, 1, \infty\}\right)$.

We claim that the monodromy $M_{\widehat{f}}$ of $\widehat{f}$ agrees with M_f, and therefore that the Belyi pair $(\mathbb{D}/K, \widehat{f})$ is equivalent to (S, f). In other words, we claim that with this identification of loops one has

$$M_f(\gamma_0)(i) = M_{\widehat{f}}(\gamma_0)(i) \quad \text{and} \quad M_f(\gamma_1)(i) = M_{\widehat{f}}(\gamma_1)(i)$$

for $i \in \{1, \ldots, N\}$.

Let $\{w_1, \ldots, w_N\}$ be a complete set of representatives of $\Gamma(2)$ modulo K_0. Then $\{\rho(w_1), \ldots, \rho(w_N)\}$ is also a complete set of representatives of Γ modulo K, and therefore

$$\{[\rho(w_1)(z)]_K, \ldots, [\rho(w_N)(z)]_K\}$$

is the $\widehat{f}$-fibre of $[z]_\Gamma$.

In terms of the natural bijection between the sets $\{1, \ldots, N\}$ and $\{K_0 w_1, \ldots, K_0 w_N\}$, the condition $M_f(\gamma_0)(i) = j$ can be rephrased (see Section 2.7.1) as $K_0 w_i \gamma_0^{-1} = K_0 w_j$.

On the other hand, if we identify also $\{1, \ldots, N\}$ to the $\widehat{f}$-fibre $\{[\rho(w_1)(z)]_K, \ldots, [\rho(w_N)(z)]_K\}$, claiming that $M_{\widehat{f}}(\gamma_0)(i) = j$ is equivalent to saying that the lift of γ_0 to the surface $\mathbb{D}/K$ starting at $[\rho(w_j)(z)]_K$ ends at $[\rho(w_i)(z)]_K$. Since this lift is nothing but the projection to $\mathbb{D}/K$ of $\rho(w_j) \circ C_0$, whose endpoint is $[\rho(w_j)(x_1(z))]_K$, we are only left to show that the identity $K_0 w_i \gamma_0^{-1} = K_0 w_j$ implies $[\rho(w_i)(z)]_K = [\rho(w_j) \circ x_1(z)]_K$, or equivalently $K\rho(w_i) = K\rho(w_j)x_1$. Now

$$
\begin{array}{ccl}
K_0 w_i \gamma_0^{-1} & = & K_0 w_j \\
& \Downarrow & \\
K\rho(w_i)\rho(\gamma_0)^{-1} & = & K\rho(w_j) \\
& \Downarrow & \\
K\rho(w_i) & = & K\rho(w_j)\rho(\gamma_0) = K\rho(w_j)x_1
\end{array}
$$

as desired.

In particular, the homomorphism M_Γ that determines the monodromy homomorphism $M_{\widehat{f}} = M_\Gamma \circ \rho$ can be described as

$$
\begin{array}{rcl}
M_\Gamma : \Gamma & \longrightarrow & \mathrm{Bij}(K\backslash\Gamma) \\
\gamma & \longmapsto & M_\Gamma(\gamma)
\end{array}
$$

where

$$
\begin{array}{rcl}
M_\Gamma(\gamma) : K\backslash\Gamma & \longrightarrow & K\backslash\Gamma \\
K\beta & \longmapsto & M_\Gamma(\gamma)(K\beta) := K\beta\gamma^{-1}
\end{array}
\tag{4.5}
$$

Notice that this is an extension of (2.27) to the case in which Γ is a triangle group. A similar discussion proves that (4.5) holds for arbitrary Fuchsian groups of genus 0.

When $\frac{1}{n} + \frac{1}{m} + \frac{1}{l} = 1$, the preceding discussion remains valid word for word if one replaces $\mathbb{D}$ with $\mathbb{C}$.

Thus we can state the following result:

Proposition 4.30 (1) *Given a Belyi pair (S, f), there exists a torsion free Fuchsian group K_0 whose inclusion in $\Gamma(2)$ induces a morphism of Riemann surfaces that gives, after compactification, a Belyi pair equivalent to (S, f).*

There also exists a (not necessarily torsion free) discrete group K whose inclusion in a triangle group $\Gamma = \Gamma_{n,m,l}$ (for some integers n, m, l) gives a Belyi pair equivalent to (S, f).

(2) The normalization $(\widetilde{S}, \widetilde{f})$ of (S, f) is induced by any of the group inclusions $\bigcap_{\gamma \in \Gamma(2)} \gamma^{-1} K_0 \gamma < \Gamma(2)$ and $\bigcap_{\gamma \in \Gamma} \gamma^{-1} K \gamma < \Gamma$.

(3) In particular, the normalization of a Belyi function is itself a Belyi function.

Proof Part (1) is the content of the comments previous to the statement of the proposition, and part (3) follows from Corollary 2.75. The statement about the inclusion $K_0 < \Gamma(2)$ in part (2) was also already proved in Proposition 2.74. Finally, the reason why the analogous statement holds for the inclusion $K < \Gamma$ is because, on the one hand, the inclusion $\bigcap_{\gamma \in \Gamma} \gamma^{-1} K \gamma < \Gamma$ is clearly normal and, on the other, the covering group $\Gamma / \bigcap_{\gamma \in \Gamma} \gamma^{-1} K \gamma$ is isomorphic to $\Gamma(2) / \bigcap_{\gamma \in \Gamma(2)} \gamma^{-1} K_0 \gamma$ via the surjection $\rho : \Gamma(2) \longrightarrow \Gamma$ defined above, hence its monodromy group has the right size. $\qquad\square$

Since an inclusion of the form $K_0 < \Gamma(2)$ (with K_0 torsion free), and also an inclusion of the form $K < \Gamma_{n,m,l}$, always induces a Belyi function in $\mathbb{D}/K_0$ and $\mathbb{D}/K$ respectively, we can state the following:

Theorem 4.31 *Let S be a compact Riemann surface different from $\mathbb{P}^1$. The following statements are equivalent:*

 (i) *S can be defined over $\overline{\mathbb{Q}}$.*
 (ii) *S admits a Belyi function.*
 (iii) *$S \simeq \dfrac{\widehat{\mathbb{H}}}{K_0}$ for some subgroup $K_0 < \Gamma(2)$.*
 (iv) *$S \simeq \dfrac{\mathbb{H}}{K}$ or $S \simeq \dfrac{\mathbb{C}}{K}$ for some subgroup K of a triangle group.*
 (v) *$S \simeq \dfrac{\widehat{\mathbb{H}}}{G}$ for some subgroup $G < \mathbb{PSL}(2, \mathbb{Z})$.*

The case $\dfrac{\mathbb{C}}{K}$ in part (iv) *can only occur in genus* 1.

Proof Only the equivalence between the last statement and the rest has not been established yet. On the one hand (iii) trivially implies (v) since $\Gamma(2) < \mathbb{PSL}(2,\mathbb{Z})$ and, on the other, an inclusion $G < \mathbb{PSL}(2,\mathbb{Z})$ induces, after compactification, a morphism to the sphere with at most three branching values, hence a Belyi function. $\qquad\square$

Example 4.32 Consider the dessin $\mathcal{D}$ with seven edges and permutation representation pair

$$\sigma_0 = (1,4,3,6,5,7,2)$$
$$\sigma_1 = (1,5,4,7,3,2,6)$$

Let (S,f) be the corresponding Belyi pair. We shall describe the Riemann surface S in the form $\mathbb{D}/K$ and the Belyi function f as an inclusion of K in a triangle group Γ. More precisely, we will exhibit a fundamental domain R and a collection of side-pairing transformations generating K. Then the dessin $\mathcal{D} = \mathcal{D}_f$ will be visible in R.

The previous discussion shows that K has to be an index 7 subgroup of the triangle group $\Gamma = \Gamma_{7,7,7}$. Let T be an equilateral hyperbolic triangle of angle $\pi/7$ and vertices v_1, v_2, v_3 in counterclockwise order. Let x_i $(i = 1,2,3)$ be a hyperbolic rotation of angle $2\pi/7$ in positive sense fixing the point v_i. The group $\langle x_1, x_2, x_3;\ x_1^7 = x_2^7 = x_3^7 = x_1 x_2 x_3 = 1 \rangle$ is our triangle group Γ, with fundamental domain $\Omega = T \cup R(T)$ where R is the reflection on the edge $[v_3, v_1]$.

The homomorphism M_Γ is determined by

$$\Gamma \xrightarrow{\quad M_\Gamma \quad} \Sigma_7$$
$$x_1 \longmapsto (1,2,7,5,6,3,4) = \sigma_0^{-1}$$
$$x_2 \longmapsto (1,6,2,3,7,4,5) = \sigma_1^{-1}$$
$$x_3 \longmapsto (1,7,6,4,2,5,3) = \sigma_1 \sigma_0$$

and therefore

$$K = M_\Gamma^{-1}(I(1)) = \{\gamma \in \Gamma \mid M_\Gamma(\gamma)(1) = 1\}$$

Thus two elements γ_1, γ_2 belong to the same coset of $K \backslash \Gamma$ if and only if $M_\Gamma(\gamma_1)(1) = M_\Gamma(\gamma_2)(1)$. We can therefore take $\{x_3, x_3^2, \ldots, x_3^7 = 1\}$ as a complete set of representatives of the seven cosets. According to Lemma 2.32

$$R = \bigcup_{i=1}^{7} x_3^i(\Omega)$$

which is a regular 14-gon with angle $2\pi/7$, is a fundamental domain for K.

We claim that K is precisely the group uniformizing Klein's Riemann surface of genus 3 (see Example 2.51). To see this, we only have to label the edges of R counterclockwise with numbers $1, \ldots, 14$, where the edge 1 is the edge $[v_1, v_2]$ of the triangle T, and check that the identification pattern is the same in both cases, namely

$$(1,6), (2,11), (3,8), (4,13), (5,10), (7,12), (9,14)$$

(see Figure 2.21). It is not difficult to find the elements of Γ realizing these identifications. For instance (see Figure 4.11), as x_1 sends edge 1 to edge 14, and x_3^3 sends edge 14 to edge 6, the identification $(1,6)$ is performed by $x_3^3 x_1$. Arguing similarly we obtain the following table:

Transformation	Action on the edges
$x_3^3 x_1$	$1 \to 6$
$x_3^5 x_2$	$2 \to 11$
$x_4^3 x_1 x_3^6$	$3 \to 8$
$x_3^6 x_2 x_3^6$	$4 \to 13$
$x_5^3 x_1 x_3^5$	$5 \to 10$
$x_3^6 x_1 x_3^4$	$7 \to 12$
$x_1 x_3^3$	$9 \to 14$

By Poincaré's polygon theorem

$$K = \langle x_3^3 x_1, x_3^5 x_2, x_4^3 x_1 x_3^6, x_3^6 x_2 x_3^6, x_5^3 x_1 x_3^5, x_3^6 x_1 x_3^4, x_1 x_3^3 \rangle$$

It is reassuring to check directly that all these generators lie in

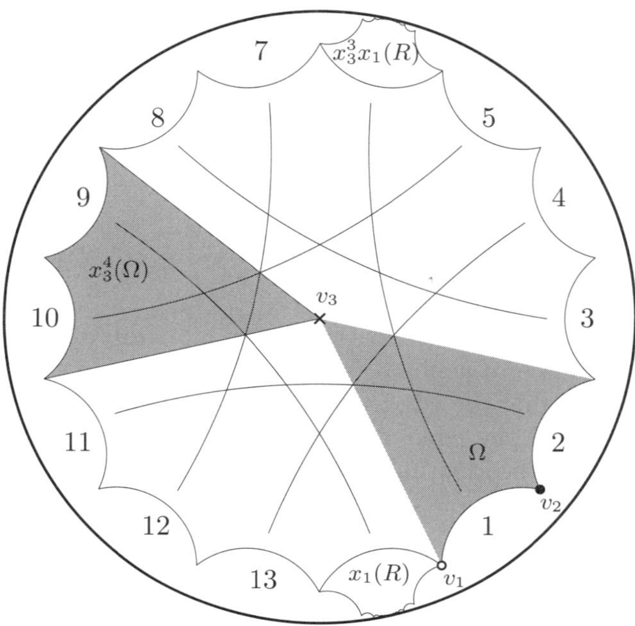

Fig. 4.11. Construction of the group K.

$K = M_\Gamma^{-1}(I(1))$ by applying M_Γ to each to find that the corresponding permutations indeed fix 1.

The resulting hyperbolic model of this dessin $\mathcal{D}$ has the boundary of R as underlying graph and the points $[v_1]_K$, $[v_2]_K$ as vertices. This is because ∂R consists of the translates of the segment $[v_1, v_2]$ by Γ, and we have identified $[v_1, v_2]$ with $[0, 1] \subset \widehat{\mathbb{C}}$.

We can think of the above construction as a hyperbolic version of the purely topological algorithm described in the proof of Proposition 4.13. In the general construction of a dessin with a prescribed permutation representation pair, one can require the faces to have a certain preferred geometric shape. For instance, in this particular case, the starting point is a single face with fourteen edges and the triangle decomposition $\mathcal{T}(\mathcal{D})$ consists of seven

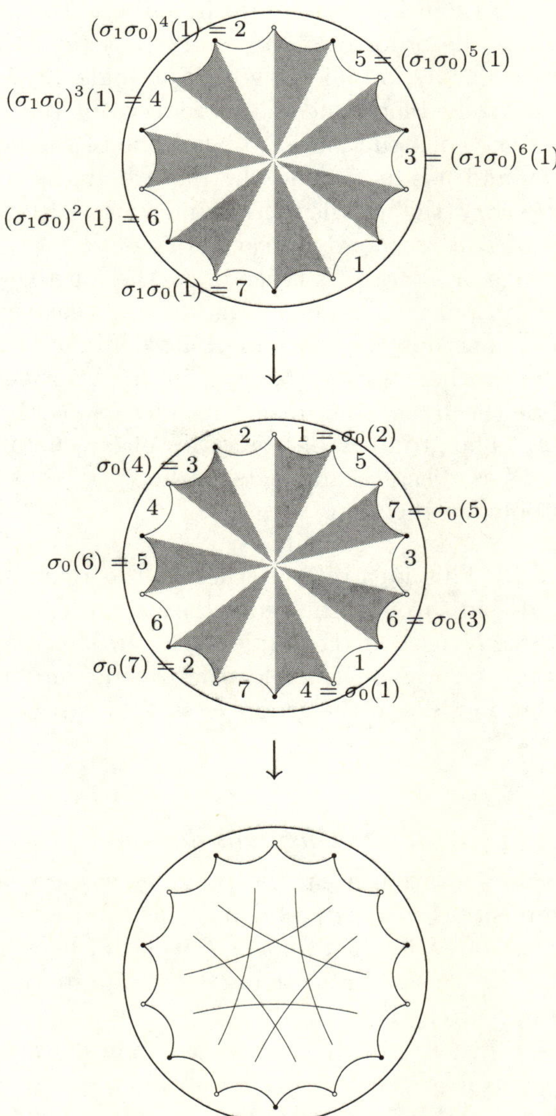

Fig. 4.12. Second description of the dessin in Example 4.32. The numbers now refer to the labels of the edges of $\mathcal{D}$ used in the permutation representation.

white triangles and seven black triangles. If we impose the condition that all of them are equilateral hyperbolic triangles of angle $\pi/7$ (since 7 is the order of σ_0, σ_1 and $\sigma_1\sigma_0$) then the face of $\mathcal{D}$ becomes a regular hyperbolic 14-gon R of angle $2\pi/7$, its vertices being alternatively black and white vertices of the dessin. The next step in the construction is to label one of the edges of R as edge 1 of $\mathcal{D}$, and use first $\sigma_1\sigma_0$ (the picture at the top of Figure 4.12) and then σ_0 (the middle picture in Figure 4.12) to determine which edge of $\mathcal{D}$ corresponds to each edge of R.

As each edge of $\mathcal{D}$ appears twice in R, the fourteen edges must be identified in pairs. The advantage now is that we can regard the identifications not only as something defining the underlying topology, but rather as side pairings in R performed by actual isometries of the hyperbolic disc (the picture at the bottom of Figure 4.12). The group K of isometries of $\mathbb{D}$ generated by these identifications is discrete and acts without fixed points due to Poincaré's polygon theorem.

Remark 4.33 This idea of constructing the Belyi surface associated to a dessin through the geometrization of Proposition 4.13 has been already used in Example 4.21. In that case, the fundamental polygon was an Euclidean hexagon, the triangle group was the group $\Gamma_{2,3,6}$, and the group K was the lattice $\mathbb{Z}\oplus\xi_6\mathbb{Z}$ (see Figure 4.6).

4.4.1 Uniform dessins

With the same notation as in the previous section, assume that the descomposition of σ_0, σ_1 and $\sigma_1\sigma_0$ as a product of disjoint cycles consists of $N/n, N/m$ and N/l cycles of length n, m and l respectively. As we saw in Section 2.4.3 the finite order elements of the triangle group $\Gamma = \Gamma_{n,m,l}$ are precisely the conjugates of powers of the generators x_1, x_2 and x_3. These are sent via the homomorphism M_Γ (see Section 4.4) to conjugates of powers of σ_0, σ_1 and σ_∞, and because of our assumptions none of them lies in the stabilizer of 1. This means that the group K is in this case torsion free, therefore it uniformizes our Belyi surface.

The following definition is then natural:

Definition 4.34 A dessin d'enfant $\mathcal{D}$ is called *uniform* if all white vertices have the same degree, and the same is true for the black vertices and the faces.

Example 4.35 The dessin $\mathcal{D}$ in Example 4.32 is uniform, hence the corresponding subgroup $K < \Gamma_{7,7,7}$ is torsion free. This group K uniformizes Klein's Riemann surface of genus 3. The dessin in Example 4.21 is uniform but, as mentioned at the end of the previous section, the group $\Gamma_{2,3,6}$ is an Euclidean triangle group and the corresponding subgroup K is in this case a lattice.

Example 4.36 Consider now the dessin $\mathcal{D}$ in Figure 4.13, already introduced at the beginning of the chapter (see Figure 4.1).

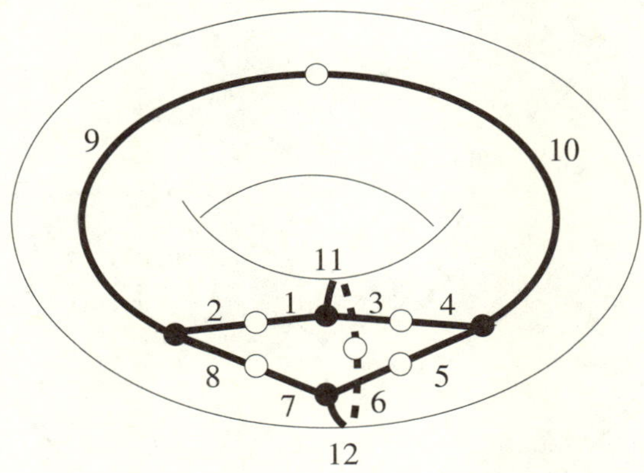

Fig. 4.13. A non-uniform dessin $\mathcal{D}$ of genus 1.

The monodromy is given by the permutations

$$\begin{aligned}
\sigma_0 &= (1,2)(3,4)(5,6)(7,8)(9,10)(11,12)\\
\sigma_1 &= (1,3,11)(2,9,8)(4,5,10)(6,7,12)\\
\sigma_1\sigma_0 &= (1,9,4,11,6,10,8,12)(2,3,5,7)
\end{aligned}$$

It follows that $\mathcal{D}$ is not uniform. It corresponds to an index 12 inclusion $K < \Gamma = \Gamma_{2,3,8}$ where K is not torsion free (of course it couldn't be since, as we know, Riemann surfaces of genus 1 are not uniformized by Fuchsian groups).

Recall that since $K = M_\Gamma^{-1}(I(1))$ the class in $K\backslash\Gamma$ of an element $x \in \Gamma$ is determined by the value $M_\Gamma(x)(1)$. We can easily check that

$$A = \{\mathrm{Id}, x_3, x_3^2, \ldots, x_3^7, x_1, x_1 x_3^7, x_1 x_3^6, x_1 x_3^5\}$$

is a complete list of representatives and therefore

$$R = \bigcup_{x \in A} x(\Omega)$$

is a fundamental domain for K (Lemma 2.32) and $\mathcal{D}$ corresponds to $\bigcup_{x \in A} x([v_1, v_2])$, see Figure 4.14.

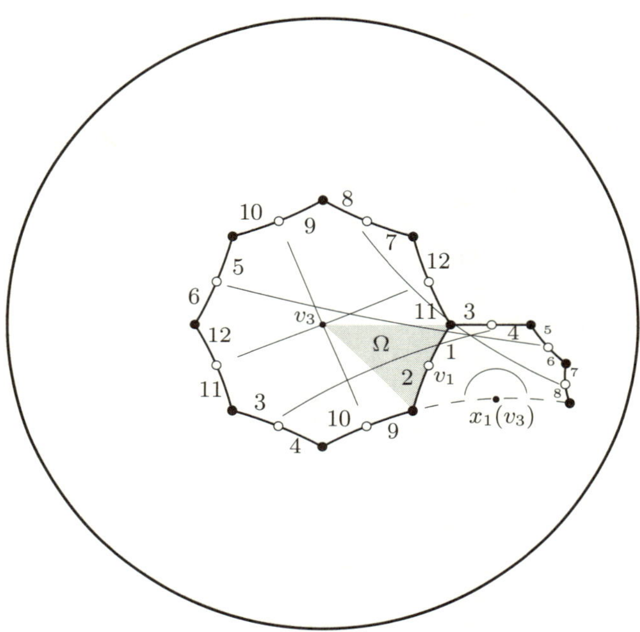

Fig. 4.14. The group K for the non-uniform dessin of Figure 4.13.

The domain R naturally decomposes into two subsets

$$R_1 = \bigcup_{j=0}^{7} x_3^j(\Omega), \quad R_2 = \bigcup_{j=0}^{3} x_1 x_3^{-j}(\Omega)$$

Here R_1 is a regular hyperbolic octogon with angle $2\pi/3$ which can be thought as the reunion of 16 triangles of angles $\pi/2, \pi/3$ and $\pi/8$ meeting at v_3. It is a geometrization of the face of $\mathcal{D}$, which, according to Proposition 4.13, corresponds to the cycle of $\sigma_1\sigma_0$ of length 8. In particular, $\partial R_1 \subset \mathcal{D}$.

On the contrary, R_2 is one half of a regular hyperbolic octogon with angle $2\pi/3$. It is the reunion of eight triangles of angles $\pi/2, \pi/3$ and $\pi/8$ meeting at $x_1(v_3)$. In the language of the proof of Proposition 4.13, this polygon corresponds to the cycle of $\sigma_1\sigma_0$ of length 4. Note that now ∂R_2 is not totally contained in $\mathcal{D}$, as the two dashed edges in Figure 4.14 do not belong to $\bigcup_{x\in A} x([v_1, v_2])$. We will see that there is an element of K which identifies these two (virtual) edges, hence in the quotient $\mathbb{D}/K$ they cancel with each other (cf. Figure 1.8). In other words, R_2 can be regarded as a topological polygon with eight edges as one expected.

The recipe given in the proof of Proposition 4.13 allows us to determine the edge-identifications in the boundary of R that give rise to $\mathcal{D}$, see Figure 4.14. The expression of the corresponding side-pairing transformations in terms of x_1, x_2 and x_3 is shown in the table below:

Transformation	Identifies the pair of edges labelled n
$x_3 x_1 x_3^3$	$n = 3$ and $n = 4$
$x_1 x_3^6 x_1 x_3^4$	$n = 5$ and $n = 6$
$x_1 x_3^5 x_1 x_3^3$	$n = 7$ and $n = 8$
$x_3^3 x_1 x_3$	$n = 9$ and $n = 10$
$x_3 x_1 x_3^3$	$n = 11$ and $n = 12$
$x_1 x_3^4 x_1$	Identifies both edges at $x_1(v_3)$

Note that, as the theory predicts, all these transformations are mapped via M_Γ into the stabilizer of 1. The element $x_1 x_3^4 x_1$ is the side-pairing transformation announced in the description of R_2. It is an order 2 hyperbolic rotation fixing $x_1(v_3)$. In particular, K is not torsion free, in accordance with the fact that $\mathcal{D}$ is not a uniform dessin.

4.4.2 Automorphisms of a dessin

Definition 4.37 Let $(X, \mathcal{D})$ be a dessin d'enfant. We denote by $\mathrm{Homeo}^+(X, \mathcal{D})$ the set of orientation-preserving homeomorphisms of X which preserve $\mathcal{D}$ as a bicoloured graph.

We define an equivalence relation in $\mathrm{Homeo}^+(X, \mathcal{D})$ by saying that $H_1 \sim H_2$ if $H_1 \circ H_2^{-1}$ preserves setwise each edge of $\mathcal{D}$. The equivalence classes by this relation will be called *automorphisms of the dessin*, and the set of all such automorphisms will be denoted by $\mathrm{Aut}(X, \mathcal{D})$ or simply as $\mathrm{Aut}(\mathcal{D})$.

Automorphisms of dessins correspond to automorphisms of the associated Belyi covers. More precisely:

Lemma 4.38 *Let $\mathcal{T} = \mathcal{T}(\mathcal{D})$ be a triangle decomposition for a dessin $(X, \mathcal{D})$ and $f_\mathcal{T}$ a continuous surjection to the sphere constructed as in Section 4.2.2. For any $H \in \mathrm{Homeo}^+(X, \mathcal{D})$ there is a unique $H_0 \in \mathrm{Homeo}^+(X, \mathcal{D})$ such that $H \sim H_0$ and $f_\mathcal{T} \circ H_0 = f_\mathcal{T}$.*

Proof We show first the existence of H_0. If, as in Remark 4.17, we denote by $H(\mathcal{T})$ the triangle decomposition obtained as the image of $\mathcal{T}$ by H, then there is an obvious choice of $f_{H(\mathcal{T})} : X \longrightarrow \widehat{\mathbb{C}}$ such that $f_{H(\mathcal{T})} \circ H = f_\mathcal{T}$. This way H induces clearly an isomorphism from $S_\mathcal{T}$ to $S_{H(\mathcal{T})}$.

Now, since $f_\mathcal{T}$ and $f_{H(\mathcal{T})}$ are obtained from the same dessin $\mathcal{D}$, the results in Section 4.2.2 show the existence of an isomorphism of Riemann surfaces $F : S_{H(\mathcal{T})} \longrightarrow S_\mathcal{T}$ preserving each edge of $\mathcal{D}$ such that $f_\mathcal{T} \circ F = f_{H(\mathcal{T})}$. Therefore we have a commutative diagram

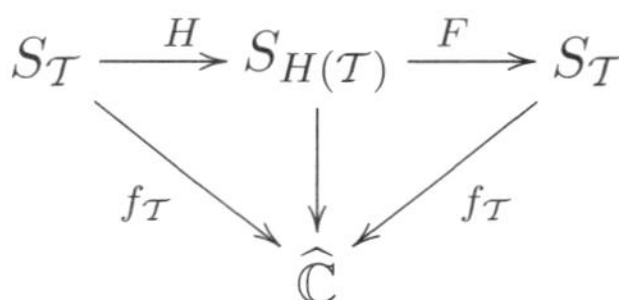

so that $f_\mathcal{T} \circ (F \circ H) = f_\mathcal{T}$. Now we can take $H_0 = F \circ H$, which, by construction, is equivalent to H.

Regarding the uniqueness of H_0, we can argue as follows. Suppose that there is another element $H_1 \in \mathrm{Homeo}^+(X, D)$ equivalent to H and satisfying $f_\mathcal{T} \circ H_1 = f_\mathcal{T}$. Then one also has the identity $f_\mathcal{T} \circ H_0 \circ H_1^{-1} = f_\mathcal{T}$. This implies firstly that $H_0 \circ H_1^{-1}$ is holomorphic with respect to the Riemann surface structure defined by $f_\mathcal{T}$, and secondly that its restriction to any edge of the dessin $e \in f_\mathcal{T}^{-1}([0, 1])$ is the identity map since the restriction of $f_\mathcal{T}$ to e is injective. Thus, by the isolated zeros principle $H_0 \circ H_1^{-1}$ equals the identity map on the whole X, as was to be seen. $\square$

Corollary 4.39 *The map*

$$\mathrm{Aut}(\mathcal{D}) \longrightarrow \mathrm{Aut}(S_{\mathcal{D}}, f_{\mathcal{D}})$$
$$H \longmapsto H_0$$

is an isomorphism of groups.

Proof Its inverse is simply given by regarding $H_0 \in \mathrm{Aut}(S_{\mathcal{D}}, f_{\mathcal{D}})$ as an element of $\mathrm{Aut}(\mathcal{D})$. $\qquad\square$

The automorphisms of a dessin can be described in a combinatorial way. Let $H \in \mathrm{Aut}(\mathcal{D})$ be an automorphism of a dessin with edges labelled e_k with $k = 1, \ldots, n$. Then H determines a permutation $\sigma_H \in \Sigma_n$ defined by $\sigma_H(i) = j$ when $H(e_i) = e_j$. As $e_{\sigma_0(i)}$ is the edge next to e_i within a given face of $\mathcal{D}$ and H is an orientation-preserving homeomorphism, $H(e_{\sigma_0(i)})$ must be the edge next to $H(e_i) = e_{\sigma_H(i)}$, that is $\sigma_H \circ \sigma_0 = \sigma_0 \circ \sigma_H$. In the same way we have $\sigma_H \circ \sigma_1 = \sigma_1 \circ \sigma_H$.

Conversely, let $(X, \mathcal{D})$ be a dessin with permutation representation pair (σ_0, σ_1). Assume there exists $\sigma \in \Sigma_n$ that commutes with σ_0 and σ_1, and let $\mathcal{T} = \{T_k^{\pm}, \ k = 1, \ldots, n\}$ be a triangle decomposition of $\mathcal{D}$.

We claim that we can choose homeomorphisms

$$H_i^{\pm} : T_i^{\pm} \longrightarrow T_{\sigma(i)}^{\pm}$$

such that all them can be glued together to form a well-defined global homeomorphism H_σ defined by $H_\sigma|_{T_i^{\pm}} = H_i^{\pm}$. For instance, glueing the three maps $\left\{ H_i^-, H_{\sigma_0(i)}^-, H_{\sigma_1^{-1}(i)}^- \right\}$ requires the compatibility conditions

$$
\begin{aligned}
H_i^+ &= H_i^- & &\text{on } T_i^+ \cap T_i^- \\
H_i^+ &= H_{\sigma_0(i)}^- & &\text{on } T_i^+ \cap T_{\sigma_0(i)}^- \\
H_i^+ &= H_{\sigma_1^{-1}(i)}^- & &\text{on } T_i^+ \cap T_{\sigma_1^{-1}(i)}^-
\end{aligned}
$$

The first condition can be achieved with no special assumption on σ, and the remaining ones can also be obviously obtained when σ commutes with σ_0 and σ_1, since this implies in particular that $T_{\sigma\sigma_0(i)}^-$ and $T_{\sigma\sigma_1^{-1}(i)}^-$ agree with $T_{\sigma_0\sigma(i)}^-$ and $T_{\sigma_1^{-1}\sigma(i)}^-$, the two neighbours of $T_{\sigma(i)}^+$ besides $T_{\sigma(i)}^-$ (see Figure 4.15).

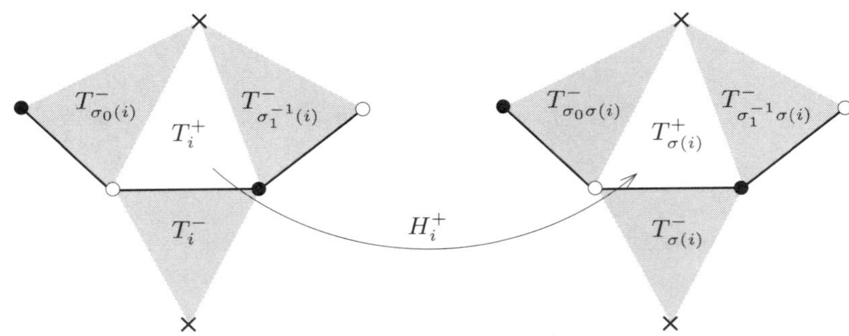

Fig. 4.15. Construction of H_σ.

We therefore have:

Theorem 4.40 Aut$(\mathcal{D})$ *is isomorphic to* $Z(\mathrm{Mon}(\mathcal{D}))$, *the centralizer of the monodromy group of* $\mathcal{D}$ *in* Σ_n.

4.4.3 Regular dessins

Definition 4.41 A dessin $(X, \mathcal{D})$ is called *regular* if Aut$(\mathcal{D})$ acts transitively on the edges of $\mathcal{D}$.

A dessin is regular if and only if the corresponding Belyi function is a Galois cover of the sphere (see Definition 2.64). This is clear since by Corollary 4.39 Aut$(\mathcal{D}) \simeq \mathrm{Aut}(S_\mathcal{D}, f_\mathcal{D})$ and the fibre $f_\mathcal{D}^{-1}(1/2)$ can be identified with the set of edges of $\mathcal{D}$.

Proposition 4.42 *Regular dessins are uniform.*

Proof Let (σ_0, σ_1) be the permutation representation pair of $\mathcal{D}$. We note that all the cycles of σ_0 must have equal length. For all j there exists $\tau_j \in \mathrm{Aut}(\mathcal{D})$ such that $\tau_j(1) = j$, and the length of the cycle of σ_0 containing 1 agrees with the length of the cycle of $\tau_j \sigma_0 \tau_j^{-1}$ containing $\tau_j(1) = j$. But, since Aut$(\mathcal{D}) = Z(\sigma_0, \sigma_1)$, we have $\tau_j \sigma_0 \tau_j^{-1} = \sigma_0$. The same argument can be applied to σ_1 and to the product $\sigma_1 \sigma_0$, therefore $\mathcal{D}$ must be uniform. $\square$

There are several equivalent ways to think of regular dessins.

Theorem 4.43 *Let $(X, \mathcal{D})$ be a dessin with corresponding Belyi pair (S, f). The following statements are equivalent:*

   (i) *$\mathcal{D}$ is regular.*

  (ii) *The Belyi function $f : S \longrightarrow \widehat{\mathbb{C}}$ is a Galois covering of the sphere.*

 (iii) *The Belyi function is induced by the inclusion of a torsion free normal subgroup inside a triangle group.*

 (iv) *The order of the monodromy group $\mathrm{Mon}(\mathcal{D})$ agrees with the number of edges of $\mathcal{D}$ or, equivalently, with the degree of f.*

Proof We have already observed that (i) $\Leftrightarrow$ (ii). On the other hand, the statement (ii) $\Leftrightarrow$ (iv) is a particular case of Proposition 2.66.

(i) $\Rightarrow$ (iii) The Fuchsian group $K < \Gamma = \Gamma_{n,m,l}$ inducing the Belyi pair is torsion free by Proposition 4.42. Now, recall that K is the preimage of the stabilizer of 1 under the homomorphism $M_\Gamma : \Gamma \longrightarrow \Sigma_N$, where N is the number of edges of $\mathcal{D}$. It remains to be shown that K is normal in Γ.

Let $x \in \Gamma$ be such that $M_\Gamma(x)(1) = 1$. By hypothesis there exists an automorphism $\tau_j \in \mathrm{Aut}(\mathcal{D}) \simeq Z(\mathrm{Mon}(\mathcal{D}))$ such that $\tau_j(1) = j$. We have

$$j = \tau_j \circ M_\Gamma(x) \circ \tau_j^{-1}(j) = M_\Gamma(x)(j)$$

where the last equality holds since τ_j commutes with $M_\Gamma(x)$. It follows that $M_\Gamma(x) = \mathrm{Id}$, that is $K = \ker(M_\Gamma) \lhd \Gamma$.

(iii) $\Rightarrow$ (ii) If $K \lhd \Gamma$ then $\Gamma \leq N(K)$ and the quotient group $\Gamma/K \simeq G$ is a subgroup of $N(K)/K \simeq \mathrm{Aut}(S)$. The Belyi function $\mathbb{D}/K \longrightarrow \mathbb{D}/\Gamma$ can be seen then as the quotient map $S \longrightarrow S/G$, and therefore it is a Galois cover. $\qquad\qquad\square$

Example 4.44 (Two dessins of genus 3 with 168 edges) The seven-edge dessin $\mathcal{D}$ we considered in Example 4.32 is obviously regular, since its monodromy group $\langle \sigma_0, \sigma_1 \rangle$ is the cyclic group of order 7. In fact $\mathrm{Aut}(\mathcal{D})$ is the group generated by the order 7 rotation around the origin τ, which acts transitively in the set of edges. According to Example 2.51 the underlying Riemann surface is $S = S_\mathcal{D} \simeq \{y^7 = x(x-1)^2\}$, $\tau(x, y) = (x, \xi_7 y)$ and the Belyi function is $\mathbf{x}(x, y) = x$.

In particular, the group K uniformizing S is normally contained in the triangle group $\Gamma_{7,7,7}$, the quotient $\Gamma_{7,7,7}/K$ being the cyclic group $\langle\tau\rangle$. We recall that $|\mathrm{Aut}(S)| = 168 > 7$, but the remaining automorphisms do not preserve the dessin $\mathcal{D}$. From the point of view of Fuchsian groups we have

$$K \lhd \Gamma_{7,7,7} < N(K)$$

the last inclusion having index $168/7 = 24$.

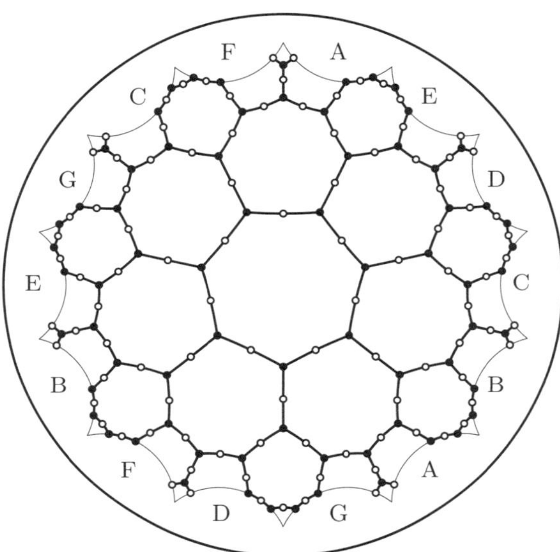

Fig. 4.16. A $(2,3,7)$-regular dessin having Klein's surface of genus 3 as the underlying Riemann surface. The side-pairing identifications defining the group K are represented here by the letters at the sides of the 14-gon.

In fact $N(K)$ has to be a triangle group $\Gamma \simeq \Gamma_{2,3,7}$ of signature $(2,3,7)$ because $|\mathrm{Aut}(S)| = [N(K) : K]$ reaches Hurwitz's bound on the number of automorphisms for a Riemann surface of genus three (see Corollary 2.42). The Galois covering of the sphere

$$\mathbb{D}/K \simeq S \longrightarrow \mathbb{D}/N(K) \simeq S/\mathrm{Aut}(S) \simeq \widehat{\mathbb{C}}$$

is the unique Belyi function on S with branching orders 2, 3 and 7. The corresponding regular $(2,3,7)$ dessin $\overline{\mathcal{D}}$ is shown in Figure 4.16. The regularity of $\overline{\mathcal{D}}$ is probably not transparent in the figure.

A tedious but safe method to confirm this is to label the edges of $\overline{\mathcal{D}}$ with numbers $1, \ldots, 168$, write down the permutation representation pair (σ_0, σ_1) and check (with the help of a computer) that the monodromy group has precisely 168 elements.

Following [Hos10] (see also [Elk99]), let us denote by $\overline{f} \in \mathcal{M}(S)$ the meromorphic function

$$\{y^7 = x(x-1)^2\} \longrightarrow \widehat{\mathbb{C}}$$
$$(x, y) \longmapsto \overline{f}(x, y) = \overline{b}(x)$$

where

$$\overline{b}(x) = \frac{\left(x^6 + 229x^5 + 270x^4 - 1695x^3 + 1430x^2 - 235x + 1\right)^3 \left(x^2 - x + 1\right)^3}{1728x(x-1)\left(x^3 - 8x^2 + 5x + 1\right)^7}$$

Clearly $\deg(\overline{f}) = 7 \cdot \deg(\overline{b}) = 7 \cdot 24 = 168$. It can be shown, again with the help of a computer, that $\overline{f}$ is invariant by the automorphism group of S. Note that it is enough to check that $\overline{f} \circ \sigma = \overline{f}$ when $\sigma \in \{\sigma_2, \sigma_3, \sigma_7\}$ is one of the generators of $\mathrm{Aut}(S)$ given in Remark 2.53. It follows that $(S, \overline{f})$ is the Belyi pair associated to $\overline{\mathcal{D}}$.

We next show that S is also the Riemann surface of a second dessin $\widetilde{\mathcal{D}}$ with the same ramification orders 2, 3 and 7 which is uniform but not regular (see [Syd97], [GTW11]). It can be constructed by rotating $\overline{\mathcal{D}}$ through an angle $2\pi/14$ (or, equivalently, through an angle π) around the origin in Figure 4.16 while leaving unchanged the fundamental 14-gon and the side-pairing identifications. The result is depicted in Figure 4.17.

The reader may find the size of the monodromy group again with the help of a computer. Since the result is $57\,624$ it follows that $\widetilde{\mathcal{D}}$ is not regular. Alternatively, the same conclusion can be obtained by the following argument regarding the Fuchsian groups involved. The dessin $\widetilde{\mathcal{D}}$ corresponds to the inclusion of the uniformizing group K in a second triangle group with signature $(2, 3, 7)$, namely the conjugate $\widetilde{\Gamma} = R\Gamma R^{-1}$ where $R(z) = \xi_{14}z$ is the rotation of order 14 around the origin. If we had $K \lhd \widetilde{\Gamma}$ then the normalizer $N(K)$ would be a an extension group of both Γ and $\widetilde{\Gamma}$. Since Γ and $\widetilde{\Gamma}$ are different groups the index $[N(K) : K]$ would be strictly larger than $[\Gamma : K] = 168 = 84 \cdot (3-1)$, which is already the maximal possible order for the automorphism group of a compact Riemann surface of genus 3. This is impossible.

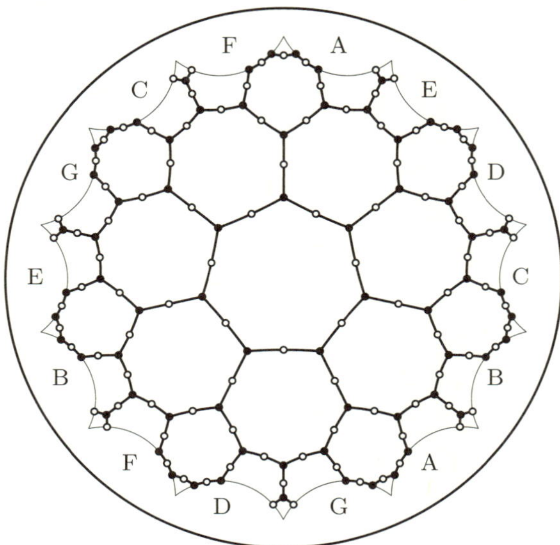

Fig. 4.17. A $(2,3,7)$-uniform (but not regular) dessin on Klein's surface of genus 3.

Let us go back to our dessin $\overline{\mathcal{D}}$. We can look at its corresponding Belyi function $\overline{f}$ from the transcendental point of view (uniformization) or from the algebraic one (algebraic curves).

The following diagram provides a decomposition of $\overline{f}$ in simpler maps connecting both points of view:

$$
\begin{array}{ccccc}
[z]_K & \longmapsto & [z]_{\Gamma_{7,7,7}} & \longmapsto & [z]_\Gamma \\
\end{array}
$$

$$
\begin{array}{ccccccc}
\mathbb{D} & \dashrightarrow & \mathbb{D}/K & \xrightarrow{\pi_1} & \mathbb{D}/\Gamma_{7,7,7} & \xrightarrow{\pi_2} & \mathbb{D}/\Gamma \\
& & \downarrow{\scriptstyle\tau} & & \downarrow{\scriptstyle\alpha} & & \downarrow{\scriptstyle\beta} \\
& & \{y^7 = x(x-1)^2\} & \xrightarrow{\mathbf{x}} & \widehat{\mathbb{C}} & \xrightarrow{\overline{b}} & \widehat{\mathbb{C}} \\
& & (x,y) & \longmapsto & x & \longmapsto & \overline{b}(x)
\end{array}
$$

where the vertical maps τ, α, β are isomorphisms making the diagram commutative. Here α, β are determined by the property that in each case the points $[v_1], [v_2], [v_3]$ are sent to $0, 1, \infty \in \widehat{\mathbb{C}}$ respectively.

Since the triangle group $\Gamma_{7,7,7}$ is also contained in $\widetilde{\Gamma}$, we have a similar diagram regarding the Belyi function $\widetilde{f}$ corresponding to $\widetilde{\mathcal{D}}$, namely

$$\mathbb{D} \dashrightarrow \mathbb{D}/K \xrightarrow{\pi_1} \mathbb{D}/\Gamma_{7,7,7} \xrightarrow{\pi_3} \mathbb{D}/\widetilde{\Gamma}$$

with vertical maps τ, α, $\widetilde{\beta}$ to

$$\{y^7 = x(x-1)^2\} \xrightarrow{\ x\ } \widehat{\mathbb{C}} \xrightarrow{\ \widetilde{b}\ } \widehat{\mathbb{C}}$$

where $\widetilde{b}$ is the function whose explicit expression we want to find out.

From the point of view of Fuchsian groups it is obvious that the functions $\overline{f}$ and $\widetilde{f}$ can be related by means of the rotation R as follows:

$$
\begin{array}{ccccc}
\mathbb{D} \dashrightarrow \mathbb{D}/K & \longrightarrow & \mathbb{D}/\Gamma_{7,7,7} & \longrightarrow & \mathbb{D}/\Gamma \\
\Big\downarrow R & & \Big\downarrow R & & \Big\downarrow R \\
\mathbb{D} \dashrightarrow \mathbb{D}/K & \longrightarrow & \mathbb{D}/\Gamma_{7,7,7} & \longrightarrow & \mathbb{D}/\widetilde{\Gamma}
\end{array}
$$

where the three vertical isomorphisms are induced by R. Note that $R \in N(\Gamma_{7,7,7})$. On the contrary, R does not belong to $N(K)$ and this is why there is no corresponding isomorphism in the second column.

The information contained in these diagrams can be assembled together as follows:

$$
\begin{array}{ccccc}
\mathbb{D} \dashrightarrow \mathbb{D}/K \simeq \{y^7 = x(x-1)^2\} & \xrightarrow{\ x\ } & \mathbb{D}/\Gamma_{7,7,7} \simeq \widehat{\mathbb{C}} & \xrightarrow{\ \overline{b}\ } & \mathbb{D}/\Gamma \simeq \widehat{\mathbb{C}} \\
\Big\downarrow R & & \Big\downarrow R \;\; M & & \Big\downarrow R \;\; \widetilde{M} \\
\mathbb{D} \dashrightarrow \mathbb{D}/K \simeq \{y^7 = x(x-1)^2\} & \xrightarrow{\ x\ } & \mathbb{D}/\Gamma_{7,7,7} \simeq \widehat{\mathbb{C}} & \xrightarrow{\ \widetilde{b}\ } & \mathbb{D}/\widetilde{\Gamma} \simeq \widehat{\mathbb{C}}
\end{array}
$$

where by construction the Möbius transformations M and $\widetilde{M}$ are given by $M = \alpha R \alpha^{-1}$ and $\widetilde{M} = \widetilde{\beta} R \beta^{-1}$. Here the isomorphism $\widetilde{\beta} : \mathbb{D}/\widetilde{\Gamma} \to \widehat{\mathbb{C}}$ is again determined the condition that $[v_1]$, $[v_2]$, $[v_3]$ are sent to $0, 1, \infty \in \widehat{\mathbb{C}}$ respectively.

Now clearly R fixes the point $[v_3]_{\Gamma_{7,7,7}}$ and permutes $[v_1]_{\Gamma_{7,7,7}}$ and $[v_2]_{\Gamma_{7,7,7}}$. It is also clear that R sends $[v_1]_\Gamma$, $[v_2]_\Gamma$ and $[v_3]_\Gamma$ to $[v_1]_{\widetilde{\Gamma}}$, $[v_2]_{\widetilde{\Gamma}}$ and $[v_3]_{\widetilde{\Gamma}}$ respectively. This means that M fixes $\infty \in \widehat{\mathbb{C}}$ and interchanges 0 and 1, while $\widetilde{M}$ fixes $0, 1$ and ∞. In other words, $M(x) = 1 - x$ and $\widetilde{M} = \mathrm{Id}$.

We find that our Belyi function $\widetilde{f} = \widetilde{b} \circ \mathbf{x} = \overline{b} \circ M^{-1} \circ \mathbf{x}$ is

$$\{y^7 = x(x-1)^2\} \longrightarrow \widehat{\mathbb{C}}$$

$$(x,y) \longmapsto \widetilde{f}(x,y) = \widetilde{b}(x)$$

where

$$\widetilde{b}(x) = -\frac{\left(x^6 - 235x^5 + 1430x^4 - 1695x^3 + 270x^2 + 229x + 1\right)^3 \left(x^2 - x + 1\right)^3}{1728x(x-1)\left(x^3 + 5x^2 - 8x + 1\right)^7}$$

Example 4.45 The genus 1 dessin considered in Example 4.21 is also a regular one. This can be checked either by computation of the monodromy group, or by direct inspection of Figure 4.6, where it becomes clear that an euclidean rotation of order 6 generates a subgroup of the automorphism group of the dessin acting transitively on the edges.

4.5 The action of $\mathrm{Gal}(\overline{\mathbb{Q}})$ on dessins d'enfants

In Section 3.3 we described how $\mathrm{Gal}(\mathbb{C}) = \mathrm{Gal}(\mathbb{C}/\mathbb{Q})$ acts on compact Riemann surfaces and morphisms by conjugation on the coefficients of the polynomials describing these objects. Since Belyi pairs are defined over $\overline{\mathbb{Q}}$ the absolute Galois group $\mathrm{Gal}(\overline{\mathbb{Q}})$ acts on them in the same way, and therefore we can make $\mathrm{Gal}(\overline{\mathbb{Q}})$ act on the dessins themselves. The transform $\mathcal{D}^\sigma$ of a dessin $\mathcal{D}$ by an element $\sigma \in \mathrm{Gal}(\overline{\mathbb{Q}})$ is defined by the rule

$$
\begin{array}{ccc}
\mathcal{D} & \dashrightarrow & \mathcal{D}^\sigma \\
\downarrow & & \uparrow \\
(S_{\mathcal{D}}, f_{\mathcal{D}}) & \xrightarrow{\;\sigma\;} & (S_{\mathcal{D}}^\sigma, f_{\mathcal{D}}^\sigma)
\end{array}
$$

that is if $(S_{\mathcal{D}}, f_{\mathcal{D}})$ is the Belyi pair corresponding to $\mathcal{D}$ (see Proposition 4.20), then we define $\mathcal{D}^\sigma$ to be the dessin corresponding to the conjugate Belyi pair $(S_{\mathcal{D}}^\sigma, f_{\mathcal{D}}^\sigma)$ (see Proposition 4.23).

Theorem 4.46 *Let $\mathcal{D}$ be a dessin. The following properties of $\mathcal{D}$ remain invariant under the action of the absolute Galois group:*

(1) *The number of edges.*
(2) *The number of white vertices, black vertices and faces.*
(3) *The degree of the white vertices, black vertices and faces.*

(4) *The genus.*

(5) *The monodromy group.*

(6) *The automorphism group.*

Proof This is a direct consequence of Theorem 3.28.   □

The number and degrees of vertices and faces are the most simple invariants of the Galois action. Sometimes these invariants do not permit to distinguish distinct Galois orbits while the monodromy group does. One can even construct more complicated invariants based on the monodromy with the help of the so-called Belyi-extending maps, a term first introduced in [Woo06].

Definition 4.47 A rational function $R : \widehat{\mathbb{C}} \longrightarrow \widehat{\mathbb{C}}$ is called *Belyi-extending* if it satisfies the following conditions:

- R is a Belyi function.
- R is defined over the rationals.
- $R\{0, 1, \infty\} \subset \{0, 1, \infty\}$.

If (S, f) is a Belyi pair and R is a Belyi-extending map, then $(S, R \circ f)$ is also a Belyi pair. Now, given $\sigma \in \mathrm{Gal}(\overline{\mathbb{Q}})$ we have $(S^\sigma, (R \circ f)^\sigma) = (S^\sigma, R \circ f^\sigma)$. Therefore, if $\mathcal{D}$ and $\mathcal{D}^R$ denote the dessins corresponding to f and $R \circ f$ respectively, the monodromy group of $\mathcal{D}^R$ is a Galois invariant of the dessin $\mathcal{D}$, and not only of the dessin $\mathcal{D}^R$. We shall refer to this group as the R-monodromy group of $\mathcal{D}$.

For example, the function $c(z) = 1 + 4z(z-1)$ is Belyi-extending. In fact, $\mathcal{D}^c$ is the clean dessin associated to $\mathcal{D}$, which is obtained from $\mathcal{D}$ by transforming all its vertices into black vertices and placing extra white vertices in the middle of the edges (see Remark 4.27). The c-monodromy group of a dessin $\mathcal{D}$ is sometimes called the *cartographic group* of $\mathcal{D}$ (see [JS97]).

There are in the literature some other invariants of the Galois action on dessins that we will not consider here. One would like to have a large enough collection of invariants in order to *separate* all orbits. Such a *complete* collection has not been constructed for the moment.

One important feature of the Galois action on dessin is the following result:

Theorem 4.48 *The restriction of the action of* $\mathrm{Gal}\left(\overline{\mathbb{Q}}\right)$ *to dessins of genus g is faithful for every g.*

In the next sections we prove this theorem for the distinct values of g.

4.5.1 Faithfulness on dessins of genus 0

Since $\mathbb{P}^1$ is the only compact Riemann surface of genus 0, the action of the absolute Galois group on a Belyi pair of genus 0 affects only the Belyi function and is trivial on the Belyi surface itself. This fact makes considerations quite special in this case.

Suppose $(X, \mathcal{D})$ is a dessin of genus 0 with only one face ($\mathcal{D}$ is therefore a tree). The corresponding Belyi function is a rational function $f : \widehat{\mathbb{C}} \longrightarrow \widehat{\mathbb{C}}$ with only one pole and, after composition with a Möbius transformation if necessary, we can assume that $f^{-1}(\infty) = \infty$, that is f is a polynomial. In the literature the term *Shabat polynomials* is often used to refer to such polynomial Belyi functions.

Note that trees (or, equivalently, Shabat polynomials) are preserved by the action of $\mathrm{Gal}\left(\overline{\mathbb{Q}}\right)$, since the number of faces is an invariant. We can now state the following:

Theorem 4.49 *The action of* $\mathrm{Gal}\left(\overline{\mathbb{Q}}\right)$ *on Shabat polynomials is faithful. In particular,* $\mathrm{Gal}\left(\overline{\mathbb{Q}}\right)$ *acts faithfully on dessins d'enfants of genus 0.*

Proof The following proof is due to Lenstra (see [Sch94a]).

Let $\alpha \in \overline{\mathbb{Q}}$ and $\sigma \in \mathrm{Gal}\left(\overline{\mathbb{Q}}\right)$ such that $\sigma(\alpha) \neq \alpha$. The starting point is a polynomial p_α ramified exactly at $\{0, 1, \alpha\}$ and such that the ramification numbers $m_0(p_\alpha)$, $m_1(p_\alpha)$ and $m_\alpha(p_\alpha)$ are pairwise distinct. For instance, if we set

$$p_\alpha(x) = \int x(x-1)^2(x-\alpha)^3 \in \mathbb{Q}(\alpha)[x]$$

the branching values of p_α are $\{p_\alpha(0), p_\alpha(1), p_\alpha(\alpha), \infty\} \subset \overline{\mathbb{Q}} \cup \{\infty\}$, hence we can apply Belyi's algorithm (cf. Section 3.1) to find a polynomial $q = q_\alpha \in \mathbb{Q}[x]$ such that $P_\alpha = q \circ p_\alpha$ is a Shabat polynomial.

Now, conjugation by σ sends P_α to $P_\alpha^\sigma = q^\sigma \circ p_\alpha^\sigma = q \circ p_{\sigma(\alpha)}$. Suppose that $(\widehat{\mathbb{C}}, P_\alpha)$ and $(\widehat{\mathbb{C}}, P_\alpha^\sigma)$ were equivalent Belyi pairs. Then there would be an automorphism T of $\widehat{\mathbb{C}}$ such that the diagram

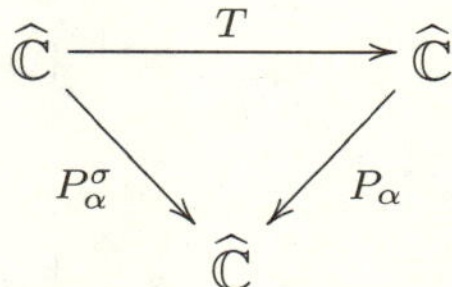

commutes. Since $T(\infty) = \infty$, it follows that T is an affine transformation $T(z) = az + b$. We therefore have

$$q\left(p_\alpha(az + b)\right) = P_\alpha(az + b) = P_\alpha^\sigma(z) = q\left(p_{\sigma(\alpha)}(z)\right)$$

From the technical Lemma 4.50 below we deduce the existence of constants c, d such that

$$p_\alpha(az + b) = c p_{\sigma(\alpha)}(z) + d \tag{4.6}$$

By construction, the ramification points of p_α are $0, 1$ and α with multiplicity 2, 3 and 4 respectively. Similarly, the ramification points of $p_{\sigma(\alpha)}$ are $0, 1$ and $\sigma(\alpha)$ with multiplicity 2, 3 and 4 respectively. Let us denote the polynomial function defined in (4.6) by P. The expression on the right-hand side of (4.6) shows that the ramification points of P of multiplicity 2, 3 and 4 agree with those of $p_{\sigma(\alpha)}$, whereas the expression on the left-hand side tells us that they are the image by $T^{-1}(z) = (z - b)/a$ of those of p_α. It follows that $b = 0$, $a = 1$ and $a\sigma(\alpha) + b = \sigma(\alpha) = \alpha$, which is a contradiction. $\square$

Lemma 4.50 (1) *Let* H_1, H_2 *be two monic polynomials of the same degree and such that* $H_1(0) = H_2(0) = 0$. *Assume that there exist polynomials* G_1, G_2 *such that* $G_1 \circ H_1 = G_2 \circ H_2$. *Then* $H_1 = H_2$.

(2) *Let* H_1 *and* H_2 *be arbitrary polynomials of the same degree such that* $G_1 \circ H_1 = G_2 \circ H_2$ *for some pair of polynomials* G_1, G_2. *Then there are constants* c, d *such that* $H_2 = cH_1 + d$.

Proof (1) Note that the two conditions $\deg(H_1) = \deg(H_2)$ and

$G_1 \circ H_1 = G_2 \circ H_2$ imply that necessarily $\deg(G_1) = \deg(G_2)$. Let

$$
\begin{aligned}
H_1(z) &= z^m + \alpha_{m-1}z^{d-1} + \cdots + \alpha_1 z \\
H_2(z) &= z^m + \beta_{m-1}z^{d-1} + \cdots + \beta_1 z \\
G_1(z) &= a_n z^n + a_{n-1}z^{n-1} + \cdots + a_1 z + a_0 \\
G_2(z) &= b_n z^n + b_{n-1}z^{n-1} + \cdots + b_1 z + b_0
\end{aligned}
$$

By hypothesis

$$
\begin{aligned}
G_1 \circ H_1(z) = \quad & a_n \left(z^m + \alpha_{m-1}z^{m-1} + \cdots + \alpha_1 z\right)^n \\
& + a_{n-1}\left(z^m + \alpha_{m-1}z^{m-1} + \cdots + \alpha_1 z\right)^{n-1} \\
& + \cdots + a_0
\end{aligned}
$$

agrees with

$$
\begin{aligned}
G_2 \circ H_2(z) = \quad & b_n \left(z^m + \beta_{m-1}z^{m-1} + \cdots + \beta_1 z\right)^n \\
& + b_{n-1}\left(z^m + \beta_{m-1}z^{m-1} + \cdots + \beta_1 z\right)^{n-1} \\
& + \cdots + b_0
\end{aligned}
$$

Comparing the terms of highest degree in both expressions we get

$$
a_n = b_n
$$

whereas the terms of degree $nm - 1$ give

$$
n a_n \alpha_{m-1} = n b_n \beta_{m-1} \quad \text{hence} \quad \alpha_{m-1} = \beta_{m-1},
$$

and those of degree $nm - 2$ produce the identity

$$
a_n \left(\binom{n}{2}\alpha_{m-1}^2 + n\alpha_{m-2}\right) = b_n \left(\binom{n}{2}\beta_{m-1}^2 + n\beta_{m-2}\right)
$$

from which we deduce $\alpha_{m-2} = \beta_{m-2}$. In general, comparison of the terms of degree $nm - j$ for $j = 1, \ldots, m - 1$ gives an identity which implies $\alpha_{m-j} = \beta_{m-j}$.

(2) Assume we have an identity $G_1 \circ H_1 = G_2 \circ H_2$ with

$$
\begin{aligned}
H_1(z) &= \alpha_m z^m + \alpha_{m-1}z^{m-1} + \cdots + \alpha_1 z + \alpha_0 \\
H_2(z) &= \beta_m z^m + \beta_{m-1}z^{m-1} + \cdots + \beta_1 z + \beta_0
\end{aligned}
$$

Denote $\widetilde{H_1} = (H_1 - \alpha_0)/\alpha_m$ and $\widetilde{H_2} = (H_2 - \beta_0)/\beta_m$. Taking $\widetilde{G_1}(x) = G_1(\alpha_m x + \alpha_0)$ and $\widetilde{G_2}(x) = G_2(\beta_m x + \beta_0)$ we have

$$
\widetilde{G_1} \circ \widetilde{H_1} = G_1 \circ H_1 = G_2 \circ H_2 = \widetilde{G_2} \circ \widetilde{H_2}
$$

Since statement (1) above implies $\widetilde{H_1} = \widetilde{H_2}$, we get

$$H_1 = \alpha_m \widetilde{H_1} + \alpha_0 = \alpha_m \widetilde{H_2} + \alpha_0 = \alpha_m (H_2 - \beta_0)/\beta_m + \alpha_0$$

and therefore the result follows by simply taking $c = \alpha_m/\beta_m$ and $d = \alpha_0 - \alpha_m\beta_0/\beta_m$. $\qquad\qquad\square$

4.5.2 Faithfulness on dessins of genus 1

The faithfullness of the action in genus 1 follows easily from the description of the moduli space in Section 2.6.1. More precisely, the key point is that the j-invariant classifies Riemann surfaces of genus 1 up to isomorphism (Corollary 2.57).

Proposition 4.51 $\mathrm{Gal}(\mathbb{C})$ *acts faithfully on the isomorphy classes of compact Riemann surfaces of genus 1.*

Proof Let $\sigma \in \mathrm{Gal}(\mathbb{C})$ and $z \in \mathbb{C}$ such that $\sigma(z) \neq z$. Take λ with $j(\lambda) = z$. Clearly $C_\lambda^\sigma = C_{\lambda^\sigma}$ has j-invariant equal to $j(\lambda^\sigma) = j(\lambda)^\sigma = \sigma(z) \neq z$, therefore it cannot be isomorphic to C_λ^σ. $\qquad\qquad\square$

Corollary 4.52 $\mathrm{Gal}(\overline{\mathbb{Q}})$ *acts faithfully on dessins d'enfants of genus 1.*

4.5.3 Faithfulness on dessins of genus g > 1

We show in this section that $\mathrm{Gal}(\overline{\mathbb{Q}})$ acts faithfully also on genus g Belyi surfaces for $g > 1$; in fact we show that the action is even faithful when restricted to hyperelliptic curves of genus g. The proof we include here was given first in [GGD07], and is a consequence of the fact that two hyperelliptic Riemann surfaces are isomorphic to each other if and only if there is a Möbius transformation relating the branching value set of the respective hyperelliptic involutions (see Corollary 2.49).

Theorem 4.53 *Let* $\sigma \in \mathrm{Gal}(\overline{\mathbb{Q}})$ *be an element of the absolute Galois group,* $\sigma \neq \mathrm{Id}$, *and* $a \in \overline{\mathbb{Q}}$ *an algebraic number such that* $\sigma(a) \neq a$. *Let* C_n *($n \in \mathbb{N}$) be the hyperelliptic curve*

$$C_n := \left\{ y^2 = (x - 1)(x - 2) \cdots (x - (2g + 1))(x - (a + n)) \right\}.$$

Then, there is an n such that C_n^σ is not isomorphic to C_n.

Proof Suppose the theorem false, so that $C_n^\sigma \simeq C_n$ for all $n \in \mathbb{N}$. Then, by Corollary 2.49, for every $n \in \mathbb{N}$ there exists some Möbius transformation $M_n \in \mathbb{PSL}(2, \mathbb{C})$ such that

$$M_n\left(\{1, 2, \ldots, 2g + 1, (a + n)\}\right) = \{1, 2, \ldots, 2g + 1, \sigma(a + n)\}$$

We have:

(1) $M_n \in \mathbb{PSL}(2, \mathbb{Q})$, since it maps three rational points to three rational points.
(2) $M_n(\{1, 2, \ldots, 2g + 1\}) = \{1, 2, \ldots, 2g + 1\}$ by (1), since $a + n \notin \mathbb{Q}$.
(3) $M_n(a + n) = \sigma(a + n) = \sigma(a) + n$, by (2).
(4) There are three distinct natural numbers p, q, r such that $M_p = M_q = M_r$. In fact, among all the transformations M_n there must be infinitely many coincidences, since by point (2) the set $\{M_n;\ n \in \mathbb{N}\}$ must contain only finitely many transformations.

We see therefore that

$$\left.\begin{array}{r} M_p(a + p) = \sigma(a) + p \\ M_p(a + q) = M_q(a + q) = \sigma(a) + q \\ M_p(a + r) = M_r(a + r) = \sigma(a) + r \end{array}\right\}$$

that is

$$\left.\begin{array}{r} M_p(a + p) - \sigma(a) = p \\ M_p(a + q) - \sigma(a) = q \\ M_p(a + r) - \sigma(a) = r \end{array}\right\}$$

Let us now consider the Möbius transformation M given by $M(z) := M_p(a+z) - \sigma(a)$. As $M(p) = p$, $M(q) = q$ and $M(r) = r$, it follows that $M = \mathrm{Id}$, and then $M_p(a + z) = z + \sigma(a)$, thus $M_p(z) = z + \sigma(a) - a$.

Now

$$1 + (\sigma(a) - a) = M_p(1) = l_1 \in \{1, 2, \ldots, 2g + 1\},$$

and similarly

$$2 + (\sigma(a) - a) = M_p(2) = l_2$$

$$\vdots$$

$$(2g + 1) + (\sigma(a) - a) = M_p(2g + 1) = l_{2g+1}$$

but then

$$(\sigma(a) - a) = (l_1 - 1) = (l_2 - 2) = \cdots = (l_{2g+1} - (2g + 1))$$

with $(l_1 - 1) \geq 0$ and $(l_{2g+1} - (2g + 1)) \leq 0$. It follows that $l_i - i = 0$ for all i, thus $l_i = M_p(i) = i$ and so $M_p = \mathrm{Id}$. Then $\sigma(a) + p = M_p(a + p) = a + p$ hence $\sigma(a) = a$, which is a contradiction. $\qquad\square$

4.6 Further examples

4.6.1 Some dessins of genus 0

Let us start with some simple remarks that will allow us to draw genus 0 dessins in a convenient manner.

(1) By the Uniformization Theorem, every genus 0 Riemann surface is isomorphic to $\widehat{\mathbb{C}}$. Thus, given a dessin $(X, \mathcal{D})$ in the (topological) surface X of genus 0 there exists an isomorphism $\tau : S_{\mathcal{D}} \longrightarrow \widehat{\mathbb{C}}$ which provides an equivalence between the Belyi pair $(S_{\mathcal{D}}, f_{\mathcal{D}})$ and a Belyi pair $(\widehat{\mathbb{C}}, R)$, where $R = f_{\mathcal{D}} \circ \tau^{-1}$ is a rational function with at most three branching values. In this way the dessin $(X, \mathcal{D})$ will be equivalent to the dessin $(\widehat{\mathbb{C}}, \tau(\mathcal{D}))$. Since the group of Möbius transformations acts transitively on triples of points of $\widehat{\mathbb{C}}$, we can choose three points of $\mathcal{D}$ (the vertices or face centres with highest degree, for instance) to correspond to three given values in $\widehat{\mathbb{C}}$. Therefore, a genus 0 dessin $\mathcal{D}$ can be always represented as a dessin $(\widehat{\mathbb{C}}, \mathcal{D})$ and, moreover, we can let any three points of $\widehat{\mathbb{C}}$ be any three points of the dessin.
(2) As we will use the extended complex plane model to represent the Riemann sphere, we will depict dessins of genus 0 as plane graphs.

The first examples we will consider are trees. We will always locate the centre of their unique face at $\infty \in \widehat{\mathbb{C}}$, hence the corresponding Belyi function will be a Shabat polynomial.

Example 4.54 Let $\mathcal{D}$ be the dessin in Figure 4.18. Let us choose the white vertex to be $0 \in \widehat{\mathbb{C}}$, and one of the black points to be $1 \in \widehat{\mathbb{C}}$. From the results of Section 4.2.2 (Summary 4.16) We deduce that $f_{\mathcal{D}}$ is a polynomial of degree 4 with a single zero of of multiplicity 4 at $z = 0$ and such that $f_{\mathcal{D}}(1) = 1$, therefore we find $f_{\mathcal{D}}(z) = z^4$. Note that ∞ is a point of ramification index 4 in agreement with the fact that $\mathcal{D}$ has four edges, each of them contained (at both sides) in the only face of $\mathcal{D}$ (see part (v) of Summary 4.16).

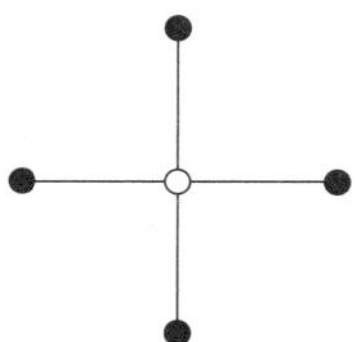

Fig. 4.18. A 4-star corresponds to $f(z) = z^4$.

In general, an n-star (with a white vertex of degree n) has $f(z) = z^n$ as associated Belyi function. A particular case is the trivial dessin in the sphere, which has only one edge (Figure 4.19), and associated Belyi function $f(z) = z$.

Fig. 4.19. The trivial dessin in the sphere.

The permutation representation pair of an n-star dessin is given by $\sigma_0 = (1, \ldots, n)$, $\sigma_1 = \mathrm{Id}$, therefore $|\mathrm{Mon}(\mathcal{D})| = n$ and so we see that these dessins are regular. Obviously, they are all defined over $\mathbb{Q}$.

Example 4.55 Besides the 4-star one there are, up to colour exchange, only two more trees with four edges (see Figure 4.20).

Fig. 4.20. Two trees with four edges.

If we place the white vertices of the dessin $\mathcal{D}$ in the left-hand side at 0 and 1, we see that the degree 4 polynomial corresponding to $\mathcal{D}$ can only be of the form $f(z) = Cz^3(z-1)$ for certain constant $C \neq 0$. Note that, since a Belyi function representing this dessin must exist, we know a priori that f has to have exactly one point α of multiplicity 2 (corresponding to the black point of degree 2). Finding α and making $f(\alpha) = 1$ will give us the constant C. To do this we compute the derivative

$$\frac{1}{C}f'(z) = 4z^3 - 3z^2 = z^2(4z - 3)$$

thus the point in question is $\alpha = \dfrac{3}{4}$. As $f\left(\dfrac{3}{4}\right) = -\dfrac{27}{256}C$, we find that

$$f_{\mathcal{D}}(z) = -\frac{256}{27}z^3(z - 1)$$

is the Belyi function associated to this dessin. The permutation representation pair is given by $\sigma_0 = (1, 2, 3)$, $\sigma_1 = (3, 4)$ and, since $Z(\mathrm{Mon}(\mathcal{D}))$ is trivial $\mathcal{D}$ has no automorphisms (see Theorem 4.40).

The tree on the right-hand side of Figure 4.20 corresponds, again up to a multiplicative constant, to $z^2(z - 1)^2$ (white vertices have been located at 0 and 1). A computation similar to the one carried out above shows that

$$f_{\mathcal{D}}(z) = 16z^2(z - 1)^2$$

is now the corresponding normalized Belyi function. The permutation representation pair is $(\sigma_0 = (1, 2)(3, 4), \sigma_1 = (2, 3))$, and we have now $Z(\mathrm{Mon}(\mathcal{D})) = \langle (1, 4)(2, 3) \rangle$, meaning that this dessin has only one non-trivial automorphism, the obvious one induced by the order two rotation around the black vertex of degree 2.

Note that both dessins in Figure 4.20 are particular cases of the configuration depicted in Figure 4.21, which we shall call the (m, n)-double star dessin, and whose corresponding Belyi function is of the form $f(z) = C_{m,n}z^m(z - 1)^n$, where the constant $C_{m,n}$ can be easily determined.

The action of the absolute Galois group is trivial on trees of less than five edges. This is a consequence of Theorem 4.46, since two different such trees must have either a different number of edges or, at least, a different collection of vertex degrees. In fact, it can

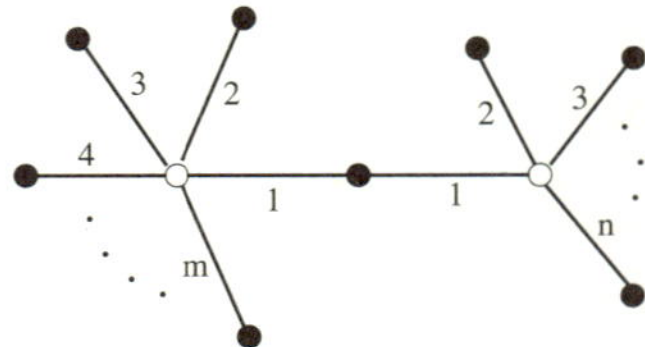

Fig. 4.21. The double star trees.

be easily seen that they are all either stars or double stars, hence their corresponding Belyi functions are defined over the rationals.

Let us consider now the case of five edges. Up to colour exchange there are six possible such trees (see Figure 4.22).

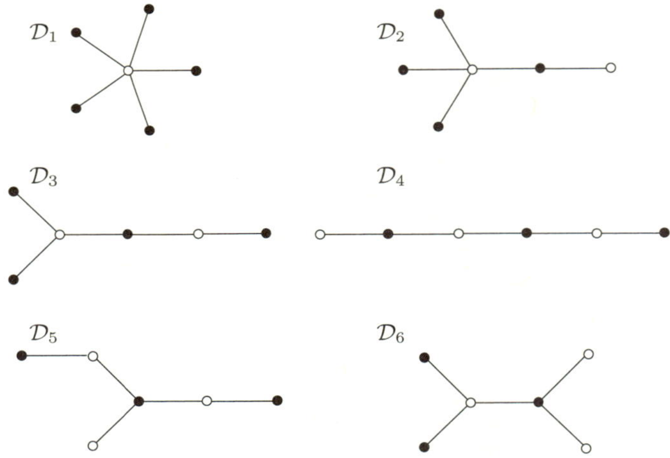

Fig. 4.22. The trees with five edges.

Note that $\mathcal{D}_1$, $\mathcal{D}_2$ and $\mathcal{D}_3$ are star or double star trees. On the other hand, $\mathcal{D}_4$ belongs also to a special family (linear trees) which we consider in the next example. As for $\mathcal{D}_5$ and $\mathcal{D}_6$, the corresponding Shabat polynomials are obtained from $z^3\left(z^2 - 2z + \frac{32}{5}\right)$ and $z^5 - \frac{10}{3}z^3 + 5z + \frac{8}{3}$ respectively after composition with a Möbius transformation sending the three branching values to 0, 1 and ∞ (see [BZ93]).

Example 4.56 (Linear trees) Let us consider the Tchebychev polynomial T_n of degree n, which is characterized by the equality $T_n(\cos z) = \cos nz$. Its explicit formula is

$$T_n(z) = \sum_{k \text{ even}} (-1)^{k/2} \binom{n}{n-k} x^{n-k}(1-x^2)^{k/2}$$

as can be seen by expanding the De Moivre's identity

$$(\cos x + i\sin x)^n = \cos nx + i\sin nx$$

and equating real parts.

Since $T_n'(x) = 0 \Leftrightarrow x = \cos\dfrac{d\pi}{n}$ for $d = 1,\ldots,n-1$, and $T_n\left(\cos\dfrac{d\pi}{n}\right) = \pm 1$, we see that T_n has three ramification values, namely $\{1, -1, \infty\}$ that can be sent to $\{0, 1, \infty\}$ by composition with the Möbius transformation $M : z \longmapsto (z+1)/2$. A straightforward computation shows that all finite ramification points of T_n have branching order 2, therefore the dessin corresponding to the Belyi function $f(z) = (T_n(z) + 1)/2$ is a linear graph in n edges. Note that since $T_n \in \mathbb{Q}[z]$, the Galois action is also trivial on linear graphs, something also predictable by consideration of the Galois invariants.

For instance, the Belyi function corresponding to the dessin $\mathcal{D}_4$ in Figure 4.22 is

$$f_{\mathcal{D}}(z) = (T_5(z) + 1)/2 = 8z^5 - 10z^3 + \frac{5}{2}z + \frac{1}{2}$$

The three fibres of the finite branching values of $f_{\mathcal{D}}$ (i.e. the vertices of $\mathcal{D}_4$) are

$$f_{\mathcal{D}}^{-1}\{0\} = \left\{-1, \cos\tfrac{\pi}{5}, \cos\tfrac{3\pi}{5}\right\}$$

$$f_{\mathcal{D}}^{-1}\{1\} = \left\{1, \cos\tfrac{2\pi}{5}, \cos\tfrac{4\pi}{5}\right\}$$

where all points are ramification points of order 2 except ± 1, which are unramified points of $f_{\mathcal{D}}$.

We remark for later use (Example 4.68) that T_5^2 is also a Belyi function. The corresponding dessin has black vertices at all points of the set

$$\left\{-1, \cos\frac{2\pi}{10}, \cos\frac{4\pi}{10}, \cos\frac{6\pi}{10}, \cos\frac{8\pi}{10}\right\}$$

and white vertices, all of them of degree 2, at the zeros of T_5 (which are all simple), namely

$$\{\cos \frac{\pi}{10}, \cos \frac{3\pi}{10}, \cos \frac{5\pi}{10}, \cos \frac{7\pi}{10}, \cos \frac{9\pi}{10}\}$$

Example 4.57 The Galois action is non-trivial already for trees with six edges. For instance, let $\mathcal{D}$ be the tree in Figure 4.23 considered in [LZ04].

Fig. 4.23. A tree $\mathcal{D}$ with six edges.

Assume that the white vertices of degree 3 and 2 are placed at $z = 0$ and $z = 1$ respectively, and let $z = a$ be the third white vertex, then the corresponding Bely function is of the form

$$f(z) = Cz^3(z-1)^2(z-a) \tag{4.7}$$

In order to determine a explicitly we compute the derivative of the function f

$$\frac{1}{C}f'(z) = 3z^2(z-1)^2(z-a) + 2z^3(z-1)(z-a) + z^3(z-1)^2$$

$$= z^2(z-1)(6z^2 + (-5a-4)z + 3a)$$

As $\mathcal{D}$ has a black vertex of degree three f must have a branch point α of order 3, distinct from 0 and 1, that occurs as double root of f'. We find out then that the discriminant of the polynomial $P(z) = 6z^2 + (-5a-4)z + 3a$ must vanish, therefore

$$25a^2 - 32a + 16 = 0 \tag{4.8}$$

hence $a = a_1 = \dfrac{4}{25}(4+3i)$ or $a = a_2 = \dfrac{4}{25}(4-3i)$.

These two values determine two Belyi functions f_j $(j = 1, 2)$. The corresponding constant C_j is deduced from the condition $f_j(\alpha_j)=1$, where $\alpha_j = \dfrac{5a_j+4}{12} = \dfrac{3 \pm i}{5}$ is the double root of the above polynomial $P(z)$. The result is $C_1 = \dfrac{3+i}{5}$, $C_2 = \dfrac{3-i}{5}$ so we get two possible candidates

$$
\begin{aligned}
f_1(z) &= \frac{3+i}{5} z^3 (z-1)^2 \left(z - \frac{4}{25}(4+3i) \right) \\
f_2(z) &= \frac{3-i}{5} z^3 (z-1)^2 \left(z - \frac{4}{25}(4-3i) \right)
\end{aligned}
$$

In principle, f_1 and f_2 could determine equivalent dessins, but this is not the case for the following reason. The dessin $\overline{\mathcal{D}}$ in Figure 4.24 has also six edges and the same collection of vertex-degrees as $\mathcal{D}$, hence the arguments above apply to it as well. In particular, its corresponding Belyi function is also either f_1 or f_2.

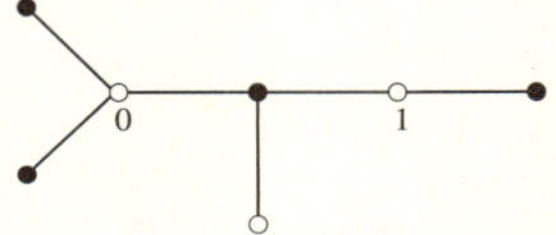

Fig. 4.24. The tree $\overline{\mathcal{D}}$ is Galois conjugate to $\mathcal{D}$.

As, clearly, there is no orientation-preserving homeomorphism of the sphere sending $\mathcal{D}$ to $\overline{\mathcal{D}}$ we may conclude that, as claimed, f_1 and f_2 correspond to the two non-equivalent dessins $\mathcal{D}$ and $\overline{\mathcal{D}}$. It follows that $\mathcal{D}$ and $\overline{\mathcal{D}}$ must lie in the same Galois, for obviously $\overline{\mathcal{D}} = \mathcal{D}^\sigma$, that is $f_2 = f_1^\sigma$, for any $\sigma \in \mathrm{Gal}(\overline{\mathbb{Q}})$ such that $\sigma(i) = -i$. In fact, since this kind of elements σ are the only ones that act non-trivially on f_1, we see that $\{\mathcal{D}, \overline{\mathcal{D}}\}$ is a complete $\mathrm{Gal}(\overline{\mathbb{Q}})$-orbit.

As a matter of fact, it can be easily seen (for instance, by directly computing the inverse image of the segment $[0,1]$) that $f_1 = f_{\mathcal{D}}$ and $f_2 = f_{\overline{\mathcal{D}}}$.

We finally observe that the graph $\overline{\mathcal{D}}$ is the image of the graph $\mathcal{D}$ by the complex conjugation $z \mapsto \overline{z}$, an orientation reversing homeomorphism which at the same time can be seen as a Galois element σ satisfying the condition $\sigma(i) = -i$ required above. We should warn the reader that this is a very special situation, since generically Galois conjugation is far from being a continuous operation, except for the case of the identity and the complex conjugation. The next example may serve as an illustration of this.

Example 4.58 We now work out the Belyi function of the tree $\mathcal{D}$ with six edges depicted in Figure 4.25 and previously considered by several authors (see [SV90], [Sch94a], [LZ04]).

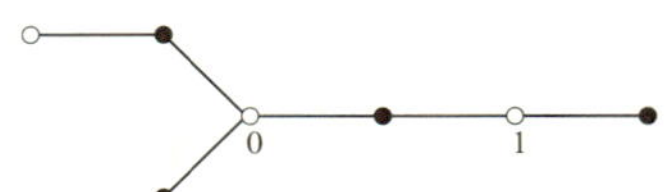

Fig. 4.25. The tree $\mathcal{D}$.

We place the vertex of degree three at $z = 0$, and the white vertex with degree 2 at $z = 1$. If we denote by $z = a$ the remaining white vertex the initial expression of the Belyi function is the same as in the previous example, i.e. $f(z) = Cz^3(z - 1)^2(z - a)$ and $(1/C)f'(z) = z^2(z - 1)P(z)$ with $P(z) = 6z^2 - (5a + 4)z + 3a$. But now $P(z)$ has to have two different roots corresponding to the two black vertices of degree 2. In particular, the discriminant of P does not vanish, hence

$$25a^2 - 32a + 16 \neq 0$$

We could now theoretically compute explicitly the two roots w_1, w_2 of P in terms of a, and then impose $f(w_1) = f(w_2)$ to determine the precise value of a. Following this method, however, would give us a complicated condition for a to meet.

We shall postpone this computation for the moment, and we first include here a beautiful and clever trick, taken from [LZ04]. Euclidean division of f by P gives $f = Q \cdot P + R$, where $\deg(R) \leq 1$ since $\deg(P) = 2$. Therefore, $R(z) = Az + B$, where A and B obviously depend on a. Now, when we evaluate f at the two values w_1, w_2 where P vanishes, we find $Aw_1 + B = Aw_2 + B$. But we have $w_1 \neq w_2$, therefore $A = 0$.

On the other hand, when we formally perform Euclidean division of f by P we get

$$A = -\frac{C}{6^5}(25a^2 - 32a + 16)(25a^3 - 12a^2 - 24a - 16)$$

and

$$B = \frac{C}{2^5 3^4}a(5a - 8)(25a^3 - 6a^2 + 8)$$

We find that the condition $A = 0$, together with the condition that the discriminant of P does not vanish, yields

$$25a^3 - 12a^2 - 24a - 16 = 0$$

We obtain three possible values of a, say $a = a_k$ $(k = 1, 2, 3)$, each one determining a Belyi function

$$f_k(z) = C_k z^3 (z-1)^2 (z-a_k)$$

where C_k can be explicitly obtained from the condition

$$1 = f_k(w_1) = R(w_1) = B$$

yielding $C_k = \dfrac{2^5 3^4}{a_k(5a_k - 8)(25a_k^3 - 6a_k^2 + 8)}.$

The two trees in Figure 4.26 have the same collection of vertex-degrees as $\mathcal{D}$, hence the above arguments apply to them and therefore their corresponding Belyi function must also be one of the functions f_k.

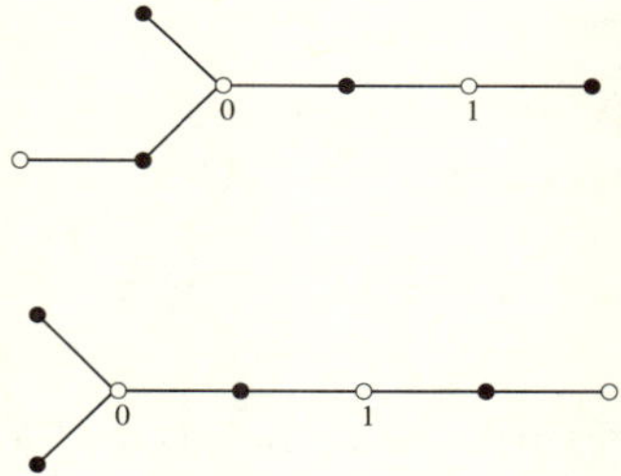

Fig. 4.26. The two Galois conjugates of the dessin in Figure 4.25.

Now, a Galois element σ permuting the roots of the polynomial $25a^3 - 12a^2 - 24a - 16$ will also permute the functions $\{f_1, f_2, f_3\}$, and this is the only kind of Galois conjugation that acts non-trivially on them. It remains to be seen that these are three non-equivalent Belyi functions. This follows from the fact that clearly there is no orientation-preserving homeomorphism between any two of the three graphs involved. We conclude that the dessins in Figures 4.25 and 4.26 form a Galois orbit with three elements.

We include next some examples of dessins on the sphere with more than one face. The reader interested in further knowledge of trees and Shabat polynomials can find much more information in the literature. In particular, there are catalogues of Belyi functions of trees up to a certain number of edges, see [BZ93].

Example 4.59 On the sphere two edges are enough to define a dessin $\mathcal{D}$ with two faces, see Figure 4.27.

Fig. 4.27. The dessin $\mathcal{D}$ on the left-hand side is the simplest possible dessin with two faces. The associated clean dessin $\mathcal{D}^c$ is shown in the right-hand side.

If we assume that the white vertex is placed at $z = 0$ and the centre of the unbounded face is placed at ∞, then f is of the form

$$f(z) = \frac{z^2}{z - \lambda}$$

for some constant λ which plays the role of the centre of the bounded face. Differentiating we get

$$f'(z) = \frac{z^2 - 2\lambda z}{(z - \lambda)^2}$$

thus the remaining ramification point of f (the black vertex) is $z = 2\lambda$. Now the value of λ that yields a normalized Belyi function is determined by the condition $1 = f(2\lambda) = 4\lambda$, hence $\lambda = 1/4$.

Note that the corresponding Belyi function f has degree 2 and only two branch values (∞ is not a ramification value of f). Note also that $\mathcal{D}$ is uniform and even regular. The associated clean dessin $\mathcal{D}^c$ in Figure 4.27 corresponds to the clean Belyi function

$$f_c(z) = 1 + 4f(z)(f(z) - 1) = 1 + \left(\frac{2z(z - 1/2)}{z - 1/4} \right)^2$$

Since $\mathcal{D}^c$ is a uniform dessin of ramification type $(2, 2, 2)$ and it has four edges, f_c must be equivalent to the Belyi function induced by an index 4 subgroup K of the triangle group $\Gamma_{2,2,2} \simeq C_2 \times C_2$ (cf. Remark 2.30), hence the trivial group. In other words, f_c is equivalent to the map

$$\widehat{\mathbb{C}} \simeq \widehat{\mathbb{C}}/\langle \mathrm{Id} \rangle \longrightarrow \widehat{\mathbb{C}}/\Gamma_{2,2,2}$$
$$z \longmapsto [z]$$

something reflected also in the fact that the triangle decomposition of $\mathcal{D}^c$ is precisely the whole tessellation of $\widehat{\mathbb{C}}$ given by the group $\Gamma_{2,2,2}$ (see Figure 2.16).

Example 4.60 The modular elliptic function

$$j(\lambda) = \frac{\left(1 - \lambda + \lambda^2\right)^3}{\lambda^2 \left(\lambda - 1\right)^2}$$

we introduced in Corollary 2.57 is another example of Belyi function. To see this, note first that two obvious ramification values of j are ∞ (which is attained at each of the points $0, 1$ and ∞ with multiplicity 2) and 0 (attained at the two roots of $\lambda^2 - \lambda + 1$ with multiplicity 3). Moreover, computation of the derivative gives

$$j'(\lambda) = \frac{(\lambda^2 - \lambda + 1)^2}{\lambda^3(\lambda - 1)^3}(2\lambda - 1)(\lambda^2 - \lambda - 2)$$

thus j has three more branching points, located at $\lambda = 1/2$ and at the two roots of $\lambda^2 - \lambda - 2$. A simple calculation shows that j maps these three points to the value $j(1/2) = 27/4$, thus the branching values of j are $\{0, 27/4, \infty\}$ and $\frac{4}{27}j$ is indeed a Belyi function.

Putting together all the relevant information about zeroes, poles, multiplicities and so on we find that the corresponding dessin $\mathcal{D}$ has six edges, two white vertices of degree 3, three black vertices of degree 2 and three faces, each of them with four edges on their boundary. One can easily get convinced that the only possible graph fulfilling all these requirements is the one shown in Figure 4.28. As j is defined over the rationals, this dessin forms a single point orbit of the Galois action.

From Figure 4.28 one sees that the permutation representation pair of $\mathcal{D}$ is $\sigma_0 = (1, 2, 3)(4, 5, 6)$, $\sigma_1 = (1, 4)(2, 6)(3, 5)$. Since these two permutations generate a group with only six elements, it follows that $\mathcal{D}$ is regular.

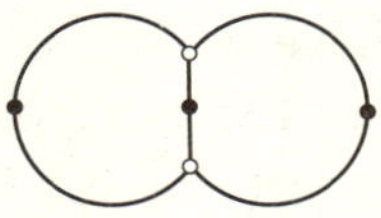

Fig. 4.28. The dessin associated to the j-function.

In fact, we know by Example 2.76 that the j-function is the normalization of the covering

$$\beta : \mathbb{P}^1 \longrightarrow \mathbb{P}^1$$

$$z \longmapsto f(z) = \frac{(1-z)^3}{z^2}$$

hence the dessin $\mathcal{D}$ associated to j is the *normalization* of the (not even uniform) dessin associated to β, see Figure 4.29.

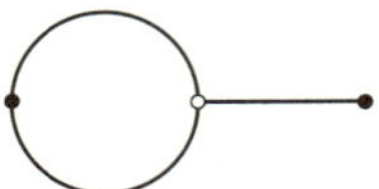

Fig. 4.29. The dessin associated to the function $\beta(z) = \dfrac{(1-z)^3}{z^2}$.

Note that by Proposition 4.30 it makes sense to speak of the normalization of a dessin $\mathcal{D}$ corresponding to a Belyi pair (S, f) as being the dessin $\widetilde{\mathcal{D}}$ corresponding to the normalization $(\widetilde{S}, \widetilde{f})$.

Remark 4.61 The function

$$j(z) = \frac{4}{27} \frac{\left(1 - z + z^2\right)^3}{z^2 (z-1)^2}$$

is Belyi-extending (see Definition 4.47).

Using the computations of Example 4.60 we can easily describe the process leading to the dessin $\mathcal{D}^j$ from a dessin $\mathcal{D}$ (Figure 4.30). In particular we deduce the following:

- Since the two points which $j^{-1}(0)$ consists of, namely $\frac{1}{2} \pm \frac{\sqrt{3}}{2} i$, lie one in the upper halfplane and one point in the lower halfplane, then $(j \circ f)^{-1}(0)$ contains one point inside each one of the triangles of the triangle decomposition $\mathcal{T} = \mathcal{T}(\mathcal{D})$ associated to $\mathcal{D}$ (see Section 4.2). These points are precisely the white vertices of $\mathcal{D}^j$, and all of them have degree 3.
- Since $j^{-1}(1) = \left\{-1, \frac{1}{2}, 2\right\}$ has one point in the interior of each of the three real segments $(-\infty, 0)$, $(0, 1)$ and $(1, \infty)$, $\mathcal{D}^j$ has one black vertex on (the interior of) each one of the sides of every triangle of $\mathcal{T}$. They all have degree 2 as vertices of $\mathcal{D}^j$.

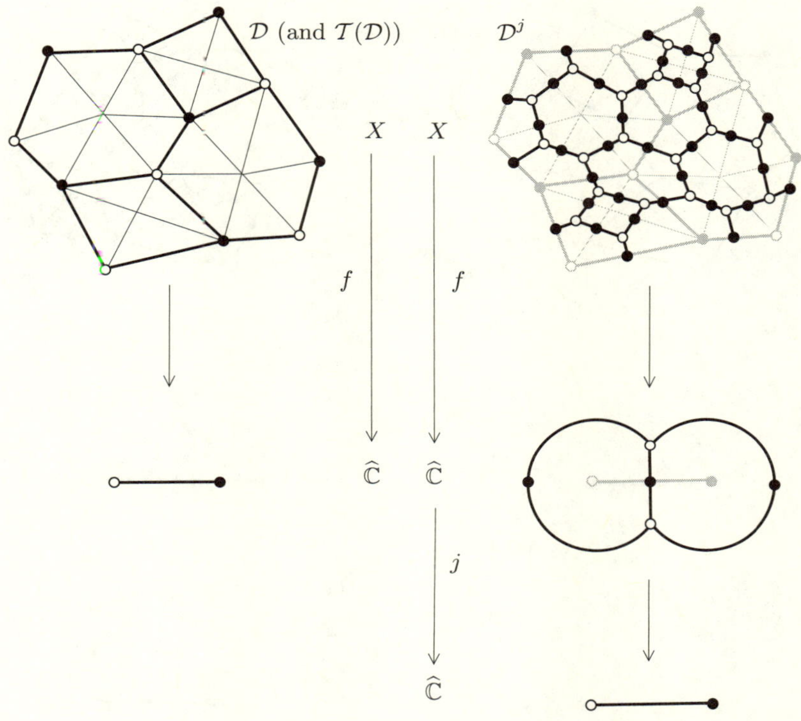

Fig. 4.30. Relation between $\mathcal{D}$ and $\mathcal{D}^j$.

- Since $j^{-1}(\infty) = \{0, 1, \infty\}$ the set of face centres of $\mathcal{D}^j$ coincides with the union of the black vertices, the white vertices and the face centres of $\mathcal{D}$.
- Since $j^{-1}\{(0, 1)\}$ has three components in $\mathbb{H}$ and three more in $\mathbb{H}^-$, any given triangle T of $\mathcal{T}$ meets three edges of $\mathcal{D}^j$. These edges connect the white vertex of $\mathcal{D}^j$ in the interior of T with the black vertices that lie on the three sides.

The relation between $\mathcal{D}$ and $\mathcal{D}^j$ is shown in Figure 4.30. It is actually easier to visualize $\mathcal{D}^{1/j}$ than $\mathcal{D}^j$, see Figure 4.31.

Example 4.62 Consider now the standard male and female symbols (see the left part of Figure 4.32), two dessins which are not Galois conjugate to each other since they have different number of edges.

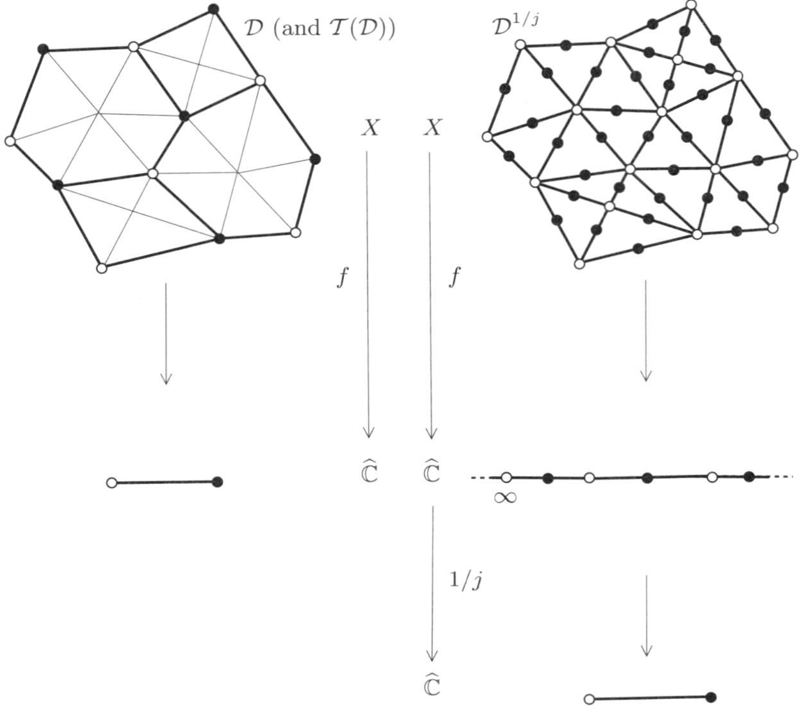

Fig. 4.31. Relation between $\mathcal{D}$ and $\mathcal{D}^{1/j}$.

Let us look for the Belyi function f associated to the female dessin. Place the white vertices at 0 and 1, as shown in the picture, and the centre of the *unbounded* face at ∞. Let b be the point in the interior of the bounded face such that $f(b) = \infty$.

Using only our knowledge of the preimages of 0 (white vertices) we deduce that, up to a multiplicative constant,

$$f(z) = \frac{z^4(z-1)^2}{P(z)}$$

where $P(z)$ is a polynomial whose degree is at most 6, since f must be a rational function of degree 6. Now, $z = b$ must be a non-ramified preimage of ∞ (it is the centre of a face with two edges, see Summary 4.16), hence $(z-b)$ divides P but $(z-b)^2$ does not. Moreover, $z = \infty$ must be a preimage of ∞ with multiplicity

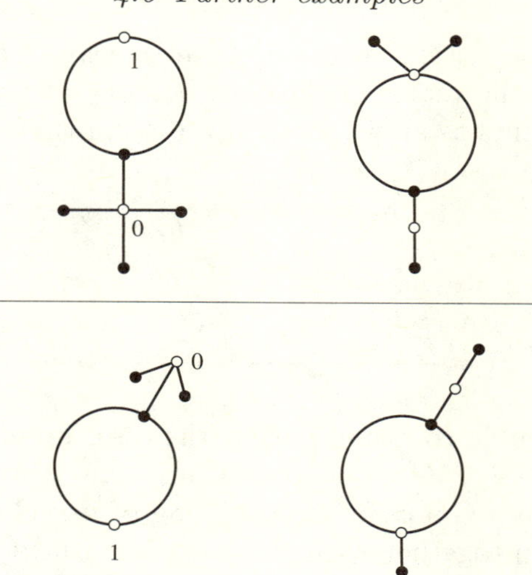

Fig. 4.32. The female and male symbols lie in two different Galois orbits, each of which has two elements.

5, as the unbounded face has ten edges if, as usual, we count twice the four of them that only bound one face. We deduce that

$$f(z) = \frac{z^4(z-1)^2}{(z-b)}$$

Differentiating we get

$$f'(z) = \frac{z^3(z-1)(5z^2 + (-3-6b)z + 4b)}{(z-b)^2}$$

and since f' must have a double root (corresponding to the black vertex of degree 3), the discriminant of $5z^2 + (-3 - 6b)z + 4b$ must vanish. Hence

$$36b^2 - 44b + 9 = 0$$

that is $b = \dfrac{-9 + 2\sqrt{10}}{18}$ or $b = \dfrac{-9 - 2\sqrt{10}}{18}$.

Once more, one of the values corresponds to the graph we started with, while the second determines the Belyi function associated to a Galois conjugate dessin, which is depicted at the top right part of Figure 4.32. By the arguments already used in Example 4.57 they constitute an orbit with only two elements.

Let now $m(z)$ denote the Belyi function for the *male dessin*.

Place the white vertices at 0 and 1, the centre of the unbounded face at ∞ and the centre of the bounded one at $z = b$. The same kind of arguments as above yield, up to a multiplicative constant,

$$m(z) = \frac{z^3(z-1)^2}{(z-b)}$$

Differentiating we get

$$m'(z) = \frac{z^2(z-1)(4z^2 + (-2-5b)z + 3b)}{(z-b)^2}$$

and since a double root must exist, the discriminant of the polynomial $4z^2 + (-2-5b)z + 3b$ must vanish, i.e. $25b^2 - 28b + 4 = 0$.

We get again a Galois orbit consisting of two elements, namely the male dessin together with the one placed next to it in Figure 4.32.

The next two examples were considered in [JS97].

Example 4.63 There exist two trees with six edges and white vertices of degrees 2,2,1,1 and black vertices of degrees 4,1,1 (see Figure 4.33).

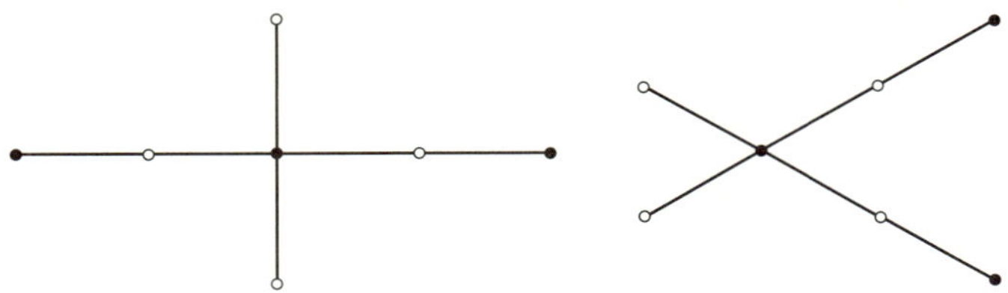

Fig. 4.33. Two dessins with different monodromy group.

We shall refer to the graphs on the left- and right-hand sides of Figure 4.33 as $\mathcal{D}_1$ and $\mathcal{D}_2$ respectively. The permutation representation pair of $\mathcal{D}_1$ is

$$\sigma_0 = (1,5)(6,3), \quad \sigma_1 = (1,2,3,4)$$

and the one of $\mathcal{D}_2$ is

$$\sigma_0 = (1,5)(6,4), \quad \sigma_1 = (1,2,3,4)$$

With the help of a computer one can easily show that the monodromy groups of $\mathcal{D}_1$ and $\mathcal{D}_2$ have order 48 and 120 respectively, hence $\mathcal{D}_1$ and $\mathcal{D}_2$ are not Galois conjugated (for the explicit description of these groups see [JS97]). The same conclusion could be reached by consideration of the automorphism groups, since the centralizer of $\langle \sigma_0, \sigma_1 \rangle$ in Σ_6 equals $\langle (1,3)(2,4)(5,6) \rangle$ and is trivial in the case of $\mathcal{D}_2$. In other words, $\mathcal{D}_2$ does not have automorphisms but the obvious order 2 rotation is an automorphism of $\mathcal{D}_1$.

Example 4.64 Let $\mathcal{D}_1$ and $\mathcal{D}_2$ be now the trees depicted in Figure 4.34.

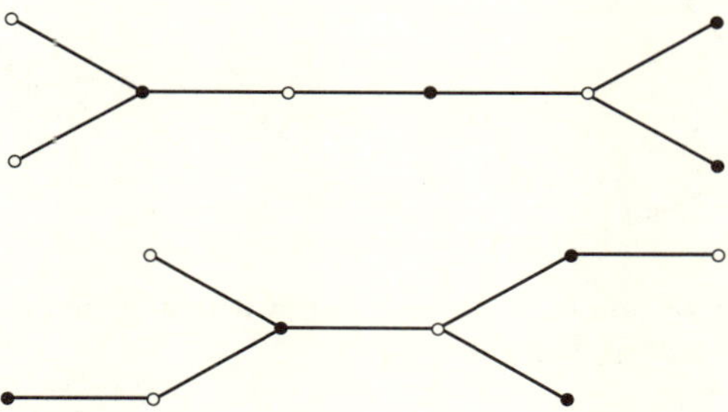

Fig. 4.34. Two dessins with the same monodromy group but different cartographic group.

Both trees have the same number of white and black vertices and the same vertex degree set. With the help of a computer it can be easily seen that their monodromy group equals Σ_7 in both cases. It can be even noticed that the automorphism group is trivial in both cases, but nevertheless they lie in different Galois orbits. One can reach this conclusion by passing to the associated clean dessins $\mathcal{D}_1^c$ and $\mathcal{D}_2^c$ (see Remark 4.27). The cartographic groups of $\mathcal{D}_1$ and $\mathcal{D}_2$, that is the monodromy groups of $\mathcal{D}_1^c$ and $\mathcal{D}_2^c$, have order $2 \cdot 7!$ in the first case an order $(7!)^2$ in the second one (see [JS97] for an explicit description of these groups), hence $\mathcal{D}_1$ and $\mathcal{D}_2$ cannot belong to the same Galois orbit.

This example illustrates also a situation in which the monodromy group is enough to show that two dessins lie in different orbits, but the automorphism group is not. Both $\mathcal{D}_1^c$ and $\mathcal{D}_2^c$ have a cyclic group of automorphisms of order 2, generated by an obvious rotation.

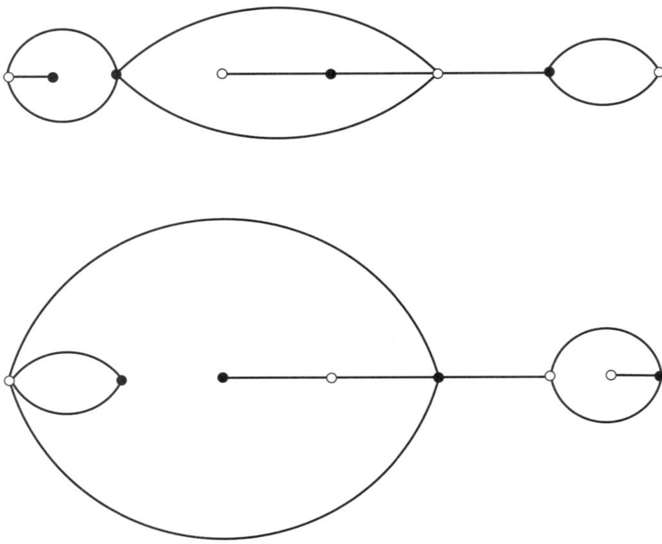

Fig. 4.35. A pair of non-conjugate dessins with the same monodromy group and the same cartographic group.

Example 4.65 The interesting case of the graphs in Figure 4.35 was studied in [Woo06].

Both dessins have trivial automorphism group, and both have the alternating group A_{10} as monodromy group. This conclusion can be easily reached with the help of a computer after labelling the edges and checking that the order of the monodromy group equals $\frac{10!}{2}$ in both cases. Moreover, their cartographic groups are isomorphic (although not coincident) subgroups of Σ_{20} of index 369 512. But computation of the j-monodromy groups, which due to the combinatorial description of the action of the Belyi-extending map j given in Remark 4.61 can be done with the help of a computer, shows that the j-monodromy groups differ even in their size. Their orders turn out to be 19 752 284 160 000 and 214 066 877 211 724 763 979 841 536 000 000 000 000.

4.6.2 Examples in genus g = 1

In the previous sections we have already encountered dessins of genus 1. For instance, the regular dessin of Example 4.21, which is defined over the rationals, and hence remains fixed by Galois conjugation.

A similar one is considered in the next example.

Example 4.66 Consider the Belyi pair (C, R) given by

$$C = \{y^2 = x^3 - x\}, \quad R(x, y) = x^2$$

We can construct the (topological) torus underlying C by glueing two copies of the Riemann sphere along suitable cuts (the grey lines in Figure 4.36) as in Example 4.24, and then finding the corresponding dessin $\mathcal{D}$ as the inverse image of the unit segment in $\widehat{\mathbb{C}}$.

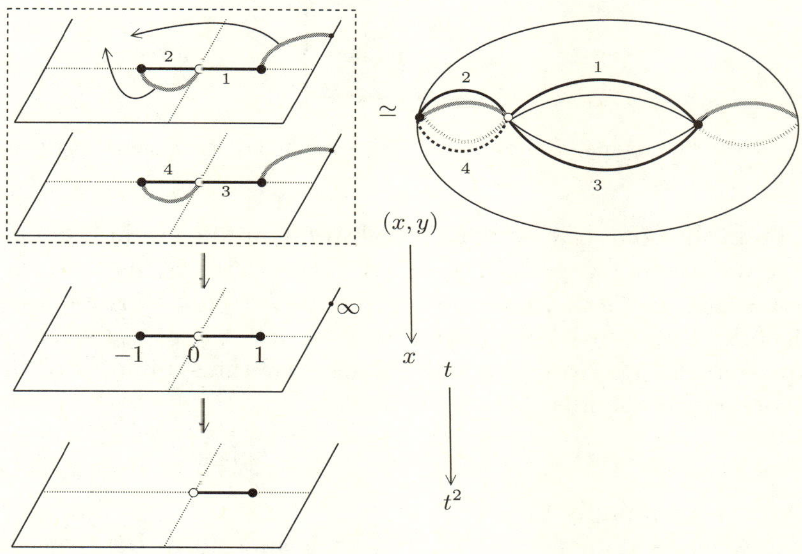

Fig. 4.36. The dessin $\mathcal{D}$ associated to the Belyi function $C(x, y) = x^2$ in the genus 1 surface with equation $y^2 = x^3 - x$.

A glance at either of the two pictures representing the dessin shows that the permutation representation pair of $\mathcal{D}$ is given by

$$\begin{aligned}
\sigma_0 &= (1, 2, 3, 4) \\
\sigma_1 &= (1, 3)(2, 4) \\
\sigma_1 \sigma_0 &= (1, 4, 3, 2) = \sigma_0 \sigma_1
\end{aligned}$$

therefore $\mathcal{D}$ is a uniform dessin of branching type $(4, 2, 4)$. Note that $\sigma_1 = \sigma_0^2$, hence $\mathrm{Mon}(\mathcal{D})$ is a cyclic group of order 4. It follows that $\mathcal{D}$ is regular and this group is nothing but $\mathrm{Aut}(X, \mathcal{D})$, the automorphism group of the dessin. Recall that this has been identified to $\mathrm{Aut}(C, R)$, the automorphism group of the Belyi pair, which is in this case obviously generated by $(x, y) \longmapsto (-x, iy)$.

Our usual geometric construction of $\mathcal{D}$ from the permutation representation pair gives now a triangle decomposition $\mathcal{T}(\mathcal{D})$ where all triangles have angles $\dfrac{\pi}{4}, \dfrac{\pi}{4}, \dfrac{\pi}{2}$, see Figure 4.37.

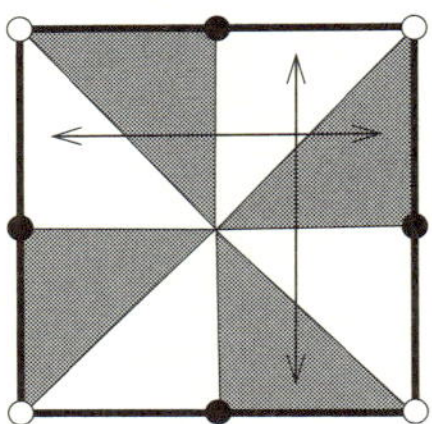

Fig. 4.37. The Riemann surface $\{y^2 = x^3 - x\}$ underlying $\mathcal{D}$ agrees with the torus $\mathbb{C}/\mathbb{Z} \oplus i\mathbb{Z}$.

From the point of view of uniformization, arguing as in Example 4.21 we see that $C \simeq \mathbb{C}/\Lambda$ where Λ is the lattice $\mathbb{Z} \oplus i\mathbb{Z}$. Recall now that there is an isomorphism from $\mathbb{C}/\Lambda$ to the corresponding algebraic curve in Weierstrass form $y^2 = 4x^3 - g_2 x - g_3$ given by equation (2.15). Note also that in this case the symmetry of the lattice yields the identities

$$\wp(iz) = -\wp(z) \quad \text{and} \quad \wp'(iz) = i\wp'(z)$$

which in turn imply $g_3 = 0$.

It is not a difficult task to find an isomorphism between the algebraic curve we started with and its Weierstrass form. In fact, the two maps in the following diagram are isomorphisms

$$\mathbb{C}/\Lambda \longrightarrow \{y^2 = 4x^3 - g_2 x\} \longrightarrow \{y^2 = x^3 - x\}$$

$$[z] \longmapsto (\wp(z), \wp'(z))$$

$$(x, y) \longmapsto \left(\frac{2}{\sqrt{g_2}} x, \frac{\sqrt{2}}{g_2^{3/4}} y \right)$$

We thus get another two equivalent descriptions of the Belyi pair corresponding to $\mathcal{D}$, namely

$$y^2 = 4x^3 - g_2 x \longrightarrow \widehat{\mathbb{C}}$$
$$(x, y) \longmapsto 4x^2/g_2$$

and

$$\mathbb{C}/\Lambda \longrightarrow \widehat{\mathbb{C}}$$
$$[z]_\Lambda \longmapsto 4\wp^2(z)/g_2$$

Obviously the rule $(x, y) \longmapsto (-x, iy)$ still defines a generator for the covering group in the Weierstrass model. On the other hand, the identity $\wp(iz) = -\wp(z)$ shows that in the transcendental model $\mathbb{C}/\Lambda$ the group is generated by $[z] \longmapsto [iz]$.

Example 4.67 The dessin in Example 4.66 above is defined over the rationals, therefore it constitutes a one point orbit of the Galois action.

On the contrary the dessin $\mathcal{D}$ in Example 4.24, whose associated Belyi pair (C, f) is given by

$$f(x, y) = \frac{-4\left(x^2 - 1\right)}{\left(x^2 - 2\right)^2}$$

and

$$C = \{y^2 = x(x - 1)(x - \sqrt{2})\}$$

is defined over $\mathbb{Q}[\sqrt{2}]$.

We claim that $\mathcal{D}$ is one of the elements of a full orbit consisting of the two dessins in Figure 4.38 ($\mathcal{D}$ is the one on the left-hand side).

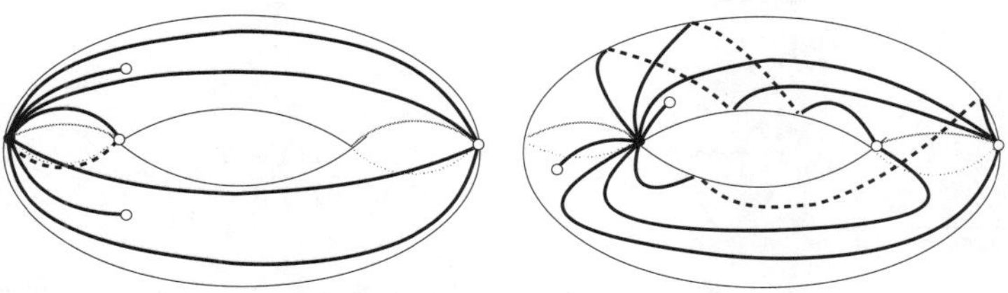

Fig. 4.38. An orbit with two elements.

Let $\sigma \in \mathrm{Gal}\left(\overline{\mathbb{Q}}\right)$ such that $\sigma(\sqrt{2}) = -\sqrt{2}$. Consider the conjugate Belyi pair (C^σ, f^σ). We find that f^σ has the same algebraic expression as f, namely

$$f^\sigma(x, y) = \frac{4(x^2 - 1)}{(x^2 - 2)^2}$$

but now it must be viewed as a function on the conjugate Riemann surface $C^\sigma = \{y^2 = x(x - 1)(x + \sqrt{2})\}$, which is not isomorphic to C since $j\left(\sqrt{2}\right) \neq j\left(-\sqrt{2}\right)$, where j denotes the classical j-invariant (see Corollary 2.57).

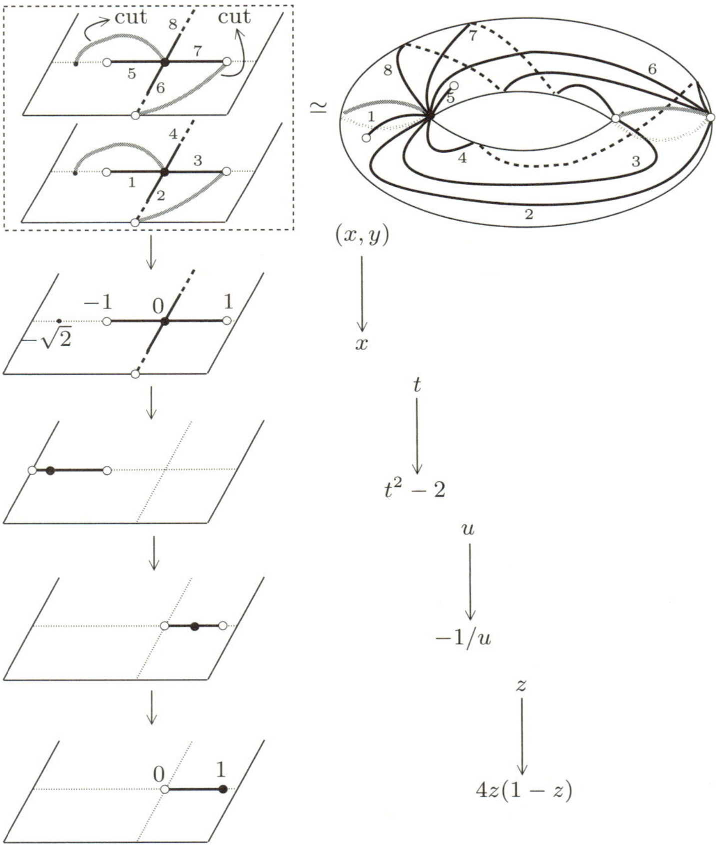

Fig. 4.39. Description of $\mathcal{D}^\sigma$ where $\sigma(\sqrt{2}) = -\sqrt{2}$ and $\mathcal{D}$ is the dessin in Example 4.24.

In order to depict $\mathcal{D}^\sigma$ we can follow the same strategy we used for $\mathcal{D}$. We find the inverse image of the unit interval $[0,1]$ by f^σ in four steps, using a description of f^σ as composition of four simpler maps, namely $f^\sigma = f_3 \circ f_2 \circ f_1 \circ \mathbf{x}$, where $\mathbf{x}$ is the x-coordinate function, $f_1(t) = t^2 - 2$, $f_2(u) = -\frac{1}{u}$ and $f_3(z) = 4z(1-z)$. The result is shown in Figure 4.39.

The permutation representation pair of $\mathcal{D}$ is

$$\sigma_0 = (1,7,5,3)(4,8), \quad \sigma_1 = (1,2,3,4,5,6,7,8)$$

while that of $\mathcal{D}^\sigma$ is

$$\sigma_0 = (2,4,6,8)(3,7), \quad \sigma_1 = (1,2,3,4,5,6,7,8)$$

The fact that these permutation representation pairs are not conjugate in Σ_8 gives us an alternative way to conclude that $\mathcal{D}$ and $\mathcal{D}^\sigma$ are not isomorphic dessins.

Example 4.68 The two dessins in Figure 4.40 were studied by F. Berg (see [Wol06]).

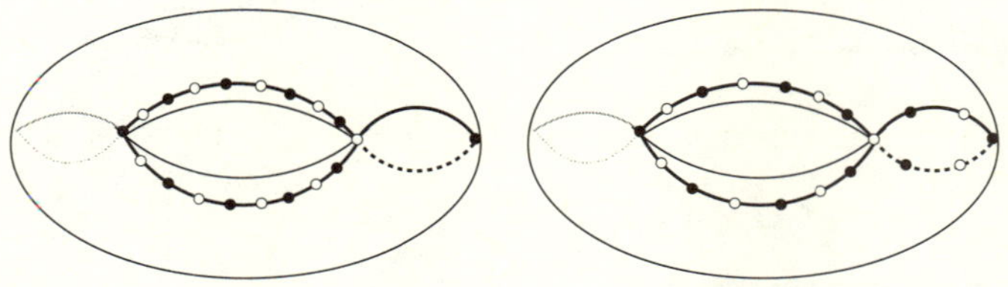

Fig. 4.40. Two $\mathrm{Gal}\left(\overline{\mathbb{Q}}\right)$-conjugate dessins.

Consider the compact Riemann surface

$$C = \left\{ y^2 = (x-1)(x+1)\left(x - \cos\frac{\pi}{10}\right) \right\}$$

and let $\sigma \in \mathrm{Gal}(\overline{\mathbb{Q}})$ be any element of the absolute Galois group whose restriction to the cyclotomic field $\mathbb{Q}(\xi)$, where

$$\xi = \xi_{20} = \cos\frac{\pi}{10} + i\sin\frac{\pi}{10}$$

is determined by $\sigma(\xi) = \xi^3$, thus

$$\sigma\left(\cos\frac{\pi}{10}\right) = \sigma\left(\frac{1}{2}\left(\xi + \xi^{-1}\right)\right) = \frac{1}{2}\left(\xi^3 + \xi^{-3}\right) = \cos\frac{3\pi}{10}$$

and therefore

$$C^\sigma = \left\{ y^2 = (x-1)(x+1)\left(x - \cos\frac{3\pi}{10}\right) \right\}$$

Let f be the meromorphic function

$$C \xrightarrow{\ f\ } \mathbb{P}^1$$

$$(x,y) \longmapsto T_5^2(x)$$

where T_5 is the Tchebychev polynomial of degree 5. Since T_5^2 is a Belyi function on $\mathbb{P}^1$ whose branching points include the branching values of the coordinate function $\mathbf{x} : C \longrightarrow \mathbb{P}^1$ (Example 4.56), it follows that f is a Belyi function on C. The dessin $\mathcal{D} = \mathcal{D}_f$ corresponding to the Belyi pair (C, f) is depicted in Figure 4.41.

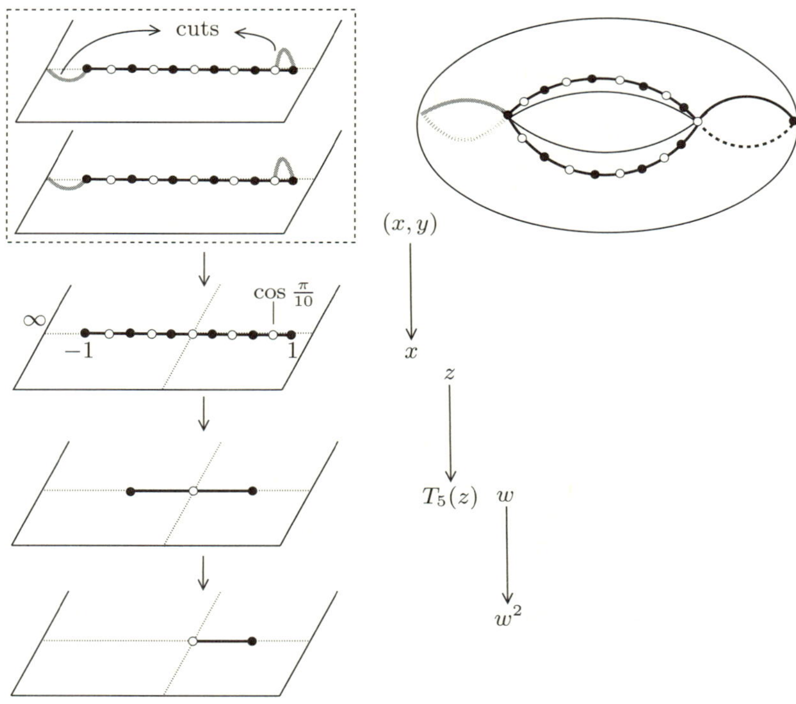

Fig. 4.41. The dessin $\mathcal{D}_f$ associated to the Belyi function $f : (x,y) \longmapsto T_5(x)^2$ defined in the Riemann surface $\{ y^2 = (x-1)(x+1)\left(x - \cos\frac{\pi}{10}\right) \}$.

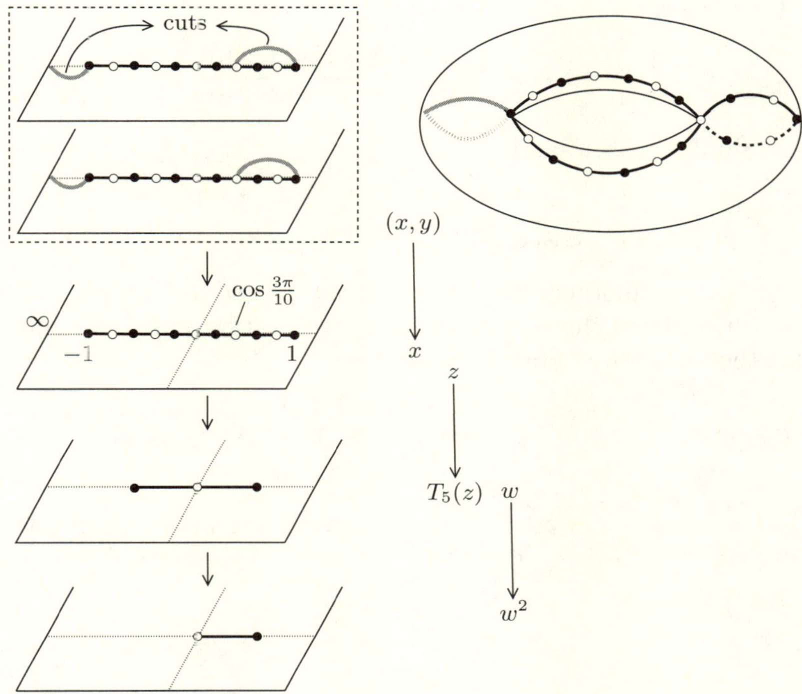

Fig. 4.42. The conjugated dessin $\mathcal{D}_f^\sigma$ is associated to the conjugated Belyi function $f^\sigma : (x, y) \longmapsto T_5(x)^2$ defined on $\left\{ y^2 = (x-1)\,(x+1)\,\left(x - \cos \frac{3\pi}{10}\right) \right\}$.

Conjugation by σ produces a Belyi pair (C^σ, f^σ), whose associated dessin $\mathcal{D}^\sigma$ is depicted in Figure 4.42.

Both dessins are not equivalent. In fact the Riemann surfaces they define are not isomorphic. The first one is isomorphic to

$$\left\{ y^2 = x\,(x-1)\left(x + \frac{1 + \cos(\pi/10)}{1 - \cos(\pi/10)}\right) \right\}$$

since $M(x) = \dfrac{x - \cos(\pi/10)}{1 - \cos(\pi/10)}$ is a Möbius transformation sending

$$\{\cos(\pi/10), 1, \infty, -1\}$$

to

$$\left\{ 0, 1, \infty, -\frac{1 + \cos(\pi/10)}{1 - \cos(\pi/10)} \right\}$$

(see Example 1.83), while the second one is isomorphic to

$$\left\{ y^2 = x\,(x-1)\left(x + \frac{1+\cos(3\pi/10)}{1-\cos(3\pi/10)}\right)\right\}$$

for a similar reason, and

$$j\left(\frac{1+\cos(\pi/10)}{1-\cos(\pi/10)}\right) \neq j\left(\frac{1+\cos(3\pi/10)}{1-\cos(3\pi/10)}\right)$$

As before, another way to show that $\mathcal{D}$ and $\mathcal{D}^\sigma$ are different is to write down the permutation representation pairs and check that they are not conjugate. These are

$$\begin{aligned}
\sigma_0 &= (1,2)(3,4)(5,6)(7,8,19,20)(9,10)(11,12)(13,14)(15,16)(17,18)\\
\sigma_1 &= (1,11)(2,3)(4,5)(6,7)(8,9)(10,20)(12,13)(14,15)(16,17)(18,19)
\end{aligned}$$

and

$$\begin{aligned}
\sigma_0 &= (1,2)(3,4)(5,6)(7,8,17,18)(9,10)(11,12)(13,14)(15,16)(19,20)\\
\sigma_1 &= (1,11)(2,3)(4,5)(6,7)(8,9)(10,20)(12,13)(14,15)(16,17)(18,19)
\end{aligned}$$

respectively.

4.6.3 Some examples in genus g ≥ 2

In Section 4.4 we studied several dessins whose underlying Riemann surface is Klein's curve, including a regular dessin of branching type $(7,7,7)$ in Example 4.32, a regular dessin of type $(2,3,7)$ in Example 4.44 and a uniform dessin of the same type also in Example 4.44.

We include in this section other examples of dessins and Belyi pairs of genus $g \geq 2$. Some more can be found in the literature, e.g. [AS05] and [AAD$^+$07].

Example 4.69 The function $f(x,y) = x^6$ is a Belyi function defined on the hyperelliptic curve C of genus 2 with equation $y^2 = x^6 - 1$. We can think of f as a composition of two maps

$$(x,y) \longmapsto x \longmapsto x^6$$

and compute $f^{-1}[0,1]$ to depict the dessin $\mathcal{D}$ corresponding to the Belyi pair (C,f) on a topological surface of genus 2, see Figure 4.43.

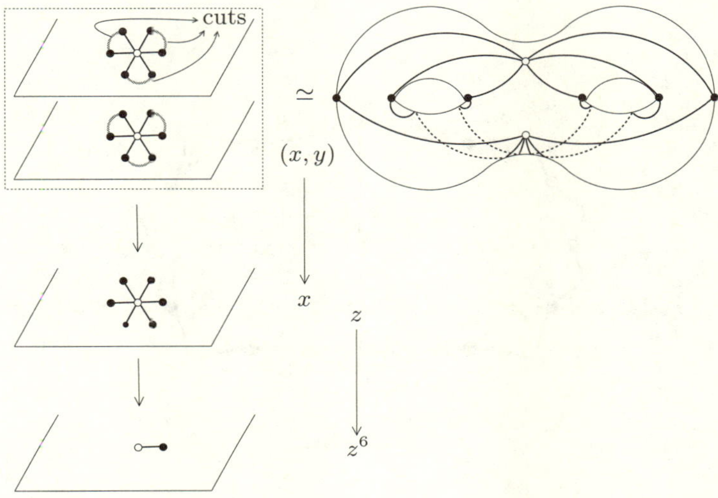

Fig. 4.43. The dessin $\mathcal{D}$ associated to the Belyi pair (C, f) where C has equation $y^2 = x^6 - 1$ and $f(x, y) = x^6$.

The permutation representation pair of $\mathcal{D}$ is given by

$$\begin{aligned}
\sigma_0 &= (1, 2, 3, 4, 5, 6)(7, 8, 9, 10, 11, 12) \\
\sigma_1 &= (1, 7)(2, 8)(3, 9)(4, 10)(5, 11)(6, 12)
\end{aligned}$$

and, since $|\langle \sigma_0, \sigma_1 \rangle| = 12$, the dessin $\mathcal{D}$ is regular.

Moreover,

$$\sigma_0 \sigma_1 = (1, 8, 3, 10, 5, 12)(2, 9, 4, 11, 6, 7)$$

hence the process described in Section 4.4 produces a Fuchsian group K uniformizing C such that $K \lhd \Gamma_{6,2,6}$ with index 12. The fundamental domain of K consists of the reunion of two regular hyperbolic hexagons with angle $\dfrac{2\pi}{6}$ (one for each of the two faces of the dessin), and $\mathcal{D}$ is represented inside the Riemann surface $\mathbb{D}/K$ as the projection of the boundary of these two hexagons (Figure 4.44).

As the pair (C, f) is defined over $\mathbb{Q}$, we know that $\mathcal{D}^\sigma = \mathcal{D}$ for any $\sigma \in \mathrm{Gal}(\overline{\mathbb{Q}})$.

Example 4.70 The function $f(x, y) = x^n$ is a Belyi function on

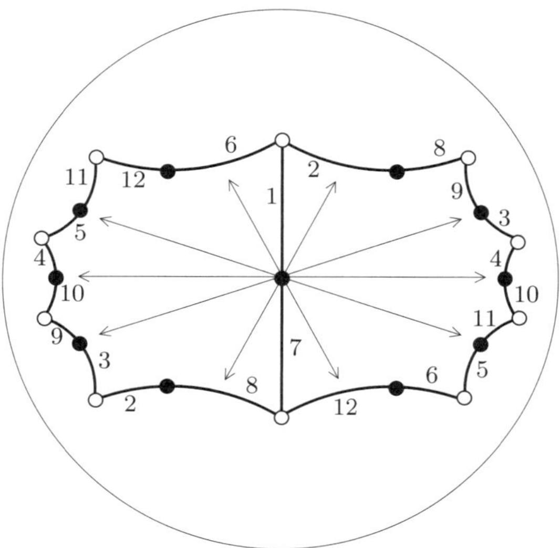

Fig. 4.44. The dessin $\mathcal{D}$ in the description $C \simeq \mathbb{D}/K$ with $K < \Gamma_{6,2,6}$. The lines indicate the side-pairing transformations generating K.

the Fermat curve $C = \{x^n + y^n = 1\}$ (see Example 1.10). The corresponding dessin $\mathcal{D}$ possesses:

- n white vertices, namely the points $(0, \xi_n^i)$ for $i = 1, \ldots, n$.
- n black vertices, namely the points $(\xi_n^i, 0)$ for $i = 1, \ldots, n$.

Thus, the only way to draw $n^2 = \deg(f)$ edges is if any white point is connected to any black point. That is, the graph underlying $\mathcal{D}$ is the complete bipartite graph $K_{n,n}$. We also note that $\mathcal{D}$ is a regular dessin with $\mathrm{Aut}(C, f) \simeq \mathbb{Z}_n^2$ generated by $\tau_1(x, y) = (\xi_n x, y)$ and $\tau_2(x, y) = (x, \xi_n y)$.

Example 4.71 The map $f(x, y) = 4x^4(1 - x^4)$ is a Belyi function on the genus 2 curve

$$C = \left\{ y^2 = (x^4 - 1)(x - \sqrt[4]{1/2}) \right\}$$

which can be described as a composition of three simpler maps, namely

$$(x, y) \longmapsto x \longmapsto x^4 \longmapsto 4x^4(1 - x^4)$$

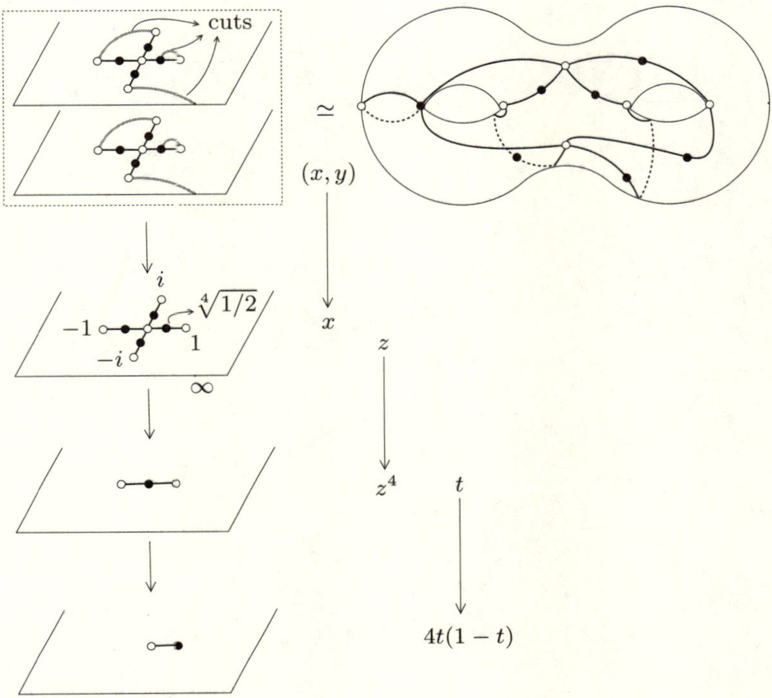

Fig. 4.45. The dessin $\mathcal{D}$ associated to the Belyi pair (C, f) where C has equation $y^2 = (x^4 - 1)(x - \sqrt[4]{1/2})$ and $f(x, y) = 4x^4(1 - x^4)$.

The inverse image of the interval $[0, 1]$ by f can be depicted as in Figure 4.45. The three cuts performed on each one of the two sheets of the **x** coordinate function are segments $[i, -1]$, $[-i, \infty]$ and $[\sqrt[4]{1/2}, 1]$.

One can easily show that, for all $\sigma \in \mathrm{Gal}(\overline{\mathbb{Q}})$, the conjugate dessin $\mathcal{D}^\sigma$ is equivalent to $\mathcal{D}$. For instance, assume $\sigma(\sqrt[4]{2}) = -\sqrt[4]{2}$. Then the function f^σ, defined in the conjugate curve

$$C^\sigma = \left\{ y^2 = (x^4 - 1)(x + \sqrt[4]{1/2}) \right\}$$

can be decomposed in the same way as f, that is

$$(x, y) \longmapsto x \longmapsto x^4 \longmapsto 4x^4(1 - x^4) = f^\sigma(x, y)$$

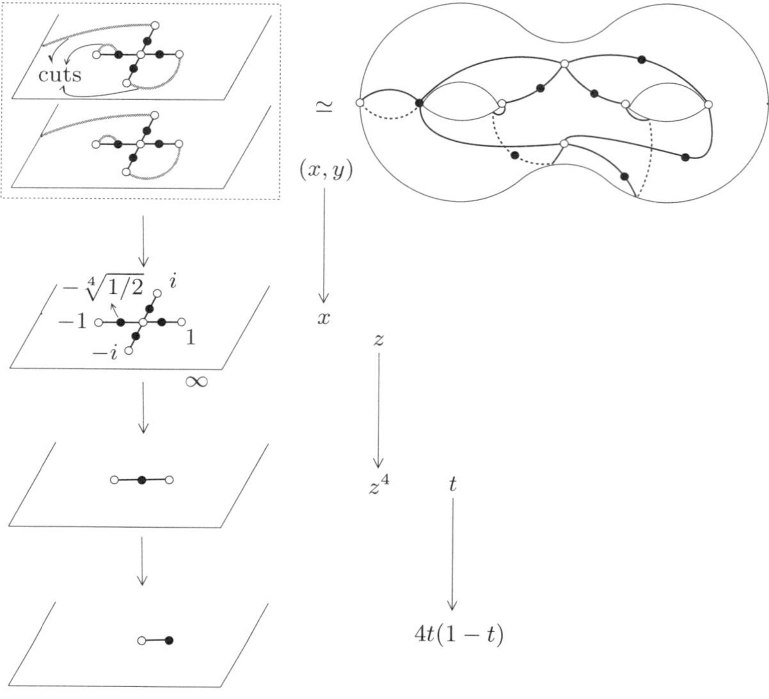

Fig. 4.46. The dessin $\mathcal{D}^\sigma$ associated to the conjugated Belyi pair (C^σ, f^σ) is equivalent to $\mathcal{D}$.

We can visualize $\mathcal{D}^\sigma$ again like we did in Figure 4.45. The result is shown in Figure 4.46.

The left-hand sides of Figures 4.45 and 4.46 are the same except for the cuts. In the second one the cuts are $[-1, -\sqrt[4]{1/2}]$, $[1, -i]$ and $[i, \infty]$, but the position of the graph with respect to the cuts is the same as in the first case. More precisely, one can easily label the edges of both graphs in a way such that both permutation representation pairs agree. This is why in the end the right-hand sides of Figures 4.45 and 4.46 are identical.

The Belyi pair (C, f) in this last example is not directly defined over the rationals but over $\mathbb{Q}[\sqrt[4]{2}]$. On the other hand, the action of the absolute Galois group is trivial on it, as it would be the case if (C, f) were defined over $\mathbb{Q}$. A natural question is whether there is a curve C_1 and a function f_1 with coefficients in $\mathbb{Q}$ such that the Belyi pair (C_1, f_1) is equivalent to (C, f). The answer is in this case affirmative, since the isomorphism

$$C \xrightarrow{\quad F \quad} C_1$$
$$(x, y) \longmapsto (\sqrt[4]{2}x, \sqrt[8]{32}y)$$

transforms the Belyi pair (C, f) into the equivalent one (C_1, f_1), where

$$f_1(x, y) = \beta \circ F^{-1}(x, y) = x^4(2 - x^4)$$

and

$$C_1 = \{y^2 = (x^4 - 2)(x - 1)\}$$

These considerations lead to the definition of the *moduli field* of a curve C (resp. of a Belyi pair (C, f)) as the fixed field $\overline{\mathbb{Q}}^{I(C)}$ (resp. $\overline{\mathbb{Q}}^{I(C,f)}$) of the subgroup of $\mathrm{Gal}(\overline{\mathbb{Q}})$

$$I(C) = \{\sigma \in \mathrm{Gal}(\overline{\mathbb{Q}}) \mid C^\sigma \text{ is isomorphic to } C\}$$

and

$$I(C, f) = \{\sigma \in \mathrm{Gal}(\overline{\mathbb{Q}}) \mid (C^\sigma, f^\sigma) \text{ is equivalent to } (C, f)\}$$

respectively.

The moduli field is obviously contained in any field of definition, but the converse is not always true, as it is shown in the following example due to Earle ([Ear71], [Ear10]). For other examples of this phenomenon see [Shi72], [Hid09].

Theorem 4.72 (C.J. Earle [Ear10]) *Let r be any real number such that $0 < r < 1$ with $r \neq \frac{-1+\sqrt{5}}{2}$. Then the Riemann surface*

$$C = \{y^2 = x(x + \xi_3^2)(x - \xi_3^2)(x - r)(x + 1/r)\}$$

cannot be defined over $\mathbb{R}$ although its field of moduli is contained in the reals.

Proof (1) It is easy to check that the formula

$$f(x, y) = \left(-\frac{1}{x}, \frac{i\xi_3^2 \overline{y}}{x^3}\right)$$

defines an anti-holomorphic automorphism of C of order 4, or equivalently an isomorphism between the Riemann surfaces C and its complex conjugate $\overline{C}$ (see Examples 1.22 and 3.13). This means that the complex conjugation lies in the stabilizer subgroup $I(C) \subset \mathrm{Gal}(\mathbb{C})$ and so the moduli field $\mathbb{C}^{I(C)}$ is contained in the field $\mathbb{R}$.

(2) The proof of the fact that C cannot be defined over the reals will be accomplished in several steps:

Step 1: if there is an isomorphism $h : C \longrightarrow S_F$ with $F(X,Y) \in \mathbb{R}[X,Y]$ then C admits an anticonformal involution s of order 2, namely the one determined by the following commutative diagram:

$$
\begin{array}{ccc}
C & \xrightarrow{\ s\ } & C \\
{\scriptstyle h}\downarrow & & \downarrow{\scriptstyle h} \\
S_F & \longrightarrow & S_F
\end{array}
$$
$$
(x,y) \longmapsto (\overline{x},\overline{y})
$$

Step 2: if there is an isomorhism $h : C \longrightarrow S_F$ as above then $\mathrm{Aut}(C)$ is strictly larger than the group generated by the hyperelliptic involution $J(x,y) = (x,-y)$.

This is easy to see. In fact in this situation $\tau = s \circ f$ would be a (holomorphic) automorphism of C different from Id (s and f have different orders) and different from J because the condition $s \circ f = J$ is equivalent to $f \circ J = s$ but, as J obviously commutes with f, the order of $f \circ J$ is the same as the order of f, that is 4.

Step 3: by Proposition 2.47 an automorphism of C different from Id or J induces a Möbius transformation $M : \widehat{\mathbb{C}} \longrightarrow \widehat{\mathbb{C}}$ preserving the branch set $\mathcal{B} = \left\{0, \infty, \xi_3^2, -\xi_3^2, r, -1/r\right\}$. This transformation must also preserve the cross-ratio of any ordered quadruplet of points in $\mathcal{B}$ (see Section 2.1.1). Direct examination of cases shows that $\left\{0, \infty, \xi_3^2, -\xi_3^2\right\}$ is the only 4-point subset of $\mathcal{B}$ such that the cross-ratio of its points is -1, $1/2$ or 2, depending on their ordering, therefore

$$
M\left(\left\{0, \infty, \xi_3^2, -\xi_3^2\right\}\right) = \left\{0, \infty, \xi_3^2, -\xi_3^2\right\}
$$

and

$$
M\left(\{r, -1/r\}\right) = \{r, -1/r\}
$$

Step 4: the group G of Möbius transformations preserving the set $\{0, \infty, \xi_3^2, -\xi_3^2\}$ is the dihedral group of order 8 generated by

$$A(z) = -z \qquad B(z) = \frac{z + \xi_3^2}{-\xi_3 z + 1}$$

To prove this, one first observe that

$$B(0) = \xi_3^2, \quad B(\xi_3^2) = \infty, \quad B(\infty) = -\xi_3^2, \quad \text{and } B(-\xi_3^2) = 0$$

Then one checks the following identities

$$B^2(z) = -\frac{\xi_3}{z}, \quad B^3(z) = \frac{\xi_3^2 z - \xi_3}{z + \xi_3^2}, \quad B^4 = \text{Id}, \quad \text{and } AB = B^3 A$$

It follows that $A, B \in G$ and that $\langle A, B \rangle$ is the dihedral group of order 8.

It is easy to show that any Möbius transformation that fixes 0 and preserves the set $\{0, \infty, \xi_3^2, -\xi_3^2\}$ is either A or the identity. From here the statement readily follows; for example, if $M \in G$ is such that $M(0) = \xi_3^2$, then $B^3 \circ M(0) = 0$, hence $B^3 \circ M = \text{Id}$ or A. In either case $M \in \langle A, B \rangle$.

Step 5: our final claim is that if $r \neq \dfrac{-1 + \sqrt{5}}{2}$ the group of Möbius transformations that preserve the whole branching set $\mathcal{B}$ is trivial. By Corollary 2.48 this is in contradiction with the claim made in Step 2 and will end the proof.

This is also easy to prove. Clearly both A and B^2 send the value r outside the branching value set $\mathcal{B}$, therefore the only possible non-trivial Möbius transformations leaving invariant the set $\mathcal{B}$, hence by Step 3 the set $\{r, -1/r\}$ are AB, AB^2 or AB^3. Consider the numbers

$$AB(r) = \frac{r - \xi_3^2}{\xi_3 r - 1}, \quad AB^2(r) = \frac{\xi_3}{r}, \quad \text{and } AB^3(r) = \frac{-r + \xi_3^2}{\xi_3 r + 1}$$

The second one is in fact not real for any real number r. A simple calculation equating $AB(r)$ to its complex conjugate $\overline{AB(r)}$ shows that the first of these three numbers is real if and only if $r^2 - r - 1 = 0$. Since $0 < r < 1$ this is impossible. Similarly, the third one is real if and only if $r^2 + r - 1 = 0$. Since $0 < r < 1$ this happens if and only if $r = \dfrac{-1 + \sqrt{5}}{2}$.

This completes the proof. $\qquad\qquad\qquad\qquad\qquad\qquad\square$

This brings to an end our introduction to the Grothendieck-Belyi theory of dessins d'enfants. Those who wish to pursue this topic are adviced to consult the book by Lando and Zvonkin [LZ04], the conference proceedings edited by Schneps and Lochak [Sch94b], [SL97], the survey articles by Jones–Singerman [JS96], Shabat–Voevodsky [SV90], Cohen–Itzykson–Wolfart [CIW94] , Jones [Jon97], Lochak [Loc04], Wolfart [Wol06] and, from a different point of view, the monograph by Bowers–Stephenson [BS04].

References

[AAD+07] N. M. Adrianov, N. Ya. Amburg, V. A. Drëmov, Yu. Yu. Kochetkov, E. M. Kreĭnes, Yu. A. Levitskaya, V. F. Nasretdinova, and G. B. Shabat. A catalogue of Belyĭ functions of dessins d'enfants with at most four edges. *Fundam. Prikl. Mat.*, 13(6):35–112, 2007.

[Abi80] W. Abikoff. *The Real Analytic Theory of Teichmüller Space*, volume 820 of *Lecture Notes in Mathematics*. Springer, Berlin, 1980.

[Ahl78] L. V. Ahlfors. *Complex Analysis*. McGraw-Hill Book Co., New York, third edition, 1978. An introduction to the theory of analytic functions of one complex variable, International Series in Pure and Applied Mathematics.

[AS05] N. M. Adrianov and G. B. Shabat. Belyĭ functions of dessins d'enfants of genus 2 with four edges. *Uspekhi Mat. Nauk*, 60(6(366)):229–230, 2005.

[Bea83] A. F. Beardon. *The Geometry of Discrete Groups*, volume 91 of *Graduate Texts in Mathematics*. Springer-Verlag, New York, 1983.

[Bea84] A. F. Beardon. *A Primer on Riemann Surfaces*, volume 78 of *London Mathematical Society Lecture Note Series*. Cambridge University Press, Cambridge, 1984.

[Bel79] G. V. Belyĭ. Galois extensions of a maximal cyclotomic field. *Izv. Akad. Nauk SSSR Ser. Mat.*, 43(2):267–276, 479, 1979.

[Bel02] G. V. Belyĭ. A new proof of the three-point theorem. *Mat. Sb.*, 193(3):21–24, 2002.

[Ber72] L. Bers. Uniformization, moduli, and Kleinian groups. *Bull. London Math. Soc.*, 4:257–300, 1972.

[BS04] P. L. Bowers and K. Stephenson. Uniformizing dessins and Belyĭ maps via circle packing. *Mem. Amer. Math. Soc.*, 170(805):xii+97, 2004.

[BT82] R. Bott and L. W. Tu. *Differential Forms in Algebraic Topology*, volume 82 of *Graduate Texts in Mathematics*. Springer-Verlag, New York, 1982.

[Bus92] P. Buser. *Geometry and Spectra of Compact Riemann Surfaces*, volume 106 of *Progress in Mathematics*. Birkhäuser Boston Inc., Boston, MA, 1992.

[BZ93] J. Betrema and A. K. Zvonkin. Plane trees and their shabat polynomials, catalogue. *Publ. LaBRI*, 75, 1993.

[Car61] H. Cartan. *Théorie Élémentaire des Fonctions Analytiques d'une ou Plusieurs Variables Complexes*. Avec le concours de Reiji Takahashi. Enseignement des Sciences. Hermann, Paris, 1961.

[CCN$^+$85] J. H. Conway, R. T. Curtis, S. P. Norton, R. A. Parker, and R. A. Wilson. *Atlas of Finite Groups.* Oxford University Press, Eynsham, 1985. Maximal subgroups and ordinary characters for simple groups, With computational assistance from J. G. Thackray.

[CIW94] P. B. Cohen, C. Itzykson, and J. Wolfart. Fuchsian triangle groups and Grothendieck dessins. Variations on a theme of Belyĭ. *Comm. Math. Phys.*, 163(3):605–627, 1994.

[Con78] J. B. Conway. *Functions of one Complex Variable*, volume 11 of *Graduate Texts in Mathematics.* Springer-Verlag, New York, second edition, 1978.

[dC76] M. P. do Carmo. *Differential Geometry of Curves and Surfaces.* Prentice-Hall Inc., Englewood Cliffs, N. J., 1976. Translated from the Portuguese.

[dR71] G. de Rham. Sur les polygones générateurs de groupes fuchsiens. *Enseignement Math.*, 17:49–61, 1971.

[Ear71] C. J. Earle. On the moduli of closed Riemann surfaces with symmetries. In *Advances in the Theory of Riemann Surfaces (Proc. Conf., Stony Brook, N.Y., 1969)*, pages 119–130. Ann. of Math. Studies, No. 66. Princeton University Press, Princeton, N. J., 1971.

[Ear10] C. J. Earle. Diffeomorphisms and automorphisms of compact hyperbolic 2-orbifolds. In *Geometry of Riemann Surfaces*, volume 368 of *London Math. Soc. Lecture Note Ser.*, pages 139–155. Cambridge University Press, Cambridge, 2010.

[Elk99] N. D. Elkies. The Klein quartic in number theory. In *The Eightfold Way*, volume 35 of *Math. Sci. Res. Inst. Publ.*, pages 51–101. Cambridge University Press, Cambridge, 1999.

[FK92] H. M. Farkas and I. Kra. *Riemann Surfaces*, volume 71 of *Graduate Texts in Mathematics.* Springer-Verlag, New York, second edition, 1992.

[For91] O. Forster. *Lectures on Riemann Surfaces*, volume 81 of *Graduate Texts in Mathematics.* Springer-Verlag, New York, 1991. Translated from the 1977 German original by B. Gilligan, Reprint of the 1981 English translation.

[Ful95] W. Fulton. *Algebraic Topology*, volume 153 of *Graduate Texts in Mathematics.* Springer-Verlag, New York, 1995. A first course.

[Gar87] F. P. Gardiner. *Teichmüller Theory and Quadratic Differentials.* Pure and Applied Mathematics (New York). John Wiley & Sons Inc., New York, 1987. A Wiley-Interscience Publication.

[GD06] G. González-Diez. Variations on Belyi's theorem. *Q. J. Math.*, 57(3):339–354, 2006.

[GD08] G. González-Diez. Belyi's theorem for complex surfaces. *Amer. J. Math.*, 130(1):59–74, 2008.

[GGD07] E. Girondo and G. González-Diez. A note on the action of the absolute Galois group on dessins. *Bull. Lond. Math. Soc.*, 39(5):721–723, 2007.

[GGD10] E. Girondo and G. González-Diez. On complex curves and complex surfaces defined over number fields. In *Teichmüller Theory and Moduli Problem*, volume 10 of *Ramanujan Math. Soc. Lect. Notes Ser.*, pages 247–280. Ramanujan Math. Soc., Mysore, 2010.

[Gro97] A. Grothendieck. Esquisse d'un programme. In *Geometric Galois Actions, 1*, volume 242 of *London Math. Soc. Lecture Note Ser.*, pages 5–48. Cambridge University Press, Cambridge, 1997. English translation on pp. 243–283.

[GTW11] E. Girondo, D. Torres, and J. Wolfart. Shimura curves with many uniform dessins. Math. Z, 2011.

[Gun62] R. C. Gunning. *Lectures on Modular Forms.* Notes by Armand Brumer. Annals of Mathematics Studies, No. 48. Princeton University Press, Princeton, N. J., 1962.

[Har74] W. Harvey. Chabauty spaces of discrete groups. In *Discontinuous Groups and Riemann Surfaces (Proc. Conf., Univ. Maryland, College Park, Md., 1973)*, pages 239–246. Ann. of Math. Studies, No. 79. Princeton University Press, Princeton, N. J., 1974.

[Hid09] R. A. Hidalgo. Non-hyperelliptic Riemann surfaces with real field of moduli but not definable over the reals. *Arch. Math. (Basel)*, 93(3):219–224, 2009.

[HM98] J. Harris and I. Morrison. *Moduli of Curves*, volume 187 of *Graduate Texts in Mathematics*. Springer-Verlag, New York, 1998.

[Hos10] K. Hoshino. The Belyi functions and dessin d'enfants corresponding to the non-normal inclusions of triangle groups. *Math. J. Okayama Univ.*, 52:45–60, 2010.

[Ive92] B. Iversen. *Hyperbolic Geometry*, volume 25 of *London Math. Soc. Student Texts*. Cambridge University Press, Cambridge, 1992.

[Jon97] G. A. Jones. Maps on surfaces and Galois groups. *Math. Slovaca*, 47(1):1–33, 1997. Graph theory (Donovaly, 1994).

[Jos02] J. Jost. *Compact Riemann surfaces*. Springer-Verlag, Berlin, second edition, 2002. An introduction to contemporary mathematics.

[JS78] G. A. Jones and D. Singerman. Theory of maps on orientable surfaces. *Proc. London Math. Soc. (3)*, 37(2):273–307, 1978.

[JS87] G. A. Jones and D. Singerman. *Complex Functions*. Cambridge University Press, Cambridge, 1987. An algebraic and geometric viewpoint.

[JS96] G. A. Jones and D. Singerman. Belyĭ functions, hypermaps and Galois groups. *Bull. London Math. Soc.*, 28(6):561–590, 1996.

[JS97] G. A. Jones and M. Streit. Galois groups, monodromy groups and cartographic groups. In *Geometric Galois Actions, 2*, volume 243 of *London Math. Soc. Lecture Note Ser.*, pages 25–65. Cambridge University Press, Cambridge, 1997.

[Kir92] F. Kirwan. *Complex Algebraic Curves*, volume 23 of *London Mathematical Society Student Texts*. Cambridge University Press, Cambridge, 1992.

[KK79] A. Kuribayashi and K. Komiya. On Weierstrass points and automorphisms of curves of genus three. In *Algebraic Geometry (Proc. Summer Meeting, Univ. Copenhagen, Copenhagen, 1978)*, volume 732 of *Lecture Notes in Math.*, pages 253–299. Springer-Verlag, Berlin, 1979.

[Kle78] F. Klein. Ueber die Transformation siebenter Ordnung der elliptischen Functionen. *Math. Ann.*, 14(3):428–471, 1878.

[Köc04] B. Köck. Belyi's theorem revisited. *Beiträge Algebra Geom.*, 45(1):253–265, 2004.

[Lan84] S. Lang. *Algebra*. Addison-Wesley, Reading, MA, second edition, 1984.

[Lan02] S. Lang. *Algebra*, volume 211 of *Graduate Texts in Mathematics*. Springer-Verlag, New York, third edition, 2002.

[Lev99] S. Levy, editor. *The Eightfold Way*, volume 35 of *Mathematical Sciences Research Institute Publications*. Cambridge University Press, Cambridge, 1999. The beauty of Klein's quartic curve.

[Loc04] P. Lochak. Fragments of nonlinear Grothendieck-Teichmüller theory. In *Woods Hole Mathematics*, volume 34 of *Ser. Knots Everything*, pages 225–262. World Sci. Publ., Hackensack, NJ, 2004.

[LZ04] S. K. Lando and A. K. Zvonkin. *Graphs on Surfaces and their Applications*, volume 141 of *Encyclopaedia of Mathematical Sciences*. Springer-Verlag, Berlin, 2004. With an appendix by Don B. Zagier, Low-Dimensional Topology, II.

[Mas71] B. Maskit. On Poincaré's theorem for fundamental polygons. *Advances in Math.*, 7:219–230, 1971.

[Mas91] W. S. Massey. *A Basic Course in Algebraic Topology*, volume 127 of *Graduate Texts in Mathematics*. Springer-Verlag, New York, 1991.

[MF82] D. Mumford and J. Fogarty. *Geometric Invariant Theory*, volume 34 of *Ergebnisse der Mathematik und ihrer Grenzgebiete [Results in Mathematics and Related Areas]*. Springer-Verlag, Berlin, second edition, 1982.

[Mir95] R. Miranda. *Algebraic Curves and Riemann Surfaces*, volume 5 of *Graduate Studies in Mathematics*. American Mathematical Society, Providence, RI, 1995.

[Nag88] S. Nag. *The Complex Analytic Theory of Teichmüller Spaces*. Canadian Mathematical Society Series of Monographs and Advanced Texts. John Wiley & Sons Inc., New York, 1988. A Wiley-Interscience Publication.

[Nar92] R. Narasimhan. *Compact Riemann Surfaces*. Lectures in Mathematics ETH Zürich. Birkhäuser Verlag, Basel, 1992.

[Sch94a] L. Schneps. Dessins d'enfants on the Riemann sphere. In *The Grothendieck Theory of Dessins d'Enfants*, volume 200 of *London Math. Soc. Lecture Note Ser.*, pages 47–77. Cambridge University Press, Cambridge, 1994.

[Sch94b] L. Schneps, editor. *The Grothendieck Theory of Dessins d'Enfants*, volume 200 of *London Mathematical Society Lecture Note Series*. Cambridge University Press, Cambridge, 1994. Papers from the Conference on Dessins d'Enfant held in Luminy, April 19–24, 1993.

[Shi72] G. Shimura. On the field of rationality for an abelian variety. *Nagoya Math. J.*, 45:167–178, 1972.

[Sie88] C. L. Siegel. *Topics in Complex Function Theory. Vol. I.* Wiley Classics Library. John Wiley & Sons Inc., New York, 1988. Translated from German by A. Shenitzer and D. Solitar, preface by W. Magnus, Reprint of the 1969 edition.

[SL97] L. Schneps and P. Lochak, editors. *Geometric Galois Actions. 1*, volume 242 of *London Mathematical Society Lecture Note Series*. Cambridge University Press, Cambridge, 1997. Around Grothendieck's "Esquisse d'un programme".

[SV90] G. B. Shabat and V. A. Voevodsky. Drawing curves over number fields. In *The Grothendieck Festschrift, Vol. III*, volume 88 of *Progr. Math.*, pages 199–227. Birkhäuser, Boston, MA, 1990.

[Syd97] R. I. Syddall. *Uniform Dessins of Low Genus*. ProQuest LLC, Ann Arbor, MI, 1997. Ph. D. thesis, University of Southampton, U. K.

[Wei56] A. Weil. The field of definition of a variety. *Amer. J. Math.*, 78:509–524, 1956.

[Wol97] J. Wolfart. The 'obvious' part of Belyi's theorem and Riemann surfaces with many automorphisms. In *Geometric Galois Actions, 1*, volume 242 of *London Math. Soc. Lecture Note Ser.*, pages 97–112. Cambridge

University Press, Cambridge, 1997.

[Wol06] J. Wolfart. *ABC* for polynomials, dessins d'enfants and uniformization–a survey. In *Elementare und Analytische Zahlentheorie*, Schr. Wiss. Ges. Johann Wolfgang Goethe Univ. Frankfurt am Main, 20, pages 313–345. Franz Steiner Verlag Stuttgart, Stuttgart, 2006.

[Woo06] M. M. Wood. Belyi-extending maps and the Galois action on dessins d'enfants. *Publ. Res. Inst. Math. Sci.*, 42(3):721–737, 2006.

Index